Progress in Molecular and Subcellular Biology 7

Progress
7
in Molecular and
Subcellular Biology

With Contributions by
T. O. Diener · H. Faulstich · F. Hutchinson
W. Köhnlein · D. J. Kopecko · A. K. Krey
R. A. Owens

Edited by: F. E. Hahn · H. Kersten
W. Kersten · W. Szybalski

Advisors: T. T. Puck · G. F. Springer
K. Wallenfels

Managing Editor: F. E. Hahn

Springer-Verlag Berlin Heidelberg New York 1980

Professor Fred E. Hahn, Ph. D.
Division of Communicable Diseases and Immunology
Walter Reed Army Institute of Research
Washington, D. C. 20012, USA

With 31 Figures

ISBN 3-540-10150-0 Springer-Verlag Berlin Heidelberg New York
ISBN 0-387-10150-0 Springer-Verlag New York Heidelberg Berlin

Library of Congress Catalog Card Number 75-79748

Offsetprinting and Bookbinding: Konrad Triltsch, Graphischer Betrieb, 8700 Würzburg.
2131/3130-543210

Contents

D.J. KOPECKO: Specialized Genetic Recombination Systems in
Bacteria: Their Involvement in Gene Expression and Evolution
(With 21 Figures)

T.O. DIENER and R.A. OWENS: Viroids (With 3 Figures)

Contributors

DIENER, T.O., Plant Virology Laboratory, Plant Protection
   Institute, AR-SEA, U.S. Department of Agriculture, Beltsville
   Agricultural Research Center, Beltsville, MD 20705, USA

FAULSTICH, HEINZ, Max-Planck-Institut für Medizinische Forschung,
   Abt. Naturstoff-Chemie, Jahnstraße 29, 6900 Heidelberg, FRG

HUTCHINSON, FRANKLIN, Department of Molecular Biophysics and
   Biochemistry, Yale University, New Haven, CT 06520, USA

KÖHNLEIN, WOLFGANG, Institut für Strahlenbiologie, Westfälische
   Wilhelms-Universität, Hittorfstraße 17, 4400 Münster, FRG

KOPECKO, DENNIS J., Department of the Army, Walter Reed Army
   Institute of Research, Walter Reed Army Medical Center,
   Washington, DC 20012, USA

KREY, ANNE K., Genetics and Teratology Program, National
   Institute of Child Health and Human Development, National
   Institutes of Health, Bethesda, MD 20205, USA
   Former address: Department of Molecular Biology, Walter Reed
   Army Institute of Research, Washington, DC 20012, USA

OWENS, R.A., Plant Virology Laboratory, Plant Protection
   Institute, AR-SEA, U.S. Department of Agriculture, Beltsville
   Agricultural Research Center, Beltsville, MD 20705, USA

The Photochemistry of 5-Bromouracil and 5-Iodouracil[1] in DNA

F. Hutchinson and W. Köhnlein

A. Introduction

Of the synthetic bases which may be incorporated in DNA, two of
the most interesting are the thymine analogs, 5-bromouracil and
5-iodouracil. DNAs containing these bases have a number of altered properties which have been used to advantage by molecular
biologists.

This review summarizes the photochemistry of these compounds.
It interprets the biological effects of ultraviolet light on
cells, viruses and DNA containing these analogs in terms of the
photochemical reactions. The biological effects have been of
considerable interest since the first observation by Greer and
Zamenhof (1957) that incorporation of bromouracil greatly sensitizes bacterial cells to the action of ultraviolet light.
This review also covers ways in which the altered photochemical
properties of DNA containing these analogs have been used as
tools to study a variety of biological properties. It also dis-
cusses other properties of DNA containing the halouracils which
may be related to the photochemical reactions.

This review is an updating of an earlier summary (Hutchinson
1973). To present the material here in the most compact form,
conclusions from this earlier review will only be summarized,
and the reader referred back to Hutchinson (1973) for the detailed documentation. Other recent reviews are by Wang (1976),
mainly the photochemistry of bromouracil and bromodeoxyuridine
in short oligonucleotides, and by Rahn and Patrick (1976),
mainly the photochemistry of DNA which contains bromouracil.

The photochemistry of 5-bromouracil and of 5-iodouracil are believed to be similar, and frequently data for the two analogs
will be combined. Most of the experimental results are for
bromouracil, since it is much easier to incorporate in DNA,
probably because the Van der Waals radius of bromine (1.95 Å)
is closer to that of the methyl group of thymine (2.0 Å) than
that of iodine (2.5 Å).

The halouracils 5-fluorouracil and 5-chlorouracil are not normally incorporated into DNA. In addition, their photochemistry
is quite different. They will not be covered in this review.

The various bases and their derivatives are referred to here in
accordance with the recommendations of the I.U.P.A.C.-I.U.B.
Commission on Biochemical Nomenclature (Eur. J. Biochem. 15,
203, 1970; J. Biol. Chem. 245, 5171, 1970): BrUra (bromouracil),
BrdUrd (bromodeoxyuridine), IUra (iodouracil), and IdUrd (iodo-
deoxyuridine).

B. Structural and Spectroscopic Information

Surprisingly, there is little definitive evidence on the structure of DNA containing BrUra. The circular dichroism of mammalian DNA in which essentially all the thymine in one strand had
been replaced by BrUra (no BrUra in the other strand) was similar to that of unsubstituted DNA, except that the positive
maximum at 275 mm (unsubstituted DNA) had been shifted to 270 nm
(Simpson and Seale 1974). This blue shift increases with the extent of BrUra incorporation (Augenlicht et al. 1974; Lapeyre and
Bekhor 1974; Nicolini and Baserga 1975). The circular dichroism
of poly(rA)·poly(rBrU), compared to that for poly(rA)·poly(rU),
suggests that the presence of the bromine atom does not change
the conformation except possibly for a slight twist of BrUra out
of the mean plane of the base pair (Bobst et al. 1976).

X-ray diffraction patterns of poly[d(A-BrU)·d(A-BrU)] are similar
to those for poly[d(A-T)·d(A-T)] (Davies and Baldwin 1963). The
pictures are of poor quality, but show unmistakably that the
structures are both similar to the DNA B form (Langridge et al.
1960). X-ray crystallographic studies suggest that the base
stacking properties of BrUra would cause it to adopt a twisted
position (compared to thymine) in DNA (Sternglanz and Bugg 1975).
Evidence to support altered stacking comes from the drastic difference between photoproducts for BrUra incorporated in DNA, and
those from synthetic polymers such as poly(dA)·poly(dBrU), or
poly[d(A-BrU)·d(A-BrU)] (Ehrlich and Riley 1974; see Sect. C,
this review).

Overall, the results fit with the common assumption that the
introduction of BrUra does not greatly change DNA from its usual
B conformation, as shown in Fig. 1.

The available spectroscopic data on the halouracils and derivatives are summarized in Hutchinson (1973). The values for the
maximum molar absorbance ε_{max} and the corresponding wavelength
λ_{max} vary with the halogen in a systematic way (Lohmann 1974).
For convenience, some useful figures are given in Table 1.

The optical absorption of DNA in which 22% thymine has been
replaced by BrUra is indistinguishable from that of unsubstituted DNA for λ between 230 nm and 285 nm (Boyce 1961). It
is presumed that for longer wavelengths DNA absorption is increased by BrUra incorporation, since such substitution increases the yield of photochemical processes. The difficulties
of purifying DNA free of small quantities of other chromophores
has so far inhibited a direct measurement.

Table 1. Some spectroscopic data on BrUra, IUra and related compounds for polar (i.e., aqueous) environments. For more extensive data, see review (Hutchinson 1973). The numbers in parenthesis give the references

Compound	Uncharged form					Singly charged form			
	Singlet Energy, cm^{-1}	Triplet Energy, cm^{-1}	λ_{max}^a nm (2,3,4)	ε_{max}^a (2,3,4)	pK_1 (2,3,4)	Singlet Energy, cm^{-1}	Triplet Energy, cm^{-1}	λ_{max}^a nm (2,3,4)	ε_{max}^a (2,3,4)
BrUra	–	25,600 (1)[b]	277.5	7,140	8.05	–	26,100 (1)[e]	302.5	6,900
IUra	–	25,300 (1)[b]	285	6,150	8.25	–	–	305	7,100
Ura	35,100 (5)	–	259.5	8,200	9.50	31,900 (5)	–	284	6,150
Thy	34,500 (6)	–	264.5	7,890	9.90	31,600 (6)	26,700 (6) / 26,300 (1)[e]	291	5,440
BrdUrd	32,400[c]	–	280	9,230	7.90	–	–	280	6,500
IdUrd	–	–	287.5	7,500	8.20	–	–	278.5	5,500
dUrd	34,900 (6)	27,400 (6)	262	10,200	9.30	35,000 (6)	28,400 (6)	262	8,630
dThd	34,100 (6)	26,300 (6)	267	9,650	9.80	34,350 (6)	27,000 (6)	267	7,380

[a] λ_{max} and ε_{max} for nucleotides are essentially the same as for the corresponding nucleosides, and presumably the energy levels are also.

[b] Using Rothman and Kearn's measurements in naphthalene crystal, and their estimate that the value would be 600 cm^{-1} higher in a polar environment (1).

[c] Calculated from the shift of 15 nm for the long wavelength absorption edge in aqueous solution of BrdUrd compared to dThd (Rapaport 1964).

[d] In ethylene glycol-water glass at 77°K, except as noted.

[e] In 9M LiCl at 77°K.

References: (1) Rothman and Kearns (1967); (2) Shugar and Fox (1952); (3) Berens and Shugar (1963); (4) Fox and Shugar (1952); (5) Longworth et al. (1966); (6) Eisinger and Lamola (1971)

4

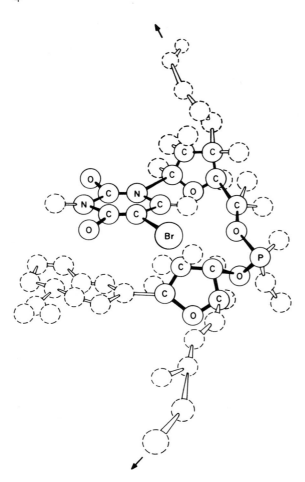

Fig. 1. The presumed configuration of one strand containing BrUra in the DNA double-helical B structure (according to Langridge et al. 1960)

Gueron et al. (1974) have discussed the transfer of electronic excitation energy in DNA. They consider that theory and experiment would suggest a transfer of excited singlet states among a few bases, extensive enough to sample the base pair distribution. The evidence for the triplet energy transfer is quite convincing, and only the thymine triplet, the lowest lying triplet state in ordinary DNA, has been detected. Various experiments suggest that triplet states can migrate 5 to 30 bases in DNA at either $77^{\circ}K$ or $300^{\circ}K$.

Since bromouracil has both singlet and triplet levels which are lower than any of the four usual DNA bases (see Table 1), it would be expected that concentration of excitation energy in BrUra could occur, but there is no direct spectroscopic evidence. Fielden et al. (1971) and Lillicrap and Fielden (1971) showed effects of small amounts of BrUra on phosphorescence of DNA after pulsed electron irradiation, suggesting some kind of "energy transfer" mechanism. However, the emission spectra they observed were almost certainly from trapped electrons or hydrogen atoms, not from BrUra triplet states.

C. Ultraviolet-induced Lesions in DNA Containing BrUra

I. Uracil

This is the major photochemical product of BrUra or IUra incorporated in DNA (Wacker 1961). It is formed by the photo-in-
duced debromination (deiodination) of BrUra (IUra), leaving
uracil with a free radical at the C5 position (review: Hutchinson 1973). This radical can rapidly extract H atoms from many
compounds to form uracil (review: Hutchinson 1973). For DNA in
the B form, the uracilyl radical would be directly adjacent to
the C2' of the deoxyribose from the nucleoside on the 5' side of
BrdUrd (Fig. 1). The assumption that uracil is formed when the
5-uracilyl radical extracts an H atom from the C2' of the nearby
deoxyribose gives a simple explanation for both the formation of
uracil and of the other photoproducts described in this section.

Irradiation of BrUra in aqueous solution also produces uracil,
but at yields one or two orders of magnitude lower (Gilbert and
Cristallini 1973; Hutchinson 1973). Campbell et al. (1974) showed
that adding various hydrogen atom donors to aqueous solutions of
BrUra increased the photoproduction of uracil very little until
the mole fraction of H donor was so high that one or more molecules would be in the hydration cage surrounding the bromine
atom. At these concentrations, the yield of uracil increased by
more than an order of magnitude.

Whether the system is described as an excited U-Br bond or as
U$^{\cdot}$ and Br$^{\cdot}$ radicals which can recombine with each other, it
usually reverts to an intact U-Br bond, unless a nearby molecule
with a sufficiently labile H atom interacts with the excited
system to form U, and presumably Br$^{\cdot}$ and a radical on the molecule donating the H atom.

For BrUra in DNA, the adjacent deoxyribose (Fig. 1) could serve
as the H-donor. Oxygen in the DNA solution does not reduce the
uracil yield (Wacker 1961; Smith 1974; Lion 1968); no evidence
was found (Hutchinson, unpublished data) for dialuric acid, the
product of the reaction of U$^{\cdot}$ with O_2 (Gilbert and Cristallini
1973). Also, the uracil yield is not affected by the presence
of -SH compounds during irradiation (Lion 1968), altough the
yields of other products are reduced, as described below. Thus,
any uracilyl radical in the DNA exists only for a very short
time, if at all.

Photodissociation of I is comparable for IUra and for IUra incorporated in DNA (Wacker 1963). The C-I bond is efficiently
broken, whereas the stronger C-Br bond is more likely to dissociate if an H atom is available to react and form Ura.

II. Single-strand Breaks and Alkali-labile Bonds

M. Lion (1966) first discovered that sedimentation of BrUra-DNA
after ultraviolet irradiation and alkaline denaturation showed
single-strand breaks. A number of investigators have confirmed
and extended these findings (review: Hutchinson 1973).

Actual single-strand breaks, chain interruptions at neutral pH, are produced by the effect of ultraviolet light on BrUra-DNA irradiated in salt solutions, measured by the altered sedimentation of supercoiled lambda DNA in neutral sucrose gradients (Hewitt and Marburger 1975). The number of such breaks is increased five-fold when the same DNA is sedimented on alkaline gradients. This result has been confirmed by others (Krasin and Hutchinson 1978b). However, when the latter experimenters irradiated the lambda supercoils inside *E. coli* host cells at ice temperature, the number of breaks in alkaline gradients was about the same as those formed on in vitro irradiation; on neutral gradients the number of breaks on in vivo irradiation was reduced by only about 30% from the number on alkaline gradients. Thus, in cells some process is converting most of the alkali-labile bonds to actual single-strand interruptions. It is not known whether this is because of a change in the chemistry following the initiating photochemical event (because of a change in the surroundings), or because enzymes acting at ice temperature are opening alkali-labile bonds.

The breaks and the alkali-labile bonds are produced when the photochemically produced uracilyl free radical extracts an H atom from the adjacent C2' from the deoxyribose of the nucleoside on the 5' side of the debrominated BrUra (Köhnlein and Hutchinson 1969; see Fig. 1). Evidence for this mechanism includes the following.

1. Deoxyribose was destroyed in irradiated BrUra-DNA (Hotz and Reuschl 1967) at yields comparable to the yield of single-strand breaks measured in alkali (see review: Hutchinson 1973).
2. Electron paramagnetic resonance of irradiated BrUra-DNA showed free radicals in the sugar (Köhnlein and Hutchinson 1969).
3. Dodson et al. (1972) found that the base on the 5' side of the break is uracil. (A small amount of an unidentified product which is not one of the normal DNA bases is also found).
4. The rate of single-strand breakage is lower in denatured DNA, in which the Br atom may not be close to a sugar, than in native DNA (Lion and Köhnlein 1972).
5. The 5-uracilyl radical is very reactive, and able to remove H atoms from a variety of materials, including deoxyribose (review: Hutchinson 1973).
6. Ogata and Gilbert (1977) isolated a 55 base pair fragment of double-helical BrUra-DNA of known base sequence and irradiated with ultraviolet light to produce much less than one break per molecule. When the DNA single strands were electrophoresed on gels, the fragment sizes corresponded to those expected from breaks at each of the bromouracil residues.

It is known that the presence of compounds containing -SH groups can greatly reduce the effect of ultraviolet light on BrUra-DNA. Table 2 summarizes the decrease in yields for some photoproducts with 0.01 M cysteamine. The first entry shows that -SH cannot compete with the reaction

U˙ + sugar ⟶ U + sugar˙

as might be expected from the previous discussion above.

Table 2. The effect of cysteamine on the yields of photoproducts

Product	Yield without cysteamine / Yield with 0.01 M cysteamine	Reference
Uracil	1.0	Lion (1968)
Destruction of deoxyribose	11	Hotz and Reuschl (1967)
Single-strand break in alkali	4	Hotz and Walser (1970)
Single-strand break in alkali	4	Lion (1970)
Single-strand break in alkali	10	Mönkehaus (1973)

The positively charged (at neutral pH) cysteamine molecule is 30-fold more effective than is neutral mercaptoethanol in media of low ionic strength (Rupp 1965). This difference is mostly erased with the addition of 1M NaCl, suggesting a high concentration of cysteamine near the negatively charged DNA in the absence of counter ions.

The chemical identity of the alkali-labile bond (four such bonds per actual single-strand break, see above) has not yet been established. Bases of all types in BrUra-DNA are released by irradiation (Lion 1972). However, the alkali-labile bond is not a deoxyribose sugar in a nucleotide chain with the base removed, because sedimentation of DNA denatured with formamide shows the same number of breaks as in the same DNA denatured with alkali (Hutchinson and Hales 1970; Lion 1972). A sugar without its base (i.e., an apurinic site made by heating DNA) is known to be stable in formamide, but to form a break in alkali (Kirtikar et al. 1975).

Two possible structures for the alkali-labile bond can be suggested:

1. From a study of the products formed by X-ray irradiation of oxygenated DNA solutions, Dizdaroglu et al. (1977) concluded that an alkali-labile site was formed by extraction of an H atom from the C2' position of deoxyribose. This would predict that extraction of an H from this same C2' carbon by a photochemically produced uracilyl radical would also form an alkali-labile bond.
2. When the uracilyl free radical extracts an H atom from the deoxyribose C2', it is assumed that the Br˙ free radical is close by, since for BrUra in solution a "cage effect" apparently causes the uracilyl and Br˙ radicals to recombine most of the time (see previous section). In DNA, it is possible

that Br· recombines with the free radical on the sugar, forming a sugar brominated in the 2 position. Such a sugar will be stable at neutral pH, but will decompose in alkali (Guthrie 1972; Szarek 1973).

The large reduction in the presence of -SH in the number of single-strand breaks (in alkali) supports the first mechanism above, as opposed to the second.

Hewitt and Marburger (1975) show that for lambda supercircular BrUra-DNA irradiated in 0.01 M Tris buffer (pH 7), 0.001 M EDTA, there is a five-fold increase in the yield of both actual single-strand breaks and alkali-labile bonds when 1M NaCl is added. The yield of uracil in BrUra-DNA irradiated with 313 nm ultraviolet light is the same for no added salt and for 1 M NaCl added (F. Hutchinson, unpublished data; M. Lion, personal communication). Perhaps at low ionic strength the charged phosphate groups on the DNA are largely neutralized by the positively charged Tris molecules (pK_A = 8.08), which might act as efficient donors of H atoms to the sugar radicals.

III. Damage to the Opposite Strand

The mechanism for the single-strand break depicted in Fig. 1 localizes the chemical changes produced by irradiation in the same strand of the double helix which contains the BrUra base which is debrominated. This conclusion is established by the presence of uracil adjacent to a single-strand break (Dodson et al. 1972).

When hybrid DNA, one strand with BrUra and the other strand with none, is irradiated with ultraviolet and sedimented in alkali, single-strand breaks are found in the strand without BrUra. The number is only 1-5% that in the strand containing BrUra (Table 3), but much larger than the number found in DNA with place is almost certainly related to those involved in the production of double-strand breaks, and will be discussed in the next section.

IV. Double-Strand Breaks

Such breaks are produced by ultraviolet irradiation of BrUra-DNA and are measured by sedimention in neutral conditions (Hotz and Walser 1970). Early experiments indicated a quadratic dependence of the number of breaks on fluence (Smets and Cornelis 1971; Lion 1972), which suggested the breaks were the result of accidental coincidence between single-strand breaks in the complementary strands. Mönkehaus and Köhnlein (1973) found that the number of double-strand breaks had a quadratic component as expected, and also a linear component at low fluences. Their DNA had considerable numbers of single-strand breaks before irradiation; it was not possible to decide conclusively whether the linear component was caused by coincidences between preexisting breaks in one strand and photoinduced breaks in the other, or by a one-photon process which broke both strands.

Table 3. The production by 313 nm light of single-strand breaks (measured
in alkali) in a DNA single strand containing only thymine, which is in
double-helical form with a complementary strand containing bromouracil.
The breaks measured are the sum of "true" single-strand breaks plus alkali-
labile bonds

DNA	Breaks in thymine strand per break in bromouracil strand	Reference
B. subtilis DNA in 0.015 M NaCl, 0.0015 M sodium citrate	0.05	Köhnlein and Mönkehaus (1972)
Mouse DNA, irradiated in the cell	∿ 0.01	Lehmann (1972)
H. influenzae DNA, irradiated in cells	0.006	Beattie (1972)
E. coli DNA, irradiated in cells	0.01	Ley (1973)
Bovine kidney DNA, irradiated in cells	0.02	Povirk (1977a)

Krasin and Hutchinson (1978a) recently measured the production
of double-strand breaks in large ($\sim 10^9$ daltons) BrUra-DNA irradiated in E. coli cells with 313 nm light. The number of double-
strand breaks was proportional to the number of single-strand
breaks measured in alkali. The possibility that these breaks are
caused by coincidence between preexisting single-strand breaks
and photoinduced breaks can be ruled out. Control experiments
make it unlikely that the breaks are artifacts produced by the
action of cellular enzymes on irradiated DNA or by shear in the
extraction process.

The number of these breaks, 0.01 per single-strand break in alkali, is equal to the number of breaks in the thymine strand of
thymine-bromouracil hybrid DNA irradiated in cells, also 0.01
per single-strand break in alkali (Table 3). This suggests that
most breaks in the thymine strand of a hybrid are actual breaks.
All measurements listed in Table 3 were made in alkali, so the
ratio of alkali-labile sites to actual breaks is not known experimentally.

It is not known if the double-strand break produced by a one-
photon event is a consequence of the only primary photochemical
event established for BrUra, debromination. The formation of
the free radicals, such as the uracilyl radical or the radical
at C2' in the sugar, is always accompanied by the production of
other free radicals. Thus, there are possibilities for free
radical damage to the strand complementary to that containing
the debrominated base. Hotz et al. (1971) irradiated BrUra-DNA
in the presence and in the absence of cysteamine. When the UV
fluences in the two cases were adjusted to yield equal numbers
of single-strand breaks measured in alkali, they found markedly

fewer double-strand breaks in the DNA irradiated in cysteamine. This suggested that some of the double-strand breaks measured in the DNA irradiated without cysteamine might not be caused by accidental coincidences between single-strand breaks, and are even more readily suppressed by -SH groups than are single-strand breaks. Similar results were reported by Mönkehaus (1973). Also, Table 3 shows there is more damage in the thymine strand of hybrid DNA irradiated in salt solution than in cells. These results suggest a free radical mechanism.

The Br˙ free radical is released in photolysis of BrUra-DNA, and could readily diffuse to the complementary strand. When DNA was exposed to Br˙ formed from the chain decomposition of N-bromosuccinimide (photolytically catalyzed), true single-strand breaks were found, with an increase of three- to five-fold under alkaline conditions (Krasin and Hutchinson 1978b). Under the conditions used, studies (e.g., Incremona and Martin 1970) show that the dominant reaction with substrate is by the bromine radical, not the succinimide radical. The predominance of alkali-labile bonds formed by Br˙ shows that more knowledge of the chemistry taking place in cells will be needed to establish the mechanism by which double-strand breaks are made.

V. Dinucleotide Products

A dinucleotide is the major photoproduct of ultraviolet irradiation of poly(rBrU) (Ehrlich and Riley 1972b), poly(dBrU) (Ehrlich and Riley 1974), and either poly(rBrU)·poly(rA) or poly(dBrU)·poly(dA) (Ehrlich and Riley 1974). This has been identified as the Ura(5-5)Ura product (Sasson et al. 1977), which is also formed on irradiation of BrUra solutions in the absence of oxygen (Ishihara and Wang 1966).

Dinucleotide photoproducts have been found on the irradiation of dBrU (3'-5') dT (Haug 1964; Peter and Drewer 1970) and dBrU (3'-5') dC (Peter and Drewer 1970). The structures of the products have not been established, but Peter and Drewer (1970) report one Br per photoproduct in both cases. The evidence suggests that much smaller amounts of dinucleotide products are formed by ultraviolet irradiation of the alternating copolymers of poly[r(BrU-A)·r(BrU-A)] or poly[d(BrU-A)·d(BrU-A)], since Ehrlich and Riley (1974) found no denaturation and absorbance changes of the magnitude which accompanied Ura (5-5) Ura formation in the homopolymers.

No firm evidence for the existence of a dinucleotide product involving BrUra has yet been found in irradiated DNA. Lion (1968) gave an upper limit of 1% (of the total photoproducts involving BrUra) for a product including both [6-^{3}H]-BrUra and [2^{14}C]-Thy in DNA in which 70% of the thymine had been replaced. Peter and Drewer (1970) set upper limits in DNA of 1% and 3.5% respectively for the photoproducts from dBrU(3'-5')dT and dBrU(3'-5')dC. Ishihara and Wang (unpublished results mentioned in Wang 1976) saw a fluorescence resembling that of Ura(5-5)Ura in acid-hydrolyzed *E. coli* DNA containing an unspecified amount of BrUra; the product stayed at the origin of the paper chromatograms and the yield did not depend on UV fluence in a consistent way, so the identification is inconclusive.

It is reasonably certain that Ura(5-5)Ura will be formed in irradiated DNA at high BrUra substitution levels. It might be expected, for example, in the DNA coding for the string of six to eight Ura residues believed to constitute part of the RNA polymerase termination signal.

The mechanism of the formation of dinucleotide products is unknown. While irradiation of BrUra in solution produces the 5-uracilyl free radical, this species is so chemically reactive that the lifetime in solution is too short to allow formation of Ura(5-5)Ura by recombination of two uracilyl radicals (summaries: Hutchinson 1973; Wang 1976).

It is possible that the dinucleotide product is formed from bases which are stacked together at the time a photon is adsorbed. The decrease at high pH in yield of the dinucleotide in irradiated dBrU (3'-5') dT (Peter and Drewer 1971) could be the result of unstacking the two bases as they become charged. The dinucleotide product might possibly result from a free radical process starting with a 5-uracilyl radical after debromination. If so, the uracilyl radical must be held in a position such that it cannot react with a sugar as shown in Fig. 1. Indeed, yields of uracil much lower than found in BrUra-DNA have been reported for poly(rBrU) (Ehrlich and Riley 1972a) and for poly[(rBrU)·(rA)] (Ehrlich and Riley 1974). The rates of strand breakage in these polymers and in poly(dBrU)·poly(dA) (Ehrlich and Riley 1974) were also lower than expected from that observed in BrUra-DNA. The dinucleotides dBrU(3'-5')dT and dBrU (3'-5')dC (see above) are such that with the bases stacked the Br atom will not be close to a sugar (see Fig. 1).

Alternatively, the absence of a suitable H-atom donor near the 5-uracilyl radical could lead to recombination with the Br· radical and a relatively long-lived triplet state which could, for example, form an excimer with a nearby base.

VI. DNA-Protein Crosslinks

Ultraviolet irradiation (254 nm) of DNA in cells or viruses leads to crosslinking to protein (review: Smith 1976). When BrUra is incorporated in the DNA, the amount of crosslinking is increased (Smith 1964; Smets and Cornelis 1971; Weintraub 1973; Lin and Riggs 1974). Irradiation with light of wavelength around 313 nm also crosslinks BrUra-DNA to protein (Weintraub 1973; Lansman and Clayton 1975).

Ultraviolet-induced crosslinking has been reported of BrUra derivatives with sulfhydryl compounds such as cysteine and glutathione (Varghese 1974), disulfides (Rupp and Prusoff 1965, for IUra) and with tryptophan derivatives (Saito et al. 1978), but not with tyrosine and histidine (Saito et al. 1978). In all cases of crosslinking, the bromine was lost and the covalent bond was made with the 5 position in the uracil ring, supporting the involvement of the 5-uracilyl free radical. However, BrUra-DNA is known to bind proteins more tightly than does normal DNA (Lin and Riggs 1972; Lin et al. 1976). This increased binding could permit more efficient crosslinking by the reactions which occur between proteins and normal DNA.

It is not known if crosslinking is initiated by photons adsorbed by DNA bases or by protein. Weintraub (1973) reports that the specificity of protein crosslinking to BrUra-DNA is increased by a factor of 5 when the irradiation wavelength is changed from 254 nm to 313 nm. This could result from the increased absorption of 313 nm radiation in BrUra-DNA. It could also occur because of relatively higher absorption of ultraviolet by protein -S-S bonds, which have a broad absorption spectrum which drops by only a factor of ten between 254 nm and 313 nm (see, e.g., Setlow and Doyle 1957). The increased crosslinking by the BrUra-DNA would then reflect the tighter binding of protein.

Unidentified Products. Several authors (e.g., Smith 1964; Lion 1968; Peter and Drewer 1970) find radioactive spots from labeled BrUra in chromatograms of hydrolyzed BrUra-DNA which suggest minor products to the extent of 10 to 20% (total) of the uracil produced. Different authors frequently find different spots in chromatograms developed with the same fluid, and different quantities of similar spots. It is unclear whether these spots vary because of differences in DNA hydrolysis techniques, because of differences in conditions surrounding the DNA at the time of irradiation, or for other reasons.

The most definite indication for another product in BrUra-DNA is given by Dodson et al. (1972). They added ^{32}p with polynucleotide kinase to the nucleoside on the 5' side of a radiation-produced single-strand break, and released the nucleotide with a 3' nuclease. Most of the ^{32}p-labeled nucleotide co-chromatographed with an authentic sample of deoxyuridine-5' monophosphate, but about 20% formed a shoulder on one side of the pdUrd peak. This material was not the 5' monophosphate of BrdUrd or of any of the natural bases, but has not otherwise been identified.

VII. Product Yields - Suitable Measures

The most appropriate measure of yields for products of the photolysis of BrUra is a quantity called the cross section, σ. For a sample containing N BrUra bases and uniformly irradiated with a small increment of fluence ΔF J/m^2, an incremental number of photoproducts ΔP are produced.

$$\Delta P = N\sigma \, \Delta F.$$

In the limit,

$$\sigma = \frac{1}{N} \frac{dP}{dF}.$$

In situations in which less than 10% of the bromouracils are photolysed, and in which the absorbance of the sample is small enough so that the fluence is essentially constant throughout, the cross section is easily calculated.

$$\sigma = \frac{\Delta P/N}{\Delta F}$$

$$= \frac{\text{fraction of BrUra converted to product}}{\text{fluence}} \frac{m^2}{J}.$$

Many experiments in photolysis express the result as a quantum yield ϕ. Let ΔP products be produced by the absorption of ΔM photons.

$$\phi = \frac{\Delta P}{\Delta M}.$$

The problem with this measure lies in applying it to BrUra incorporated in DNA. The number of photoproducts is, in principle, unambiguous; the number of absorbed photons is not. Should one use the number of photons adsorbed by all the bases in the DNA, or only those adsorbed by BrUra residues? If the latter, by what factor should the absorption coefficient of free BrUra be multiplied to account for the hypochromic decrease in absorption which occurs on base stacking?

In this review, we will generally use cross sections.

VIII. Product Yields - Quantitative Values

In Table 4 are summarized some cross sections for the photoproduction of two products, uracil and DNA single-strand breaks measured in alkali.

It is clear that incorporation of BrUra into DNA enhances debromination by a factor of 25-50, and that the yield of single-strand breaks measured in alkali is comparable to, but somewhat less than, the yield of uracil. At 313 nm, the absorption of BrdUrd is 50 - fold lower than at its maximum at 280 nm (Rapaport 1964), and Table 4 shows the 313 nm cross sections are a factor of 50-150 less than those at 254 nm. These ratios are compatible, bearing in mind the following:

1. Any results at a nominal "313 nm" are critically dependent on the effective wavelength of the irradiating beam, and hence on the way the beam was prepared; the absorption coefficient of BrdUrd decreases by a factor of 10 between 310 nm and 320 nm (Rapaport 1964).

Table 4. Some cross sections σ and quantum yields ϕ

Material	λ, nm	Product	σ m²/J · 10^6	ϕ	References
BrUra in solution	254	uracil	2.8	$2.0 \cdot 10^{-3}$	Hutchinson (1973)
		uracil	2.5	$1.8 \cdot 10^{-3}$	Campbell et al. (1974)
	313	uracil	0.04	$2.4 \cdot 10^{-3}$	Hutchinson (1973
BrUra-DNA 6-100% replacement of thymine	254	uracil	60-115	–	Summary: Hutchinson (1973)
	313	uracil	0.9	–	Lion (unpublished data 1977)
BrUra-DNA in cells	313	uracil	0.8±0.2	–	Krasin and Hutchinson (1978)
	254	single-strand break in alkali	30-90	–	Summary: Hutchinson (1973)
	313		0.32	–	Ley (1973)
	313		0.5	–	Krasin and Hutchinson (1978)
IUra in solution	254	uracil	~100-200	$\sim 100 \cdot 10^{-3}$	Gilbert and Schulte-Frohlinde (1970)
IUra-DNA	254	uracil	~ 60	–	Wacker (1963)

2. For BrUra incorporated in DNA, the involvement of other bases
   in absorbing photons and transferring the energy to incorporated BrUra is unknown (see Sect. B of this review).

If the transfer of energy from other DNA bases to BrUra is important, there should be higher yields of photoproducts at low
concentrations of BrUra in the DNA than at high. The three
available bits of information are contradictory:

1. Carrier and Setlow (1972) determined the relative efficiency
   of producing single-strand breaks (measured in alkali) in DNA
   preparations which had 14, 35, and 100% of Thy replaced by
   BrUra; at 265 nm, the cross sections were factors of 160,
   130, and 70, respectively, larger than at 313 nm. The cross
   section at 313 nm was approximately constant. The higher efficiency for 265 nm irradiation of the DNA with low BrUra
   is most easily explained as energy funneled into the BrUra
   from the other bases.
2. The fraction of uracil produced by a given fluence of 254 nm
   radiation was approximately the same for DNAs having 6, 28,
   and 100% of Thy replaced by BrUra (Hutchinson 1973).
3. For both 254 nm and 313 nm irradiation, the cross section for
   loss of bacterial colony-forming ability was a linear fraction
   of the extent of thymine replacement by BrUra (measured by
   banding in CsCl density gradients) from a few percent to 100%
   (Köhnlein et al., unpublished data).

   These last two points argue that transfer of energy is not
   important. More experiments are needed to resolve the dis-
   crepancy.

When Ogata and Gilbert (1977) determined the fragmentation by
ultraviolet light of BrUra-DNA of known sequence (see above),
they observed that the cross section varied greatly from one
BrUra to another. Binding of proteins also can change the probability of scission at particular BrUra residues (Simpson 1979).
Thus, the cross sections in Table 4 are only averages.

Table 5 tabulates the relative yields of a number of products
involving BrUra incorporated in DNA. Normalizing the yields to
that of single-strand breaks is convenient because these ratios
probably stay roughly constant (except where noted) even when
the yields change with different wavelengths of irradiating
light, varying concentrations of sulfhydryl compound, etc. The
yield of single-strand breaks in alkali is also the easiest to
measure.

D. Biological Effects of BrUra and IUra Photoproducts

To maintain a unified point of view in this review, we will
first discuss what is known of the ways in which cells can
repair or modify BrUra and IUra photoproducts in DNA. We will

Table 5. Yields of various photoproducts in BrUra-DNA irradiated in cells, relative to the yield of single-strand breaks measured in alkali

Product	Yield relative to single-strand break in alkali	References	Comments
Uracil	1.3-2	Summary: Hutchinson (1973)	The yield of uracil is not changed by adding -SH groups, whereas that of other products decreases (see Table 2). For DNA irradiated in salt solutions, the relative yield is probably close to 1
Single-strand breaks at neutral pH	0.7-0.9	Krasin and Hutchinson (1978)	For DNA irradiated is salt solutions, yield is 0.2 breaks (neutral) per break in alkali (Hewitt and Marburger 1975)
Destruction of purine deoxyribose	~1	Hotz and Reuschl (1967)	
Double-strand break	0.01	Krasin and Hutchinson (1978)	This excludes the breaks formed by coincidence between random single-strand breaks in the complementary strands. The ratio given will probably drop with increasing -SH concentration (Hotz et al. 1971)
Protein-DNA crosslink	~0.1	Smets and Cornelis (1971) (calculation in Hutchinson 1973)	The accuracy of the calculations is questionable, but the high yield would suggest that the crosslink is a significant lesion
	~1	Weintraub (1973)	
Dinucleotide product			
BrU-T	< .013	Lion (1968)	
BrU-T	< .01	Peter and Drewer (1970)	
BrU-C	< .035	Peter and Drewer (1970)	

then discuss the various phenomena observed on ultraviolet irradiation of cells or viruses whose DNA includes the analogs.
The extent to which the phenomena can be interpreted in terms
of molecular biology will be pointed out, and also those aspects
for which no reasonably solid explanations have yet been advanced.

I. Cellular Repair of BrUra Photoproducts

Uracil. Rapid excision of the photochemically produced uracil from
BrUra-DNA irradiated in cells has been reported in mammalian
cells (Smets and Cornelis 1971) and in bacteria (Dennis and Hutchinson 1972). From the previous section, we can estimate that
20-50% of the uracils were in an intact double helix where an H
atom (from -SH compounds?) had been donated to the sugar which
the uracilyl radical had attacked (the possibility of a rear-
ranged sugar, or one with an -SR group added, cannot be ruled
out). Ten to 20% of the uracils would be next to an alkali-labile bond, and 40-60% next to a single-strand break. The excision is complete enough so that probably all three types of
uracil are excised.

Uracil is also produced in DNA by deamination of cytosine at
elevated temperatures (Hurst and Kuksis 1958; Ullman and McCarthy
1973; Lindahl and Nyberg 1974). These uracils will be mispaired
with guanine in the double-helical structure, whereas the ones
photochemically produced in BrUra-DNA can form a Watson-Crick
pair with adenine.

The heat-induced uracils are excised by DNA glycosylases which
remove uracil from its sugar (Lindahl 1974). Makino and Munakata
(1977) have evidence for a glycosylase from *B. subtilis* which
produces single-strand breaks (assayed in alkali) in BrUra-DNA
irradiated in the presence of cysteamine in cells or in solution.
This shows that this enzyme excised uracil even when it can be
paired. It is presumably the glycosylase found by Duncan et al.
(1976), which produces an apyrimidinic site (Cone et al. 1977)
which is then attacked by a specific endonuclease (reviews:
Lindahl and Ljungquist 1975; Verly 1975). Repair of the break
then probably includes excision and resynthesis, which might involve the Kornberg DNA polymerase I (Negishi et al. 1976).

Single-Strand Breaks and Alkali-Labile Bonds. The single-strand breaks
measured in alkali sedimentation, including both actual breaks
and alkali-labile bonds, are rapidly repaired in both bacterial
(Ley and Setlow 1972) and mammalian cells (Ben-Hur and Elkind
1972). The actual single-strand breaks probably do not differ
essentially from those produced by ionizing radiation; their
repair probably proceeds by the same pathways, about which surprisingly little is known except that a high percentage are
rapidly repaired, and that little resynthesis accompanies the
process (see various articles in Hanawalt and Setlow 1975).
Nothing is known about the pathways by which the alkali-labile
bonds are repaired.

Double-Strand Breaks. The double-strand breaks are presumably repaired by a process similar to that for such breaks produced
by ionizing radiation; in *E. coli*, this requires both an active
recA gene and the presence of an intact DNA double helix with
homologous base sequences (Krasin and Hutchinson 1977).

Other Lesions. Nothing is known of the repair of other lesions.
The reported yields of DNA-protein crosslinking, particularly,
are so high that one tends to presume repair, but there are no
data.

II. Effects of UV Repair Systems on BrUra-DNA

Irradiated BrUra-DNA contains lesions which are not repaired by
photoreactivation enzymes or by the enzymes associated with the
E. coli uvr repair systems (review: Hutchinson 1973). These systems are associated mainly with repair of pyrimidine dimers,
and apparently are not closely involved in the BrUra photoproducts. The rate of excision of pyrimidine dimers from DNA is not
affected by BrUra incorporated in bacterial (Lion 1968) or mammalian cells (Cornelis 1978).

One might expect that cell strains deficient in various DNA repair systems might react to BrUra (or IUra) incorporation plus
ultraviolet light in much the same way as to ionizing radiations
or other strand-breaking agents.

III. Sensitization of Biological Effects to UV Light by BrUra or IUra Incorporation

Incorporation of either BrUra or IUra usually sensitizes DNA,
viruses, and cells to the effects of ultraviolet light. The effect increases with the extent of replacement of thymine by
BrUra, and depends on the wavelength of the irradiating light,
being much larger at 313 nm than at 240-280 nm.

In normal DNA, the photoproducts produced by ultraviolet light
involve thymine and cytosine - largely dimers, photoadducts,
etc. (review: Rahn and Patrick 1976); on the incorporation of
BrUra (and in all probability IUra) the predominant lesions become those described above. With this point in mind, many significant elements in the sensitization can be interpreted in a
reasonable way.

The quantitative measure of sensitization is complicated by the
fact that curves relating, e.g., colony-forming ability of cells
to UV fluence, can change shape when BrUra is incorporated. In
the following, we will discuss the effect of BrUra or IUra incorporation plus ultraviolet on colony-forming ability of cells,
plaque-forming ability of viruses, or the ability of DNA to transform bacterial cells. The quantitative measure of sensitization
will be the ratio of the fluence required to reduce the measured
parameter to 37% with unsubstituted DNA, to the fluence needed

for the same response when BrUra has been incorporated. Substitution of high ($\geq$ 50%) levels of BrUra gives sensitizations of the order of ten for λ = 240-280 nm, hundreds for λ = 313 nm (review: Hutchinson 1973).

This sensitization to 254 nm radiation is not caused by increased *numbers* of lesions. For example, 1 J/m^2 produces 65 pyrimidine dimers in a genome of *E. coli* (Rupp and Howard-Flanders 1968) as well as smaller numbers of other photoproducts. Substitution of 22% BrUra will decrease the number of dimers to perhaps 50 (see Lion 1968), but 1 J/m^2 will form about 25 single-strand breaks measured in alkali, and 0.25 double-strand breaks (see Tables 4 and 5). The colony-forming ability of wild-type cells will be sensitized by a factor of about 7 (Boyce and Setlow 1963).

The primary factor in the sensitization to 254 nm radiation is almost certainly the change in photoproducts to ones which are more difficult for wild-type cells to repair. Krasin and Hutchinson (1978) have determined that three to five DNA double-strand breaks, produced by UV irradiation of *E. coli* cells with incorporated BrUra, reduces colony-forming ability to 37%; in the same strain, six to eight double-strand breaks from gamma rays reduces colony-forming ability to the same level. Repair of DNA double-strand breaks does not seem to be very extensive, and the data are certainly consistent with the concept that a major cause for the loss of colony-forming ability is an unrepaired double-strand break. (See also the section below on the effect of BrUra incorporated in one and in two strands of double helical DNA.)

The large increase in sensitization at 313 nm by BrUra incorporation is readily explained. From Table 4, the cross section at 313 nm for production of BrUra photoproducts is two orders of magnitude less than at 254 nm, whereas various action spectra involving unsubstituted DNA show a decrease in effectiveness of three to four orders of magnitude (Setlow 1974). Thus, the sensitization produced by BrUra incorporation should be one to two orders of magnitude greater for 313 nm irradiation than for 254 nm, as is found.

An interesting case is the effect of BrUra incorporation in the single-strand DNA virus, ϕX174. For close to 100% replacement of thymine by BrUra, the plaque-forming ability of the virus is sensitized by a factor less than 2 for radiations of wavelength between 240 and 280 nm (Denhardt and Sinsheimer 1965). This presumably represents a small increase after BrUra incorporation in numbers of UV-induced lesions in the single-strand DNA for which there is little or no repair. At 302 nm, the sensitization increases to 10, reflecting a relatively larger number of BrUra photoproducts.

We have seen in the preceding section that the presence of -SH groups greatly reduces the number of all photoproducts from BrUra (and those from IUra) except the formation of uracil. T1 phage with IUra are actually *less* affected by 254 nm radiation in the presence of cysteamine than phage with normal DNA when both are

assayed on *uvrA*⁻ cells (Rupp and Prusoff 1964); equivalent observations have been made on λ phage with BrUra (Radman et al.
1969a,b). This shows that uracil introduced in the DNA (some of
it probably excised) does not markedly affect plaque-forming
ability; the presence of the halouracil would also reduce the
formation of pyrimidine dimers (Lion 1968), lesions whose effects would be greatest in *uvr*⁻ cells which cannot excise them.

The considerable sensitization of DNA by incorporation of BrUra
(and IUra) seems to be general, with the exception of the two
special circumstances just mentioned (single-strand viruses,
irradiation of DNA in the presence of high -SH concentrations).
In addition to the cases previously reviewed (Hutchinson 1973),
sensitization to ultraviolet is also reported for: ability to
replicate for T4 phage (Byrd et al. 1975), for chlorella (Nečas
1976; the negative result reported by Nečas 1974, might be caused
by failure to incorporate BrUra), and algal virus (Singh 1975);
induction of transformed mammalian cells (Fogel 1972; Teich et
al. 1973; Barrett et al. 1975); increased numbers of mutations
in Neurospora (Nemcrofsky 1975) and *E. coli* (Kondratev and Skavronskaya 1971).

A particularly interesting case is the great sensitization by
BrUra and IUra of ultraviolet-induced chromosomal aberrations in
mammalian cells (Chu 1965; Puck and Kao 1967; Pontecorvo 1971;
Bender et al. 1973; Ikushima and Wolff 1974; Thiron 1976) and
in plant cells (Verma et al. 1977). Bender et al. (1973) suggested that the observed aberration would best be explained
by a DNA double-strand break made by a single-strand nuclease
acting on the strand opposite to a radiation-induced single-
strand break; it now seems more likely that a double-strand
break directly formed by irradiation might be responsible.

IV. BrUra Incorporated in One Strand and in Two

By appropriate control of the time of BrUra exposure, it is possible to incorporate the analog either in one strand of a DNA
double helix, or in both strands. By the discussion above, uracil, and 99% of the single-strand breaks and the alkali-labile
bonds, will be only in the strand with BrUra. Double-strand
breaks will, of course, affect both strands. Since the mechanisms of DNA-protein crosslinking is undescribed, it is not
known if this would be limited to the strand with BrUra, although there would seem to be a possibility that this is so.

E. coli cells with 1-20% BrUra in one strand only of the DNA
double helix were only a little less sensitive to 313 nm light
than cells with the same amount of BrUra uniformly distributed
in the DNA (Krasin and Hutchinson 1978a). However, Chinese
hamster cells with >50% BrUra in one strand were very much
less sensitive than those with similar BrUra levels in both
strands (Roufa 1976). Two factors probably explain the difference. Hagan and Elkind (1979) have shown that when ultraviolet
irradiation is soon after a pulse of BrUra incorporation in one
strand, as in Roufa's experiments, considerable repair can take

place before DNA synthesis occurs on the template containing the analog; the effects of the lesions are thus reduced from those in cultures grown several generations in BrUra, where some cells will be in DNA synthesis at the time of irradiation and thus will be strongly affected. Also, at high levels of BrUra in both strands in Roufa's experiments, photochemical events involving nearby portions of both strands of the double helix could form lesions such as double-strand breaks. The number of such lesions will decrease with the inverse square of the degree of BrUra substitution, so the contribution from this source could be negligible in the *E. coli* experiments.

Fox and Meselson (1963) made lambda phage containing BrUra in only one strand. These phages were irradiated with long wavelength ultraviolet in 5% AET (2-aminoethylisothiouronium bromide, a compound which decomposes in solution to form -SH groups). Half of these phages were inactivated reasonably rapidly, and the other half were much more resistant. One interpretation of their result was that one specific strand is required for lambda DNA replication. However, Ihler (1975) incorporated BrUra specifically either into the *r* strand or the *l* strand of lambda DNA, and showed that the two kinds of phage were equally sensitive to long wavelength ultraviolet when irradiated in solution in the absence of -SH compounds. Under these conditions, neither phage showed a resistant fraction at any level close to 50%. It could be that the Fox and Meselson result is the consequence of high -SH concentrations which suppressed processes leading to damage to both strands (i.e., double-strand breaks), which could complicate the interpretation of Ihler's experiments; alternatively, there could be some unknown artifact in one of the experiments.

E. Uses of the Photochemistry of Bromouracil in Biology

The purpose of this section is to illustrate some ways in which the photochemistry of BrUra has been exploited by molecular biologists to gain information about biological processes.

I. Sensitization to Ultraviolet Light by BrUra Incorporation of Processes Which Depend on DNA

We have seen in Sect. D that DNA is sensitized to ultraviolet light by incorporation of BrUra. (Two exceptions are single-strand DNA, and DNA irradiated in the presence of sufficiently high concentrations of -SH compounds and assayed under conditions in which pyrimidine dimers are not excised.) This sensitization is particularly large for 313 nm radiation, of which significant quantities are emitted by ordinary fluorescent lights. This sensitization has been exploited on many ways.

Some examples of early experiments are given in the earlier review (Hutchinson 1973). Some more recent uses are mentioned here.

Köhnlein (1974) was able to separate (in equilibrium density gradients) B. subtilis transforming DNA with BrUra in one specific strand of the double helix from DNA with BrUra in the other specific strand. When the activity of the BrUra strands was destroyed with 313 nm light, the transforming activity of the remaining strand in each of the two preparations was equal, showing that both DNA strands can transform. Matsumoto et al. (1974) mapped various markers in B. subtilis grown in BrUra by observing the order in which the transforming activity of various markers became sensitized to ultraviolet. Kimball and Setlow (1974) showed that the number of mutations found in Haemophilus influenzae transforming DNA after treatment of the cells with nitrosoguanidine was about the same when cells were incubated in media with dThd and with BrdUrd. This suggests that extensive DNA replication is not required to "fix" such mutations.

Laemmli et al. (1974) used the sensitization of phage DNA by BrUra to show that DNA synthesized after the activation of specific genes was packaged into preformed heads. Since incorporation of BrUra sensitised the growth of F-pili in E. coli cells to ultraviolet, Fives-Taylor and Novotny (1974) suggested DNA as the sensitive target. When BrUra uptake and DNA synthesis in lymphocytes was stimulated by one antigen, Lazda and Baram (1974) found that the stimulation by a second antigen was sensitized to long wavelength ultraviolet, showing that the same subpopulation of cells was responding to both antigens.

An interesting use of sensitization by BrUra was by Newman and Kubitschek (1978). E. coli cells were exposed to BrUra for a short time, then incubated in growth medium without BrUra. The sensitivity of these cells to 313 nm light showed a cyclic variation with time after the BrUra pulse, which clearly defined the time required for the DNA replication fork to move around the circular E. coli genome.

Puck and Kao (1967) introduced a particularly useful technique for greatly enriching a culture for a specific mutation. A population of cells is grown in a medium containing BrdUrd, the medium being designed to permit DNA synthesis except for cells which have the desired mutation. After prolonged incubation, the culture is irradiated with long wavelength ultraviolet to kill off the cells incorporating BrUra, leaving a resistant population which will include those cells with the desired mutation. Some uses of this technique include the isolation of mutants in mammalian cells (Roufa et al. 1973), in Physarum polycephalum (Haugli and Dove 1972), and in Dictyostelium discoidium (Kessin et al. 1974).

Povirk (1977b) showed that initiation of DNA replicons was inhibited by 313 nm irradiation in mammalian cell DNA containing BrUra, but not in DNA within the same cell that did not contain the analog and therefore was not damaged. Lansman and Clayton (1975) took advantage of the fact that mammalian cells which lack a thymidine kinase for chromosomal DNA will have a mitochondrial-specific thymidine kinase. Since these cells will selectively incorporate exogenous BrdUrd in mitochondrial DNA, long wavelength ultraviolet will selectively break the covalently closed circular mitochondrial DNA. Kao et al. (1973) could follow the fate of homologous and heterologous DNAs incorporated in skin fibroblasts by observing the fragmenting effect of ultraviolet light, as well as the buoyant density, for DNAs in the recipient cells.

II. Light-Induced Crosslinking Between BrUra-DNA and Proteins

A significant new application of the photochemistry of BrUra has been its use in forming light-induced crosslinks between DNA which contains BrUra and proteins. A notable early experiment was that of Lin and Riggs (1974), who were able to form covalent bonds between the *lac* repressor protein molecules and the specific DNA sequence in the *E. coli* genome to which the repressor binds. Weintraub (1973) had previously shown covalent crosslinking on irradiation of BrUra-DNA with attached histones.

Crosslinking of proteins also occurs with DNA which does not contain BrUra (Weintraub 1973; Lin and Riggs 1974; Smith 1976), but the magnitude of the effect becomes large enough for a useful tool with BrUra in the DNA. The specificity of crosslinking to BrUra-DNA, compared to crosslinking with normal DNA, is greatly improved by using 313 nm light (Weintraub 1973). As explained in Sect. C of this review, the details of the crosslinking mechanism are not yet clear.

Ogata and Gilbert (1977) have recently determined in detail which BrUra residues in the *lac* operator have their photochemistry altered by the presence of *lac* repressor. Because of the current interest in DNA-protein interactions, it may be anticipated that this technique will be much used in the near future.

A special case involves photochemical reactions between halouracils and enzymes which form complexes with these thymine analogs. IdUrd sensitizes the enzyme thymidine kinase to ultraviolet light (Voytek et al. 1972), and can be used as a photoaffinity label (Cysyk and Prusoff 1972). For such purposes, IdUrd is better than BrdUrd (Chen and Prusoff 1972), a feature perhaps related to the higher yields of photoproducts for IdUrd irradiated in solution (see Sect. C).

III. Measurement of Repair Replication

Regan et al. (1971) introduced an ingenious way of detecting a low level of DNA repair. After cells had been irradiated with ultraviolet light, one fraction was incubated for repair replication in medium containing BrdUrd, another fraction with dThd. Each fraction was then irradiated with 313 nm light, and the cell lysates sedimented in alkaline gradients to measure the DNA single-strand length. DNA containing patches of BrdUrd incorporated during repair replication is photolysed to shorter pieces than is the DNA from cells incubated with dThd. Detection of a break in a DNA strand of $6 \cdot 10^8$ daltons is equivalent to measuring the repair of one lesion per $2 \cdot 10^6$ bases. Further, the size of the repair "patch" may be estimated from the 313 nm fluence required to photolyse half the patches opened at limiting high levels, and the cross sections per BrUra for strand breakage, measured under the same conditions used to photolyse the patches.

This assay has been used to show that some agents acting on mammalian cells (X-rays, ethyl methanesulfonate, methyl methanesulfonate) produce lesions in which repair synthesis introduces only three to four nucleotides, whereas other agents (ultraviolet light, N-acetoxy-2-acetylaminofluorene) cause lesions whose repair causes the insertion of 100 bases (Regan and Setlow 1974, 1976).

When cells which have been irradiated with ultraviolet light are incubated, newly synthesized single DNA strands have average lengths which are roughly equal to the average distance between radiation-produced dimers; this is interpreted as gaps in the DNA caused by the inability of the polymerase to synthesize past a photoproduct (Rupp and Howard-Flanders 1968). On further incubation, these gaps are filled, a process referred to as "post-replication repair". Use of the BrUra-photolysis method suggests that filling these gaps involves the insertion of the order of 10^3 nucelotides in mammalian cells (Buhl et al. 1972), and 10^4 nucleotides in E. coli cells (Ley 1973). The involvement of "gap filling" in postreplication repair has also been indicated by the BrUra-photolysis method for X-irradiated cells (Korner and Malz 1975) and for cells treated with alkylating agents (Fujiwara 1975).

When the method is used, allowance must be made for breaks which will be produced by 313 nm photolysis in strands which do not themselves contain BrUra, but on which a BrUra-strand has been synthesized (see Table 3). Under some conditions, such breaks can mask those introduced by photolysis of BrUra patches (Beattie 1972).

F. Differential Staining of Sister Chromatids Containing BrUra

Latt (1973) observed that in mammalian cells which had been grown for two generations with BrdUrd, some chromatids fluoresced more strongly after staining with Hoechst 33258 dye than did others. The strongly fluorescent chromatids were those containing hybrid DNA with thymine in one strand and BrUra in the other, and the weakly fluorescent ones had BrUra in both DNA strands. This provides a method of studying physical recombination between sister chromatids which is capable of visualizing smaller pieces of exchanged material than the use of [^{3}H]-thymidine and radio-autography. Latt and Wohlleb (1975) showed that incorporation of BrUra in DNA or chromatin increases dye binding, but decreases fluorescence. Their data suggest increased intersystem crossing from the fluorescent singlet state of the dye to a non-radiating triplet state because of the presence of the heavy bromine atom. Similar selective staining has also been observed with acridine orange (Kato 1974; Dutrillaux et al. 1974; Perry and Wolff 1974).

This method has the disadvantage that the fluorescence observed with either dye decreases rather rapidly with exposure to the exciting light beam. It was therefore an advance when Perry and Wolff (1974) observed that when preparations were treated with Hoechst 33258 and heated, the strongly fluorescing chromatids stained darkly with Giemsa whereas the weakly fluorescing chromatids stained much less intensely (see also Kihlman and Kronborg 1975). Indeed, Korenberg and Freedlender (1974) found differential Giemsa staining with heating to 88°C without pretreatment with fluorescent dyes.

The method most commonly used now involves treating with fluorescent dyes such as Hoechst 33258 or acridine orange, then before staining with Giemsa, irradiating with sunlight (Goto et al. 1975; Sugiyama et al. 1976) or near ultraviolet light (Lambert et al. 1976; Pera and Mattias 1976; Scheid 1976; Schwartzman and Cortes 1977).

It is hypothesized that irradiation before Giemsa staining might sufficiently depolymerize DNA with both strands incorporating BrUra so as to minimize staining. From the data in Table 4 of this review, one can estimate 0.1-0.9 breaks per BrUra base in the DNA at the ultraviolet exposures used by Pera and Mattias (1976) and by Scheid (1976). This should cause extensive disintegration in DNA with BrUra in both strands. Chromatin with both DNA strands containing BrUra was broken more drastically by 313 nm light than when one strand was still unsubstituted (Taichman and Freedlander 1976). Mouse chromosomes containing BrUra were readily shattered by light from fluorescent lamps, under conditions where unsubstituted human chromosomes in the same man-mouse fused cell were unaffected (Thiron 1976).

Fluorescent dyes applied before irradiation would increase the number of breaks; for example, acridine orange plus visible light is known to cause breaks in normal DNA (Freifelder et al. 1961). No direct data exist to show whether the fluorescent dye increases the differential staining, or acts by giving differential staining at a lower level of irradiation (see Sugiyama et al. 1976). However, Stetten et al. (1977) have shown that pretreatment of cells with Hoechst 33258 increases the differential effect of near ultraviolet light on loss of colony-forming ability between cells with BrUra in the DNA and those without. Rosenstein et al. (1980) have shown that the dye increases the differential effect of near ultraviolet in forming strand breaks in DNA with and without BrUra. The presence of compounds such as cysteamine or KI seems to inhibit differential staining (Scheid and Traupe 1977).

Hoechst 33258 and 360 nm irradiation produced Ura in the DNA of *E. coli* cells, and in BrdUrd in solution; in the latter case, smaller quantities of an unidentified acid-labile product were also formed (Ben-Hur et al. 1978).

Chromatids treated with Hoechst 33258 and sunlight so as to
stain weakly with Giemsa have also a much reduced reaction to
the Feulgen stain for DNA (Sugiyama et al. 1976). Chromosomes
containing BrUra show a more pronounced "paling", as observed
in the light microscope after treatment with ethidium bromide
plus visible light, than do normal chromosomes (Berns et al.
1976). They also stain less well by the Feulgen reaction.
Ethidium bromide plus visible light is known to produce strand
breaks in DNA (Martens and Clayton 1977).

Conceivably, the heating used by some before Giemsa staining
is required to produce breaks in DNA containing BrUra by the
process described by Grigg (1977).

Finally, it must be emphasized that incorporated BrUra leads
to an increased frequency of recombinational events without
irradiation in mammalian (Wolff and Perry 1974; Mazrimas and
Stetka 1978) and plant cells (Kihlman and Kronborg 1975) and
in lambda phage (Little 1976).

G. Incorporation of Halouracils in DNA

Organisms synthesize the pyrimidine nucleotides by the pathways
which are shown in simplified form in Fig. 2. BrUra and IUra can
be incorporated in cellular DNA when the various derivatives of
the analogs can compete with the pools of the thymine derivatives
shown in Fig. 2.

To compete, the analogs must first enter the cell. The entry of
thymine and thymidine is known to involve an active transport
system in Chinese hamster cells (Breslow and Goldsby 1969) and
in _E. coli_ cells (McKeown et al. 1976). Presumably, this would
also be true for the analogs.

The extent of thymine replacement in the DNA depends on the molar
ratio of 5-halodeoxyuridylic acid to thymidylic acid available to
the polymerases. The amount of incorporated analog may be increased in the following ways:

1. Use of mutants blocked in some step of the synthesis of thymidine triphosphate (i.e., thymine or thymidine requiring). Note
   that yeasts such as _Saccharomyces cerevisiae_ lack thymidine kinase
   and thus must be supplied with the 5'-nucleotide analog (Brendel and Haynes 1973).
2. Selective inhibition of thymidylate synthetase by fluorodeoxyuridylic acid (circled number 1 in Fig. 2), usually done by
   adding FUra or FdUrd (Cohen et al. 1958). This method is widely used in mammalian cells and even in plants (e.g., Kihlman
   and Kronborg 1975). This inhibition is made more effective by
   the addition of uracil or uridine to suppress catabolism of
   fluorodeoxyuridylic acid (Szybalski 1962).
3. Inhibition of the methylation of pdUrd (circled number 1 in
   Fig. 2) by addition of folic acid antagonists, such as amin-
   opterin or sulfanilamide (Zamenhof et al. 1958).

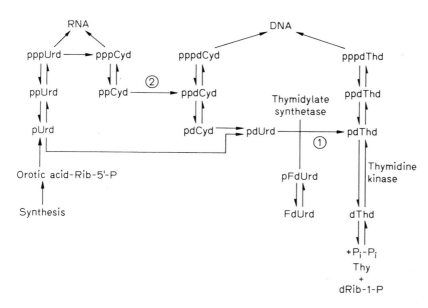

Fig. 2. A simplified diagram of the metabolic and catabolic pathways involving the pyrimidines. Inhibition of thymidylic acid synthesis (circle 1) and of the conversion of ribonucleotide to deoxyribonucleotide (circle 2) are the two chief methods of enhancing the incorporation of halouracils (adapted from Szybalski 1974)

4. Feedback inhibition of the conversion of ppCyd to ppdCyd (circled number 2 in Fig. 2) (Szybalski 1962).

Halouracil incorporation by (2) and (3) above may be improved by addition of dCyd, since halouracil triphosphates inhibit conversion of ppCyd to ppdCyd (Reichert et al. 1960; Meuth and Green 1974). Addition of dCyd can increase the supply of pppdCyd needed for DNA synthesis; it could also tend to increase thymidine concentrations even in the presence of inhibitors of step (1) (Fig. 2), which might explain Horn and Davidson's observations (1976) that adding dCyd decreased BrdUrd incorporation.

Incorporation of the halouracils in mammals (not usually of importance in photochemical research) is reviewed by Szybalski (1974).

I. Measurement of Incorporation

For cells which cannot make their own thymidine, the simplest method is to use a growth medium with a known ratio of BrdUrd/dThd (or the bases), each carrying a different radioisotope label. The ratio of the disintegrations per minute of the two isotopes in a sample of purified DNA, compared to that in the growth medium, gives the BrUra/Thy ratio in the DNA (Rydberg 1977a). A similar method uses [^{32}P]-phosphate and labeled thymine (Lion and Doerner 1972).

Bromouracil incorporation increases the density at which DNA bands in a neutral CsCl density gradient. According to Luk and Bick (1977),

$$\% \text{ BrUra} = 100 \, \frac{(\text{density})_{\text{sub}} - (\text{density})_{\text{unsub}}}{0.1730 \cdot (\% \text{ thymine in DNA}/50)}$$

where $(\text{density})_{\text{sub}}$ and $(\text{density})_{\text{unsub}}$ are the banding densities in neutral CsCl for DNA with BrUra incorporated and for homologous DNA without bromouracil.

The extent of BrUra incorporation can always be measured by hydrolysis and chromatography. Antibodies with reasonable specificity for BrUra bases can be made (Gratzner et al. 1975; Zamchuk and Braude 1975), and qualitative estimates of BrUra incorporation are possible from the sensitization of some biological property of DNA to ultraviolet light.

II. Discrimination in BrUra Incorporation

When bacteria which depend on exogenous thymine for growth are incubated in media with various ratios of BrUra/thymine (or nucleosides), the ratio BrUra/thymine in the DNA can range from about that in the medium to as little as 1/15 that ratio (review: Rydberg 1977b).

In *E. coli*, Rydberg (1977b) has shown that various levels of this discrimination can be correlated with various mutations affecting enzymes involved in the catabolism of thymine, mutants which have the property that they can grow on low levels of exogenous thymidine.

$$\text{dThd} + \text{P}_i \xrightarrow{} \text{dRib-1-P} \xrightarrow[\text{(drm)}]{\text{deoxyribo-mutase}} \text{dRib-5-P} \xrightarrow{\text{deoxyribo-aldolase}} \begin{array}{c} \text{glyceralde-} \\ \text{hyde-3-P} \\ + \\ \text{acetaldehyde} \end{array}$$

with a "+ Thy" below dRib-1-P.

A *dra*$^-$ mutant, deficient in the enzyme deoxyriboaldolase, discriminates strongly against BrUra incorporation. A *drm*$^-$ mutant discriminates very little, with BrUra/Thy ratios in the DNA equal to 0.5-1 those in the growth medium. *Dra*$^-$*drm*$^-$ double mutants incorporate BrUra only a little less readily than do *drm*$^-$ mutants. These results, and the dependence of bromouracil incorporation on thymidine concentration, are consistent with the idea that high concentrations of dRib-5-P inhibit BrUra incorporation. While dRib-5-P is known to be an efficient inducer of enzymes involved in the degradation of deoxy sugars, it is not known how the incorporation of BrUra is affected.

Kanner and Hanawalt (1968) noted that repair replication discriminated against BrUra more than did normal DNA synthesis, suggesting discrimination by the DNA polymerases. However, Bick (1977) finds that pppBrUrd and pppdThd are equally well utilized by the *E. coli* polymerase I in copying poly d(A-T) in solution.

III. Effects of High Levels of BrUra in DNA

It has been found in many laboratories that the growth of bacterial or mammalian cells is greatly inhibited under conditions in which most of the thymine in the DNA has been replaced by BrUra. The reasons for this are not known, but are probably related to some of the non-photochemical effects of BrUra incorporation detailed in the last section of this review. It may be noted that numerous workers have been able to get high levels of BrUra substitution in viruses. In fact, reasonable substitution of IUra has been obtained in bacteriophage (e.g., Rupp and Prusoff 1964; Byrd et al. 1975).

Cell strains have been reported which will grow for many generations with nearly all thymine replaced by BrUra for: *Escherichia coli* (Hewitt et al. 1967): *Bacillus subtilis* (Bishop and Sueoka 1972); Syrian hamster cells (Davidson and Bick 1973; Bick and Davidson 1974). In fact, this last cell line *requires* some BrUra to grow. The properties of these strains which make them viable at high BrUra substitution are not known.

H. Relationship Between Other Biological Effects of the Halouracils and Photochemical Effects

The incorporation of BrUra and IUra in DNA produces a number of biological effects in addition to those already discussed. This section points out what connections may exist between these effects and the photochemical processes covered in this review.

I. Sensitization of DNA to Ionizing Radiations

Incorporation of BrUra sensitizes DNA to ionizing radiations (review: Szybalski 1974). Typically, a high degree of incorporation reduces the dose needed to produce a given effect by perhaps a factor of two, compared to an order of magnitude change in response to 254 nm ultraviolet light.

On the basis of pulse and steady state radiolysis studies on BrUra and BrUra-DNA, Adams (1967) and Zimbrick et al. (1969a,b) suggested that the hydrated electron e^-_{aq} produced by the action of ionizing radiations reacted with Br of BrUra.

$$BrUra + e^-_{aq} \longrightarrow Br^- + Ura\cdot$$

The evidence for this step is very strong (see also: Reuschl 1966; Gilbert et al. 1967; Danziger et al. 1968; Langmuir and Hayon 1969; Ward and Kuo 1970; Kourim et al. 1971; Adams and Willson 1972; Bansel et al. 1972). The uracilyl radical thus produced might then react with the nearby sugar (Fig. 1) to produce a single-strand break or an alkali-labile bond. Thus, the incorporation of BrUra might result in strand breaks caused by the hydrated electron.

Some workers have reported an increase with BrUra incorporation in the yield produced by ionizing radiations of DNA single-strand breaks (measured in alkali) and/or double-strand breaks (Kaplan 1966; Sawada and Okada 1972; Bonura and Smith 1977). However, certain experiments, which by their design would be expected to provide the most definitive evidence, have shown *no* increase in breaks with BrUra incorporation (Freifelder and Freifelder 1966; Lett et al. 1970).

Rahn and Stafford (1979) formed radicals in the 5-pyrimidine position by ultraviolet irradiation of DNA containing 5-ICyt. Only 0.17 breaks (in alkali) were produced per I dissociated. Clearly such radicals, which probably act similarly to those formed by electron capture by Br, only occasionally extract H atoms from the adjacent sugars and cause breaks. This contrasts with photolysis of DNA with BrUra, in which most Br atoms are released by the intervention of a nearby H-donor (see Sect. C).

II. Lesions in Unirradiated BrUra-DNA

There are extensive data showing that DNA containing BrUra has a lower average single-strand molecular weight when extracted from cells than does DNA extracted from the same cells grown in thymine (Szybalski and Opara-Kubinska 1965; Baker and Case 1974; Grivell et al. 1975; Pietrzykowska et al. 1975; Pietrzykowska and Krych 1977). In addition, cellular DNA which contains BrUra will incorporate new (radioactively labeled) bases in short stretches (presumably repair replication) at two- to threefold greater rate than will unsubstituted DNA (Grivell et al. 1975). Exposure of BrUra-DNA to fluorescent lights could give rise to these effects (Stahl et al. 1961), but it is reasonably certain that in the cases cited there were other factors. Possibilities include the following:

1. Halogen substitution at the 5 position increases the likelihood of thermal depyrimidination (Grigg 1977) by opening the N-glycosylic bond (Olafsson and Bryan 1974).
2. There are enzymes in *E. coli* cells which correct mismatched base pairs which apparently occur during BrUra incorporation in DNA (Rydberg 1977a). The action of these enzymes presumably includes incision in one strand near the mismatch giving rise to a single-strand break.
3. Compounds containing -SH groups debrominate BrUra in solution (Sedor and Sander 1973; review: Bradshaw and Hutchinson 1977; Wataga et al. 1973; Chikuma et al. 1978). If -SH compounds in cells did this to BrUra incorporated in DNA, single-strand

breaks could be produced either as a consequence of the reaction, or as part of the excision of the uracil from the DNA
by glycosylases (see the section on repair of BrUra lesions).

III. Mutagenesis

BrUra is a potent mutagen without light (review: Drake 1970).
In addition, the effect is enhanced by ultraviolet irradiation
(e.g., Kondratev and Skavronskaya 1971). In the dark, BrUra
probably mutagenizes by promoting mispairing of bases during
DNA synthesis, although there are some awkward bits of information which do not fit into a simple model (Drake 1970). The
possibility that mutation is caused by the lesions in unirradiated BrUra-DNA (see preceding section) is unlikely for lambda
phage; mutagenesis by BrUra incorporation in the phage is not
increased by ultraviolet irradiation of the host cell before
adsorption (Hutchinson and Stein 1977), a treatment which increases the observed mutagenic effect of most physical and
chemical agents on lambda phage (review: Witkin 1976).

IV. Effects on Development

The effects of the halouracils on the development of embryos
are striking (reviews: Rutter et al. 1975; Skalko and Packard
1975), but, by the conditions of the experiments, are not caused
by photochemical events. It is not known how the effects are
produced. Possibilities include the lesions in unirradiated DNA
(see above), the increased binding of proteins to DNA (Lin and
Riggs 1972; Rutter et al. 1975), and possible changes in methylation patterns induced by the presence of BrUra (Holliday and
Pugh 1975; but see the negative results reported by Singer et
al. 1977).

It is possible that similar processes are responsible for induction by BrUra and IUra of tumor viruses (e.g., Teich et al.
1973; Besmer et al. 1975).

V. Clinical Use of IUra Derivatives in Viral Infections

IUra derivatives are used in clinical medicine to control certain
surface viral infections (review: Prusoff and Goz 1975). The effect is, with reasonable certainty, independent of photochemical
reactions, and may be related to the loss of plaque-forming
ability without ultraviolet exposure for, e.g., T4 phage (Byrd
et al. 1975). An augmentation of the effects of IdUrd by near
ultraviolet exposure has been reported (Thiel and Wacker 1962),
but has not been found useful in practice.

Acknowledgments. The authors thank a number of individuals for discussion,
communication of results before publication, and comments: Dr. Frank Krasin,
Dr. M. Lion, Prof. D. Schulte-Frohlinde, Dr. Herbert Kubitschek, Prof. John
Ward, Mr. M. Hagan, Dr. M. M. Elkind, Prof. W. H. Prusoff, and Prof. S. Wolff.

The preparation of this review was supported in part by contract from the U.S. Department of Energy and by a grant from the Deutsche Forschungsgemeinschaft. During the preliminary planning of this review, one of us (W.K.) was the recipient of a travel grant from the Deutsche Forschungsgemeinschaft.

References

Adams, G.E.: The general application of pulse radiolysis to current problems in radiobiology. Curr. Top. Radiat. Res. 3, 35 (1967)

Adams, G.E., Willson, R.L.: Mechanism of 5-BrUdr sensitization. Pulse radiolysis study of one electron transfer in nucleic acid derivatives. Int. J. Radiat. Biol. 22, 589 (1972)

Augenlicht, L., Nicolini, C., Baserga, R.: Circular dichroism and thermal denaturation studies of chromatin and DNA from BrdU-treated mouse fibroblasts. Biochem. Biophys. Res. Commun. 59, 920 (1974)

Baker, R.F., Case, S.T.: Effect of 5-bromodeoxyuridine on the size distribution of DNAs isolated from sea urchin embryos. Nature (London) 249, 350 (1974)

Bansel, K.M., Patterson, L.K., Schuler, R.H.: The production of halide ion in the radiolysis of aqueous solutions of the 5-halouracils. J. Phys. Chem. 76, 2386 (1972)

Barrett, J.C., Schechtman, L., Ts'o, P.: An investigation of the DNA involvement in neoplastic transformation in vitro transformation of hamster fibroblasts induced by BrdU incorporation coupled with irradiation of near ultraviolet light. Abstr. 41 (15th Annu. Meet. Am. Soc. Cell Biol.) (1975)

Beattie, K.L.: Breakage of parental strand in *Haemophilus influenzae* by 313 nm radiation after replication in the presence of 5-bromodeoxyuridine. Biophys. J. 12, 1573 (1972)

Bender, M.A., Bedford, J.S., Mitchell, J.B.: Mechanisms of chromosomal aberration production. II. Aberrations induced by 5-bromodeoxyuridine and visible light. Mutat. Res. 20, 403 (1973)

Ben-Hur, E., Elkind, M.M.: Damage and repair of DNA in 5-BrdU labeled Chinese hamster cells exposed to fluorescent light. Biophys. J. 12, 636 (1972)

Ben-Hur, E., Prager, A., Riklis, E.: Photochemistry of the bisbenzimidazole dye 33258 Hoechst with bromodeoxyuridine and its biological effects on BrdUrd-substituted *E. coli*. Photochem. Photobiol. 27, 559 (1978)

Berens, K., Shugar, D.: Ultraviolet absorption spectra and structure of halogenated uracils and their glycosides. Acta Biochim. Pol. 10, 25 (1963)

Berns, M.W., Leonardson, K., Winter, M.: Laser microbeam irradiation of rat kangaroo cells (PTK$_2$) following selective sensitization with bromodeoxyuridine and ethidium bromide. J. Morphol. 149, 327 (1976)

Besmer, P., Smotkin, D., Haseltine, W., Fan, H., Wilson, A.T., Paskind, M., Weinberg, R., Baltimore, D.: Mechanism of induction of RNA tumor viruses by halogenated pyrimidines. Cold Spring. Harbor Symp. Quant. Biol. 39, 1103 (1975)

Bick, M.D.: A quantitative method for distinguishing BrdUTP and dTTP in soluble pools. Anal. Biochem. 78, 582 (1977)

Bick, M.D., Davidson, R.L.: Total substitution of bromodeoxyuridine for thymidine in the DNA of a bromodeoxyuridine dependent cell line. Proc. Natl. Acad. Sci. USA 71, 2082 (1974)

Bishop, R.J., Sueoka, N.: 5-bromouracil tolerant mutants of *Bacillus subtilis*. J. Bacteriol. 112, 870 (1972)

Bobst, A.M., Torrence, P.F., Kouidou, S., Witkop, B.: Dependence of inter-
feron induction on nucleic acid conformation. Proc. Natl. Acad. Sci. USA
73, 3788 (1976)

Bonura, T., Smith, K.C.: Sensitization of E. coli to gamma-radiation by
5-bromouracil incorporation. Int. J. Radiat. Biol. 32, 457 (1977)

Boyce, R.P.: Ultraviolet light inactivation of E. coli and bacteriophage
containing 5-bromouracil-substituted DNA. Ph.D. Thesis, Yale Univ. (1961)

Boyce, R.P., Setlow, R.B.: The action spectra for ultraviolet light inactivation of systems containing 5-bromouracil-substituted DNA. Biochim.
Biophys. Acta 68, 446 (1963)

Bradshaw, T.K., Hutchinson, D.W.: 5-substituted pyrimidine nucleosides and
nucelotides. Chem. Soc. Rev. 6, 43 (1977)

Brendel, M., Haynes, R.H.: Exogenous thymidine 5'-monophosphate as a precursor for DNA synthesis in yeast. Genetics 126, 337 (1973)

Breslow, R., Goldsby, R.: Isolation and characterization of thymidine transport mutants of Chinese hamster cells. Exp. Cell Res. 55, 339 (1969)

Buhl, S.N., Setlow, R.B., Regan, J.D.: Steps in DNA chain elongation and
joining after ultraviolet irradiation of human cells. Int. J. Radiat.
Biol. 22, 417 (1972)

Byrd, D.M., Goz, B., Prusoff, W.H.: Comparison of the lethal effect of
5-iodouracil incorporated into T4 ø in the presence and absence of near-
visible light. Photochem. Photobiol. 21, 407 (1975)

Campbell, J.M., Schulte-Frohlinde, D., von Sonntag, C.: Quantum yields in
the ultraviolet photolysis of 5-bromouracil in the presence of hydrogen
donors. Photochem. Photobiol. 20, 465 (1974)

Carrier, W.L., Setlow, R.B.: Ultraviolet sensitivity of DNA containing bromodeoxyuridine. VI Int. Congr. Photobiol., Bochum, Abstr. 96 (1972)

Chen, M.S., Prusoff, W.H.: Kinetic and photochemical studies and alteration
of ultraviolet sensitivity of E. coli thymidine kinase by halogenated
allosteric regulators and substrate analogs. Biochemistry 16, 3310 (1977)

Chikuma, T., Negishi, K., Hayatsu, H.: Formation of S-[5-(2'-deoxyuridyl)]
thiol compounds in the dehalogenation of 5-bromo and 5-iodo-2' deoxyuridine
with cysteine derivatives. Chem. Pharmaceut. Bull. 26, 1746 (1978)

Chu, E.H.Y.: Effects of ultraviolet radiations on mammalian cells. Mutat.
Res. 2, 75 (1965)

Cohen, S.S., Flaks, J.G., Barner, H.D., Loeb, M.R., Lichtenstein, J.: The
mode of action of 5-fluorouracil and its derivatives. Proc. Natl. Acad.
Sci. USA 44, 1004 (1958)

Cone, R., Duncan, J., Hamilton, L., Friedberg, E.C.: Partial purification
and characterization of a uracil DNA N-glycosidase from B. subtilis.
Biochemistry 16, 3194 (1977)

Cornelis, J.J.: The influence of inhibitors on dimer removal and repair of
single-strand breaks in normal and bromodeoxyuridine substituted DNA of
HeLa cells. Biochim. Biophys. Acta 521, 134 (1978)

Cysyk, R., Prusoff, W.H.: Alteration of ultraviolet sensitivity of thymidine
kinase by allosteric regulators, normal substrates and a photoaffinity
label, 5-iodo-2-deoxyuridine, a metabolic analog of thymidine. J. Biol.
Chem. 247, 2522 (1972)

Danziger, R.M., Hayon, E., Langmuir, M.E.: Pulse radiolysis and flash
photolysis study of aqueous solutions of simple pyrimidines, uracil and
bromouracil. J. Phys. Chem. 72, 3842 (1968)

Davidson, R.L., Bick, M.D.: Bromodeoxyuridine dependence - A new mutation
in mammalian cells. Proc. Natl. Acad. Sci. USA 70, 138 (1973)

Davies, D.R., Baldwin, R.L.: X-ray studies of two synthetic DNA copolymers.
J. Mol. Biol. 6, 251 (1963)

Denhardt, D.J., Sinsheimer, R.L.: The process of infection with bacteriophage øX174. VI. Inactivation of infected complexes by ultraviolet irradiation. J. Mol. Biol. 12, 674 (1965)

Dennis, W.S., Hutchinson, F.: Repair of single-strand breaks induced by ultraviolet light in *E. coli* DNA containing bromouracil. VI. Int. Congr. Photobiol., Bochum, Abstr. 108 (1972)

Dizdaroglu, M., Schulte-Frohlinde, D., von Sonntag, C.: γ-radiolyses of DNA in oxygenated aqueous solution. Structure of an alkali-labile site. Z. Naturforsch. 32c, 1021 (1977)

Dodson, M.L., Hewitt, R., Mandel, M.: Nature of ultraviolet light induced strand breakage in DNA containing bromouracil. Photochem. Photobiol. 16, 15 (1972)

Drake, J.W.: The Molecular Basis of Mutation. San Francisco: Holden-Day 1970

Duncan, J., Hamilton, L., Friedberg, E.C.: Degradation of uracil-containing DNA. II. Evidence for N-glycosidase and nuclease activities in unfractionated extracts of *B. subtilis*. J. Virol. 19, 338 (1976)

Dutrillaux, B., Fosse, A.M., Prieur, M., Jejeune, J.: Chromatid exchanges in human mitotic cells. BUDR treatment and bichromatic fluorescence by acridine orange. Chromosoma 48, 327 (1974)

Ehrlich, M., Riley, M.: Photolysis of polyribobromouridylic acid. Photochem. Photobiol. 16, 385 (1972a)

Ehrlich, M., Riley, M.: Oligonucleotide photoproducts formed by photolysis of polyribobromouridylic acid. Photochem. Photobiol. 16, 397 (1972b)

Ehrlich, M., Riley, M.: Effect of base sequence on the ultraviolet irradiation products of double-stranded polynucleotides containing bromouracil and adenine. Photochem. Photobiol. 20, 159 (1974)

Eisinger, J., Lamola, A.A.: Luminescence spectroscopy of nucleic acids. Methods Enzymol. 21, 24 (1971)

Fielden, E.M., Lillicrap, S.C., Robins, A.B.: The effects of 5-bromouracil on energy transfer in DNA and related model systems: DNA with incorporated 5-BUdR. Radiat. Res. 48, 421 (1971)

Fives-Taylor, P., Novotny, C.P.: Effect of thymine-5-bromouracil substitution on F pili. J. Bacteriol. 118, 175 (1974)

Fogel, M.: Induction of virus synthesis in polyoma transformed cells by DNA anti-metabolites and by irradiation after pretreatment with 5-bromodeoxyuridine. Virology 49, 12 (1972)

Fox, E., Meselson, M.: Unequal photosensitivity of the two strands of DNA in bacteriophage lambda. J. Mol. Biol. 7, 583 (1963)

Fox, J.J., Shugar, D.: Spectrophotometric studies of nucleic acid derivatives and related compounds as a function of pH. II. Natural and synthetic pyrimidine nucleosides. Biochim. Biophys. Acta 9, 369 (1952)

Freifelder, D., Freifelder, D.R.: Mechanism of X-ray sensitization of bacteriophage T7 by 5-bromouracil. Mutat. Res. 3, 177 (1966)

Freifelder, D., Davison, P.F., Guiduschek, E.P.: Damage by visible light to the acridine orange-DNA complex. Biophys. J. 1, 389 (1961)

Fujiwara, Y.: Postreplication repair of alkylation damage to DNA of mammalian cells in culture. Cancer Res. 35, 2780 (1975)

Gilbert, E., Cristallini, C.: Ultraviolet photolysis of 5-bromouracil in aqueous solution. Influence of oxygen and deoxy-D-ribose. Z. Naturforsch. 28B, 615 (1973)

Gilbert, E., Schulte-Frohlinde, D.: Photolysis of 5-iodouracil in aqueous oxygen saturated solution. Z. Naturforsch. 25B, 492 (1970)

Gilbert, E., Volkert, O., Schulte-Frohlinde, D.: Radiochemistry of aqueous oxygen containing solutions of 5-bromouracil. Identification of radiolysis products. Z. Naturforsch. 22b, 477 (1967)

Goto, K., Akematsu, T., Shimazu, H., Sugiyama, T.: Simple differential Giemsa staining of sister chromatids after treatment with photosensitive dyes and exposure to light and the mechanisms of staining. Chromosoma 53, 223 (1975)

Gratzner, H.G., Leif, R.C., Ingram, D.J., Castro, A.: The use of antibody specific for bromodeoxyuridine for the immunofluorescent determination of DNA replication in single cells and chromosomes. Exp. Cell Res. 95, 88 (1975)

Greer, S., Zamenhof, S.: Effect of 5-bromouracil in DNA of *E. coli* on sen-
    sitivity to ultraviolet irradiation. Abstr. Am. Chem. Soc. 131st Meet.
    p3C (1957)
Grigg, G.W.: Selective breakage of DNA alongside 5-bromodeoxyuridine nucleo-
    tide residues by high temperature hydrolysis. Nucleic Acids Res. **4**, 969
    (1977)
Grivell, A.R., Grivell, M.B., Hanawalt, P.C.: Turnover in bacterial DNA
    containing thymine or 5-bromouracil. J. Mol. Biol. **98**, 219 (1975)
Gueron, M., Eisinger, J., Lamola, A.A.: Excited states of nucleic acid
    bases. In: Principles in Nucleic Acid chemistry, Vol. I, p. 312. New
    York: Academic Press 1974
Guthrie, R.D.: Glycosans and anhydro sugars. In: The carbohydrates (eds.
    W. Pigman, D. Horton). Vol. I A. New York: Academic Press 1972
Hagan, M.P., Elkind, M.M.: Changes in repair competency after 5-bromo-
    deoxyuridine pulse labeling and near-ultraviolet light. Biophys. J. **27**,
    75 (1979)
Hanawalt, P.C., Setlow, R.B.: Molecular Mechanisms for Repair of DNA. Vol.
    A, B. New York: Plenum Press 1975
Haug, A.: Photochemical decomposition of TdBU. Z. Naturforsch. **19B**, 143
    (1964)
Haugli, F.B., Dove, W.F.: Mutagenesis and mutant selection in *Physarum
    polycephalum*. Mol. Gen. Genet. **118**, 109 (1972)
Hewitt, R., Marburger, K.: The photolability of DNA containing 5-bromo-
    uracil. I. Single-strand breaks and alkali-labile bonds. Photochem.
    Photobiol. **21**, 431 (1975)
Hewitt, R., Suit, J C., Billen, D.: Utilization of 5-bromouracil by thymine-
    less bacteria. J. Bacteriol. **93**, 86 (1967)
Holliday, R., Pugh, J.E.: DNA modification mechanisms and gene activity
    during development. Science **187**, 226 (1975)
Horn, D., Davidson, R.L.: Inhibition of biological effects of bromodeoxy-
    uridine by deoxycytidine-correlation with decreased incorporation of
    bromodeoxyuridine into DNA. Somat. Cell Genet. **2**, 469 (1976)
Hotz, G., Reuschl, H.: Damage to deoxyribose molecules and to U-gene re-
    activation in ultraviolet-irradiated 5-bromouracil DNA of phage T4. Mol.
    Gen. Genet. **99**, 5 (1967)
Hotz, G. Walser, R.: On the mechanism of radiosensitization by 5-bromo-
    uracil. The occurrence of DNA single-strand breaks in ultraviolet-ir-
    radiated phage T4 as influenced by cysteamine. Photochem. Photobiol. **12**,
    207 (1970)
Hotz, G., Mauser, R., Walser, R.: Infectious DNA from coliphage T1. III. The
    occurrence of single-strand breaks in stored, thermally treated, and
    ultraviolet irradiated molecules. Int. J. Radiat. Biol. **19**, 519 (1971)
Hurst, R.O., Kuksis, A.: Degradation of deoxyribonucleic acid by hot alkali.
    Can. J. Biochem. Physiol. **36**, 919 (1958)
Hutchinson, F.: The lesions produced by ultraviolet light in DNA containing
    5-bromouracil. Q. Rev. Biophys. **6**, 201 (1973)
Hutchinson, F., Hales, H.: Mechanism of the sensitization of bacterial
    transforming DNA to ultraviolet light by the incorporation of 5-bromo-
    uracil. J. Mol. Biol. **50**, 59 (1970)
Hutchinson, F., Stein, J.: Mutagenesis of lambda phage: 5-bromouracil and
    hydroxylamine. Mol. Gen. Genet. **152**, 29 (1977)
Ihler, G.: Preparation and photochemical properties of strand-specific
    5-bromouracil substituted lambda phage. Radiat. Res. **61**, 298 (1975)
Ikushima, T., Wolff, S.: Sister chromatid exchanges induced by light
    flashes to 5-bromodeoxyuridine and 5-iododeoxyuridine substituted Chinese
    hamster chromosomes. Exp. Cell Res. **87**, 15 (1974)

Incremona, J.H., Martin, J.C.: N-bromosuccinimide, mechanisms of allylic bromination and related reactions. J. Am. Chem. Soc. 92, 627 (1970)

Ishihara, H., Wang, S.Y.: Photochemistry of 5-bromouracils: Isolation of 5-5' diuracils. Nature (London) 210, 1222 (1966)

Kanner, L., Hanawalt, P.C.: Efficiency of utilization of thymine and 5-bromouracil for normal and repair DNA synthesis in bacteria. Biochim. Biophys. Acta 157, 532 (1968)

Kao, P.C., Regan, J.D., Volkin, E.: Fate of homologous and heterologous DNAs after incorporation into human skin fibroblasts. Biophys. Biochim. Acta 324, 1 (1973)

Kaplan, H.S.: DNA strand scission and loss of viability after X-irradiation of normal and sensitized bacterial cells. Proc. Natl. Acad. Sci. USA 55, 1442 (1966)

Kato, H.: Spontaneous sister chromatid exchanges detected by a BUdR labeling method. Nature (London) 251, 70 (1974)

Kessin, R.H., Williams, K.L., Newell, P.C.: Linkage analysis in *Dictyostelium discoidium* using temperature-sensitive growth mutants selected with bromo-deoxyuridine. J. Bacteriol. 119, 776 (1974)

Kihlman, B.A., Kronborg, D.: Sister chromatid exchanges in *Vicia faba*. Demonstration of a modified fluorescence plus Giemsa (FPG) technique. Chromosoma 51, 1 (1975)

Kimball, R.F., Setlow, J.K.: Mutation fixation in MNNG-treated *H. influenzae* as determined by transformation. Mutat. Res. 22, 1 (1974)

Kirtikar, D.M., Slaughter, J., Goldthwait, D.A.: Endonuclease II of *E. coli*: degradation of gamma-irradiated DNA. Biochemistry 14, 1235 (1975)

Köhnlein, W.: Transforming activity in both complementary strands of *B. subtilis* DNA. Z. Naturforsch. 29c, 63 (1974)

Köhnlein, W., Hutchinson, F.: ESR-studies of normal and 5-bromouracil-substituted DNA of *Bacillus subtilis* after irradiation with ultraviolet light. Radiat. Res. 39, 745 (1969)

Köhnlein, W., Mönkehaus, F.: Experimental evidence for intramolecular energy transfer in hybrid DNA of *B. subtilis* after irradiation with long wavelength Uv. Z. Naturforsch. 27b, 708 (1972)

Kondratev, Y.S., Skavronskaya, A.G.: The effect of 5-bromouracil on the sensitivity of Hcr^+ and Hcr^- bacteria to the lethal and mutagenic action of ultraviolet light. Sov. Genet. 7, 1218 (1971)

Korenberg, J.R., Freedlender, E.F.: Giemsa technique for the detection of sister chromatid exchanges. Chromosoma 48, 355 (1974)

Korner, I., Malz, W.: Postreplication gap filling in the DNA of X-ray damaged Chinese hamster cells. Stud. Biophys. 51, 115 (1975)

Kourim, P., Bors, W., Schulte-Frohlinde, D.: Gamma radiolysis of aqueous solutions of 5-bromo-2-deoxyuridine in the presence of oxygen. Z. Naturforsch. 26b, 308 (1971)

Krasin, F., Hutchinson, F.: Repair of DNA double-strand breaks in *E. coli*, which requires *recA* function and the presence of a duplicate genome. J. Mol. Biol. 116, 81 (1977)

Krasin, F., Hutchinson, F.: Double-strand breaks from single photochemical events in DNA containing 5-bromouracil. Biophys. J. 24, 645 (1978a)

Krasin, F., Hutchinson, F.: Strand breaks and alkali-labile bonds induced by ultraviolet light in DNA with 5-bromouracil in vivo. Biophys. J. 24, 657 (1978b).

Laemmli, U.K., Teaff, N., D'Ambrosia, J.: Maturation of the head of bacteriophage T4. III. DNA packaging into preformed heads. J. Mol. Biol. 88, 749 (1974)

Lambert, B., Harrison, K., Lindsten, J., Sten, M., Werelius, B.: Bromodeoxy-uridine induced sister chromatid exchanges in human lymphocytes. Heredi-tas 83, 163 (1976)

Langmuir, M.E., Hayon, E.: Transient species produced in the photochemistry of 5-bromouracil and its N-methyl derivatives. J. Chem. Phys. 51, 4893 (1969)

Langridge, R., Marvin, D.A., Seeds, W.E., Wilson, H.R., Hooper, C.W., Wilkins, M.H.F., Hamilton, L.D.: The molecular configuration of deoxy-ribonucleic acid. II. Molecular models and their Fourier transforms. J. Mol. Biol. 2, 38 (1960)

Lansman, R.A., Clayton, D.A.: Selective nicking of mammalian mitochondrial DNA in vivo: Photosensitization by incorporation of 5-bromodeoxyuridine. J. Mol. Biol. 99, 761 (1975)

Lapeyre, J.-N., Bekhor, I.: Effect of 5-Bromo 2' deoxyuridine and dimethyl sulfoxide on properties and structure of chomatin. J. Mol. Biol. 89, 137 (1974)

Latt, S.A.,: Microfluorometric detection of DNA replication in human meta-phase chromosomes. Proc. Natl. Acad. Sci. USA 70, 3395 (1973)

Latt, S.A., Wohlleb, J.C.: Optical studies of the interaction of 33258 Hoechst with DNA, chromatin and metaphase chromosomes. Chromosoma 52, 297 (1975)

Lazda, V.A., Baram, P.: Participation of different cell populations in antigen- and mitogen-induced lymphocyte proliferation. J. Immunol. 112, 1705 (1974)

Lehmann, A.R.: Postreplication repair of DNA in ultraviolet-irradiated mammalian cells. J. Mol. Biol. 66, 319 (1972)

Lett, J.T., Caldwell, I., Little, J.G.: Repair of X-ray damage to the DNA in *Micrococcus radiodurans:* The effect of 5-bromodeoxyuridine. J. Mol. Biol. 48, 395 (1970)

Ley, R.D.: Postreplication repair in an excision-defective mutant *E. coli.* Ultraviolet light-induced incorporation of bromodeoxyuridine into paren-tal DNA. Photochem. Photobiol. 18, 87 (1973)

Ley, R.D., Setlow, R.B.: Rapid repair of lesions induced by 313 nm light in bromouracil-substituted DNA of *E. coli.* Biochem. Biophys. Res. Commun. 46, 1089 (1972)

Lillicrap, S.C., Fielden, E.M.: The effect of 5-bromouracil on energy trans-fer in DNA and related model systems. Radiat. Res. 48, 432 (1971)

Lin, S.Y., Riggs, A.D.: *Lac* operator analogues: Bromodeoxyuridine substi-tution in the *lac* operator affects the rate of dissociation of the *lac* repressor. Proc. Natl. Acad. Sci. USA 69, 2574 (1972)

Lin, S.Y., Riggs, A.D.: Photochemical attachment of *lac* repressor to bromo-deoxyuridine-substituted *lac* operator by ultraviolet irradiation. Proc. Natl. Acad. Sci. USA 71, 947 (1974)

Lin, S.Y., Lin, D., Riggs, A.D.: Histones bind more tightly to bromodeoxy-uridine-substituted DNA than to normal DNA. Nucleic Acids Res. 3, 2183 (1976)

Lindahl, T.: An N-glycosidase from *Escherichia coli* that releases free uracil from DNA containing deaminated cytosine residues. Proc. Natl. Acad. Sci. USA 71, 3649 (1974)

Lindahl, T., Ljungquist, S.: Apurinic and apyrimidinic sites in DNA. In: Molecular Mechanisms for Repair of DNA (eds. P.C. Hanawalt, R.B. Setlow) Vol. A, p. 31. New York: Academic Press 1975

Lindahl, T., Nyberg, B.: Heat-induced deamination of cytosine residues in DNA. Biochemistry 13, 3405 (1974)

Lion, M.B.: Search for a mechanism to explain the high ultraviolet sen-sitivity of 5-bromouracil-substituted DNA. 3rd Int. Congr. Radiat. Res. (Cortina) Abstr. p. 142 (1966)

Lion, M.B.: Search for a mechanism for the increased sensitivity of bromouracil-substituted DNA to ultraviolet radiation. Biochim. Biophys. Acta
155, 505 (1968)

Lion, M.B.: Single-strand breaks in the DNA of irradiated 5-bromouracil-
substituted T3 coliphage. Biochim. Biophys. Acta 209, 24 (1970)

Lion, M.B.: Mechanism of sensitization of ultraviolet radiation by 5-bromo-
uracil-substituted DNA. Isr. J. Chem. 10, 1151 (1972)

Lion, M.B., Doerner, T.: Determination of the distribution of 5-bromo-
uracil and 5-iodouracil in the DNA of viable and total phage populations.
Biochim. Biophys. Acta 277, 25 (1972)

Lion, M.B., Köhnlein, W.: Effect of DNA conformation on the ultraviolet
damage in 5-bromouracil substituted DNA of T3 coliphage. VI Int. Congr.
Photobiol., Bochum, Abstr. 107 (1972)

Little, J.W.: The effect of 5-bromouracil on recombination of phage lambda.
Virology 72, 530 (1976)

Lohmann, W.: Halogen substitution effect on the optical adsorption bands
of uracil. Z. Naturforsch. 29c, 493 (1974)

Longworth, J.W., Rahn, R.O., Shulman, R.G.: Luminescence of pyrimidines,
purines, nucleosides and nucleotides at 77°K. The effect of ionization
and tautomerization. J. Chem. Phys. 45, 2930 (1966)

Luk, D.C., Bick, M.D.: Determination of 5-bromodeoxyuridine in DNA by
buoyant density. Anal. Biochem. 77, 346 (1977)

Makino, F., Munakata, N.: Isolation and characterization of a B. subtilis
mutant with a defective N-glycosidase activity for uracil-containing DNA.
J. Bacteriol. 131, 438 (1977)

Martens, P.A., Clayton, D.A.: Strand breakage in solution of DNA and
ethidium bromide exposed to visible light. Nucleic Acids Res. 4, 1393
(1977)

Matsumoto, K., Shibata, T., Saito, H.: Genetic mapping in Bacillus subtilis by 5-bromouracil sensitization to ultraviolet inactivation of
transforming activities. J. Bacteriol. 119, 666 (1974)

Mazrimas, J.A., Stetka, D.G.: Direct evidence for the role of incorporated
BudR in the induction of sister chromatid exchanges. Exp. Cell Res. 117,
23 (1978)

McKeown, M., Kahn, M., Hanawalt, P.: Thymidine uptake and utilization in
E. coli: A new gene controlling nucleoside transport. J. Bacteriol. 126,
814 (1976)

Meuth, M., Green, H.: Induction of a deoxycytidineless state in cultured
mammalian cells by bromodeoxyuridine. Cell 2, 109 (1974)

Mönkehaus, F.: Influence of cysteamine on intramolecular energy transfer in
5-bromouracil-substituted phage DNA. Int. J. Radiat. Biol. 24, 517 (1973)

Mönkehaus, F., Köhnlein, W.: Single- and double-strand breaks in 5-BrU substituted DNA of B. subtilis and phage PBSH after irradiation with long
wavelength ultraviolet and their correlation to intramolecular energy
transfer. Biopolymers 12, 329 (1973)

Nečas, J.: Attempts to sensitize some chlorococcal algae using 5-bromo-
uracil for the induction of mutations by ultraviolet light. Biochem.
Physiol. Pflanz. 166, 115 (1974)

Nečas, J.: Sensitization dependence of ultraviolet irradiation effects on
concentration of 5-bromodexoyuridine in a precultivation medium for a
chlorococcal alga. Biochem. Physiol. Pflanz. 170, 487 (1976)

Negishi, K., Hayatsu, H., Tanooka, H.: Pol A dependent repair of 5-bromo-
uracil labeled Bacillus subtilis transforming DNA irradiated with ultraviolet in the presence of cysteamine. Int. J. Radiat. Biol. 30, 491 (1976)

Nemcrofsky, A.: The interaction effect of ultraviolet irradiation and
5-bromouracil at the rib 1 locus in Neurospora crassa. Can. J. Genet.
Cytol. 17, 275 (1975)

Newman, C.N., Kubitschek, H.E.: Variation in periodic replication of the chromosome in *Escherichia coli* B/r TT. J. Mol. Biol. <u>121</u>, 461 (1978)

Nicolini, C., Baserga, R.: Circular dichroism spectra and ethidium bromide binding of 5-deoxybromouridine-substituted chromatin. Biochem. Biophys. Res. Commun. <u>64</u>, 189 (1975)

Ogata, R., Gilbert, W.: Contacts between the *lac* repressor and thymines in the *lac* operator. Proc. Natl. Acad, Sci. USA <u>74</u>, 4973 (1977)

Olafsson, P.G., Bryan, A.M.: The influence of 5-halo substituants on the thermal depyrimidination of the glycosidic bond in 2'-deoxyuridines. Arch. Biochem. Biophys. <u>165</u>, 46 (1974)

Pera, F., Mattias, P.: Labeling of DNA and differential sister chromatid staining after BrdU treatment in vivo. Chromosoma <u>57</u>, 13 (1976)

Perry, P., Wolff, S.: New method for the differential staining of sister chromatids. Nature (London) <u>251</u>, 156 (1974)

Peter, H., Drewer, R.: Photoproducts of bromouracil-labeled DNA and the structure of 5-bromodeoxyuridyl-thymidine photoproduct. Photochem. Photobiol. <u>12</u>, 269 (1970)

Peter, H., Drewer, R.: Photochemistry of ^{14}C-labeled 5-bromo-2-deoxyuridylyl (3'-5') thymidine. Determination of quantum yields as a function of pH. Photochem. Photobiol. <u>14</u>, 561 (1971)

Pietrzykowska, I., Krych, M.: Lethal and mutagenic BU-induced lesions in DNA and their repair. Stud. Biophys. <u>61</u>, 17 (1977)

Pietrzykowska, I., Lewandowska, K., Shugar, D.: Liquid holding recovery of bromouracil-induced lesions in DNA of *Escherichia coli* CR34 and its possible relation to dark repair mechanisms. Mutat. Res. <u>30</u>, 21 (1975)

Pontecorvo, G.: Induction of directional chromosome elimination in somatic cell hybrids. Nature (London) <u>230</u>, 367 (1971)

Povirk, L.F.: Radiation-induced depression of DNA synthesis in cultured mammalian cells. Ph. D. Thesis, Univ. California, Berkeley (1977a)

Povirk, L.F.: Localization of inhibition of replicon initiation to damaged regions of DNA. J. Mol. Biol. <u>114</u>, 141 (1977b)

Prusoff, W.H., Goz, B.: Halogenated pyrimidine deoxyribonucleosides. In: Antimetabolites and Immunosuppressive Agents (eds. A.C. Sartorelli, D. Johns). Vol. II, Chap. 5. Berlin, Heidelberg, New York: Springer 1975

Puck, T.T., Kao, F.-T.: Genetics of somatic mammalian cells. V. Treatment of 5-bromodeoxyuridine and visible light for isolation of nutritionally deficient mutants. Proc. Natl. Acad. Sci. USA <u>58</u>, 1227 (1967)

Radman, M., Roller, A., Errera, M.: Protection and host cell repair of irradiated lambda phage. II. Irradiation of 5-bromouracil-substituted phage with near visible light. Mol. Gen. Genet. <u>104</u>, 147 (1969a)

Radman, M., Roller, A., Errera, M.: Protection and host cell repair of irradiated lambda phage. III. Ultraviolet irradiation of 5-bromo-uracil-substituted phage. Mol. Gen. Genet. <u>104</u>, 152 (1969b)

Rahn, R.O., Patrick, M.H.: Photochemistry of DNA. In: Photochemistry and Photobiology of Nucleic Acids (ed. S.Y. Wang), Vol. II, p. 97. New York: Academic Press 1976

Rahn, R.O., Stafford, R.S.: Photochemistry of DNA containing iodonated cyto-sine. Photochem. Photobiol. <u>30</u>, 449 (1979)

Rapaport, S.A.: Action spectrum for inactivation by ultraviolet light of bacteriophage T4 substituted with 5-bromodeoxyuridine. Virology <u>22</u>, 125 (1964)

Regan, J.D., Setlow, R.B.: Two forms of repair in the DNA of human cells damaged by chemical carcinogens and mutagens. Cancer Res. <u>34</u>, 3318 (1974)

Regan, J.D., Setlow, R.B.: Repair of human DNA: Radiation and chemical damage in normal and *Xeroderma pigmentosum* cells. In: Biology of Radiation Carcinogenesis (eds. J.M. Yuhas, R.W. Tennant, J.D. Regan), pp. 103. New York: Raven Press 1976

Regan, J.D., Setlow, R.B., Ley, R.D.: Normal and defective repair of damaged DNA in human cells: A sensitive assay utilizing the photolysis of bromo-deoxyuridine. Proc. Natl. Acad. Sci. USA 68, 708 (1971)

Reichert, P., Canellakis, Z.N., Canellakis, E.S.: Regulatory mechanisms in the synthesis of deoxyribonucleic acid in vitro. Biochim. Biophys. Acta 41, 558 (1960)

Reuschl, H.: Kinetic studies of gamma radiolysis of 5-bromouracil in aqueous solution. Z. Naturforsch. 21b, 643 (1966)

Rosenstein, B.S., Setlow, R.B., Ahmed, F.E.: Use of the dye Hoechst 33258 in a modification of the bromodeoxyuridine photolysis technique for the analysis of DNA repair. Photochem. Photobiol. 31, 215 (1980)

Rothman, W., Kearns, D.R.: Triplet states of bromouracil and iodouracil. Photochem. Photobiol. 6, 775 (1967)

Roufa, D.J.: Bromodeoxyuridine strand symmetry and the repair of photolytic breaks in Chinese hamster cell chromosomes. Proc. Natl. Acad. Sci. USA 73, 3905 (1976)

Roufa, D.J., Sadow, B., Caskey, C.T.: Derivation of TK⁻ clones from re-vertant TK⁺ mammalian cells. Genetics 75, 515 (1973)

Rupp, W.D.: The photochemistry of iodouracil as related to the survival of ultraviolet-irradiated T1 bacteriophage substituted with 5-iodo-2'-deoxyuridine. Ph. D. Thesis, Yale Univ. (1965)

Rupp, W.D., Howard-Flanders, P.: Discontinuities in the DNA synthesized in an excision-defective strain of *Escherichia coli* following ultraviolet irradiation. J. Mol. Biol. 31, 291 (1968)

Rupp, W.D., Prusoff, W.H.: Incorporation of 5-iodo 2'-deoxyuridine into bacteriophage T1 as related to ultraviolet sensitization or protection. Nature (London) 202, 1288 (1964)

Rupp, W.D., Prusoff, W.H.: Photochemistry of iodouracil. II. Effects of sulfur compounds, ethanol and oxygen. Biochem. Biophys. Res. Commun. 18, 158 (1965)

Rutter, W.J., Pictet, R.L., Githins, S., III, Gordon, J.S.: The mode of action of the thymidine analogue, 5-bromodeoxyuridine, a model terato-genic agent. In: New Approaches to the Evaluation of Abnormal Embryonic Development (eds. D. Neubert, H.J. Merker), pp. 804. Stuttgart: Thieme 1975

Rydberg, B.: Bromouracil mutagenesis in *E. coli*: Evidence for involvement of mismatch repair. Mol. Gen. Genet. 152, 19 (1977a)

Rydberg, B.: Discrimination between bromouracil and thymine for uptake into DNA in *drm⁻* and *dra⁻* mutants of *E. coli* K12. Biochim. Biophys. Acta 476, 32 (1977b)

Saito, I., Ito, S., Matsumura, T.: Photoinduced coupling reaction of 5-bromouridine to tryptophan derivatives. JACS 100, 2901 (1978)

Sasson, S., Wang, S.Y., Ehrlich, M.: 5-5' diuridinyl, a major photoproduct from ultraviolet-irradiation of polynucelotides containing bromouracil. Photochem. Photobiol. 25, 11 (1977)

Sawada, S., Okada, S.: Effects of 5-BrUdR labeling on radiation-induced DNA breakage and subsequent rejoining in cultured mammalian cells. Int. J. Radiat. Biol. 21, 599 (1972)

Scheid, W.: Mechanism of differential staining of BrUdR-substituted *Vicia faba* chromosomes. Exp. Cell Res. 101, 55 (1976)

Scheid, W., Traupe, H.: Further studies on the mechanism involved in dif-ferential staining of BUdR-substituted *Vicia faba* chromosomes. Exp. Cell Res. 108, 440 (1977)

Schwartzman, J.B., Cortes, F.: Sister chromatid exchanges in *Allium cepa*. Chromosoma 62, 119 (1977)

Sedor, F.A., Sander, E.G.: Effect of thiols on the dehalogenation of 5-iodo and 5-bromouracil. Biochem. Biophys. Res. Commun. 50, 328 (1973)

Setlow, R.B.: The wavelengths in sunlight effective in producing skin
 cancer, a theoretical analysis. Proc. Natl. Acad. Sci. USA 71, 3363 (1974)
Setlow, R.B., Doyle, B.: The action of monochromatic ultraviolet light on
 proteins. Biochim. Biophys. Acta 24, 27 (1957)
Shugar, D., Fox, J.J.: Spectrophotometric studies on nucleic acid derivatives
 and related compounds as a function of pH. I. Pyrimidines. Biochem. Biophys.
 Acta 9, 199 (1952)
Simpson, R.T., Seale, R.L.: Characterization of chromatin extensively sub-
 stituted with 5-bromodeoxyuridine. Biochemistry 13, 4609 (1974)
Simpson, R.B.: Contact between *E. coli* RNA polymerase and thymines in the
 lac UV5 promotor. Proc. Natl. Acad. Sci. USA 76, 3233 (1979)
Singer, J., Stellwagen, R.H., Roberts-Ems, J., Riggs, A.D.: 5-methylcytosine
 content of rat hepatoma DNA substituted with bromodeoxyuridine. J. Biol.
 Chem. 252, 5509 (1977)
Singh, P.K.: Sensitization of algal virus to UV by the incorporation of
 5-bromouracil and mutations of host alga *Plectonema broyanum*. Z. Allg.
 Mikrobiol. 15, 547 (1975)
Skalko, R.G., Packard, D.S., Jr.: Mechanisms of halogenated nucleoside
 embryotoxicity. Ann. N. Y. Acad. Sci. 255, 552 (1975)
Smets, L.A., Cornelis, J.J.: Repairable and irrepairable damage in 5-bromo-
 uracil-substituted DNA exposed to ultraviolet radiation. Int. J. Radiat.
 Biol. 19, 445 (1971)
Smith, K.C.: The photochemistry of thymine and bromouracil in vivo. Photochem. Photobiol. 3, 1 (1964)
Smith, K.C.: The radiation-induced addition of proteins and other molecules
 to nucleic acids. In: Photochemistry and Photobiology of Nucleid Acids.
 (ed. S.Y. Wang), Vol. II, p. 187. New York: Academic Press 1976
Stahl, F.W., Crasemann, J.M.K., Okun, L., Fox, E., Laird, C.: Radiation-
 sensitivity of bacteriophage containing 5-bromodeoxyuridine. Virology
 13, 98 (1961)
Sternglanz, H., Bugg, C.E.: Relationship between the mutagenic and base
 stacking properties of halogenated uracil derivatives. The crystal
 structures of 5-chloro- and 5-bromouracil. Biochim. Biophys. Acta 378,
 1 (1975)
Stetten, G., Davidson, R.L., Latt, S.A.: 33258 Hoechst enhances the selectivity of the bromodeoxyuridine-light method of isolating conditional
 lethal mutants. Exp. Cell Res. 108, 447 (1977)
Sugiyama, T., Goto, K., Kano, Y.: Mechanism of differential Giemsa method
 for sister chromatids. Nature (London) 259, 59 (1976)
Szarek, W.A.: Deosyhalogeno-sugars. Adv. Carbohydr. Chem. Biochem. 28, 225
 (1973)
Szybalski, W.: Properties and applications of halogenated deoxyribonucleic
 acids. In: The Molecular Basis of Neoplasia, pp. 147. Austin: Un. Texas
 Press 1962
Szybalski, W.: X-ray sensitization by halopyrimidines. Cancer Chemother.
 Rep. 58, 539 (1974)
Szybalski, W., Opara-Kubinska, Z.: Radiobiological and physiochemical
 properties of 5-bromodeoxyuridine-labeled transforming DNA as related
 to the nature of the critical radiosensitive structures. In: Cellular
 Radiation Biology, pp. 223. Baltimore: Williams and Wilkins 1965
Taichman, L., Freedlender, E.F.: Separation of chromatins containing BrU
 in one or both strands of the DNA. Biochemistry 15, 447 (1976)
Teich, N., Lowy, D.R., Hartley, J.W., Rowe, W.P.: Studies of the mechanism
 of induction of infectious murine leukemia virus from AKR mouse embryo
 cell lines by 5-iododeoxyuridine and 5-bromodeoxyuridine. Virology 51,
 163 (1973)

Thiel, R., Wacker, A., Treatment of herpetic keratitis with thymine-analogous compounds. Klin. Monatsbl. Augenheilkd. 141, 94 (1962)

Thiron, J.P.: Chromosome damage in mouse-human hybrid cells after BUdR treatment and light irradiation. Mutat. Res. 35, 479 (1976)

Ullman, J.S., McCarthy, B.J.: Alkali deamination of cytosine residues in DNA. Biochim. Biophys. Acta 294, 396 (1973)

Varghese, A.J.: Photoreactions of 5-bromouracil in the presence of cysteine and glutathione. Photochem. Photobiol. 20, 461 (1974)

Verly, W.G.: Maintenance of DNA and repair of apurinic sites. In: Molecular Mechanisms for Repair of DNA (eds. P.C. Hanawalt, R.B. Setlow), Vol. A, p. 39. New York: Plenum Press 1975

Verma, R.S., Cummins, J.E., Walden, D.B.: Chromosome aberrations produced by 5-bromodeoxyuridine with concurrent exposure to long wavelength UV in Zea mays root tip cells. Can. J. Genet. Cytol. 19, 447 (1977)

Voytek, P., Chang, P.K., Prusoff, W.H.: Kinetic and photochemical studies of 3-N-methyl-5-iodo-2'-deoxyuridine. J. Biol. Chem. 247, 367 (1972)

Wacker, A.: Strahlenchemische Veränderungen von Pyrimidinen in vivo und in vitro. J. Chim. Phys. 58, 1041 (1961)

Wacker, A.: Molecular mechanisms of radiation effects. Prog. Nucleic Acid Res. 1, 369 (1963)

Wang, S.Y.: Pyrimidine biomolecular photoproducts. In: Photochemistry and Photobiology of Nucleic Acids (ed. S.Y. Wang), Vol. I, p. 296. New York: Academic Press 1976

Ward, J.F.: Molecular mechanisms of radiation-induced damage to nucleic acids. Adv. Radiat. Biol. 5, 181 (1975)

Ward, J.F., Kuo, I.: The effects of radiation modifiers on sugar-phosphate bond breakage in deoxynucleotides irradiated in aqueous solution. IV. Int. Congr. Radiat. Res., Evian, France (1970)

Wataga, Y., Negishi, K., Hayatsu, H.: Debromination of 5-bromo-2-deoxy-uridine by cysteine. Formation of deoxyuridine and S-(5-2-deoxyuridyl) cysteine. Biochemistry 12, 3992 (1973)

Weintraub, H.: The assembly of newly replicated DNA into chromatin. Cold Spring Harbor Symp. Quant. Biol. 38, 247 (1973)

Witkin, E.M.: Ultraviolet mutagenesis and inducible DNA repair in E. coli. Bacteriol. Rev. 40, 869 (1976)

Wolff, S., Perry, P.: Differential giemsa staining of sister chromatids and the study of sister chromatid exchanges without autoradiography. Chromosoma 48, 341 (1974)

Zamchuk, L.A., Braude, N.A.: Immunogenic properties of Escherichia coli and T4 DNA containing 5-bromodeoxyuridine. Mol. Biol. USSR 9, 565 (1975)

Zamenhof, S., Rich, K., DeGiovanni, R.: Further studies on the introduction of pyrimidines into deoxyribonucleic acids of E. coli. J. Biol. Chem. 232, 651 (1958)

Zimbrick, J.D., Ward, J.F., Myers, L.S., Jr.: Studies on the chemical basis of cellular radiosensitization by 5-bromouracil substitution in DNA. I. Pulse and steady state radiolysis of 5-bromouracil and thymine. Int. J. Radiobiol. 16, 505 (1969a)

Zimbrick, J.D., Ward, J.F., Myers, L.S., Jr.: Studies on the chemical basis of cellular radiosensitization by 5-bromouracil substitution in DNA. II. Pulse and steady state radiolysis of regular and bromouracil-substituted DNA. Int. J. Radiobiol. 16, 525 (1969b)

Non-Intercalative Binding to DNA

A. K. Krey

A. Introduction

Many biologically active substances produce their genetic, biochemical, pharmacological, and tinctorial effects as consequences
of their interactions with DNA. Binding to the nucleic acid has,
therefore, been widely investigated. Two types of structural
models describe the attachment reactions.

The intercalation model of Lerman (1961, 1963, 1964a,b) pertains
to flat heterocyclic ring systems which are inserted between the
levels of base pairs in DNA. Intercalation, although anticipated
earlier (Michaelis 1947), could only be envisaged after the proposal of the double-helical structure of DNA (Watson and Crick
1953). Intercalation was offered as an explanation of the frameshift mutagenesis of aminoacridines (Lerman 1961) and of their
strong mode of attachment to DNA (Peacocke and Skerrett 1956).
Subsequently, the intercalation theory has been extended to many
other DNA-complexing compounds.

A second, external, model of binding to DNA and other macromolecules was derived for acridine orange (Bradley and Wolf 1959;
Stone and Bradley 1961) and has now been recognized to describe
the secondary or weaker mode of attachment of intercalating substances to DNA (Blake and Peacocke 1968). Bimodal binding of
intercalators (Blake and Peacocke 1968) is, however, not the
topic of this review.

There exists a number of biologically active compounds which
for structural reasons cannot be intercalated but bind exclusively to the periphery of the double helix. This review dis-
cusses these substances which bind to DNA only according to the
second, non-intercalative, structural model.

In general, DNA binding studies begin with determinations of
changes in the optical properties of ligands when their complexes with DNA are formed. The obtained optical changes are
customarily converted into adsorption isotherms, known as
Scatchard plots (Scatchard 1949), which yield information on
stoichiometries and apparent association constants of the
ligands' binding reactions. Derivation of the parameters for
the attachment of a series of DNA-complexing compounds from
equilibrium dialyses, monitored optically, suggested a trend
that intercalating substances may preferentially interact with
G·C pairs in DNA (Müller and Crothers 1975; Müller et al. 1975)
while non-insertive ligands bind to A·T pairs (Müller and Gautier
1975). The present review will bring out the A·T preferences
of non-intercalative compounds. Some substances (berenil,
Hoechst 33258, and perhaps anthramycin) appear to prefer A·T

pairs per se, while others (polylysine, distamycin A, netropsin, and methyl green) seem to require for external attachment
the B-form which is imposed on DNA (review: Bram 1976; Arnott
1977) by an abundance of A·T pairs.

There are no direct criteria for peripheral binding to DNA,
unless a ligand has a chemical structure which excludes the
possibility of intercalation, such as spermine or polylysine.
To decide in favor of outside attachment, i.e. against intercalation, one must show that the compounds in question fail
to satisfy the criteria which have been established for the
intercalation reaction.

Intercalation places the chromophoric ring system of DNA-complexing substances into the hydrophobic environment between
the base pairs of DNA. Such insertion results in hypochromic
effects on the ligands' absorption spectra, while absence of
such an effect carries a strong suggestion of an external binding
process. The insertion of molecules between base pairs of DNA
causes increases in the length of DNA. These increases are recognized hydrodynamically by increases in DNA's viscosity and
decreases in its rate of sedimentation (Lerman 1961; Lerman
1964b), customarily observed for low-molecular-weight, rod-like
pieces of the macromolecule used to separate stiffening effects
(Müller and Crothers 1968; Cohen and Eisenberg 1969). Length
increases of DNA produced by intercalation can also be studied
by optical methods such as light scattering (Mauss et al. 1967),
small angle X-ray scattering (Luzzati et al. 1961), X-ray fiber
diffraction (Neville and Davies 1966), electron microscope observations (Freifelder 1971), or autoradiographic techniques
(Cairns 1962). Absence of these diverse hydrodynamic and optical effects argues against intercalation binding.

Conclusive evidence for or against intercalation comes from
hydrodynamic effects of substances on closed circular DNA
(Waring 1970a,b; 1971). Insertion produces through extension
and local unwinding of the duplex structure a removal and reversal of supercoils of closed circular DNA, which is conveniently
monitored by decreases and subsequent increases in its sedimentation coefficient. Absence of the hydrodynamic changes of super-
coiled circular DNA provides evidence against intercalation
binding (Waring 1970a,b, 1971), and supports the suggestion of
an external attachment of the ligands to be discussed here to
the periphery of the DNA helix.

B. Poly-L-Lysine

One of these externally binding substances, polylysine, has been
widely studied as model for histones in association with chromatin DNA. Histones are basic proteins which assist in the formation of chromatin structure by folding DNA into a compact state
with a secondary structure in the B-conformation (Bram 1971a)
and a higher-order arrangement in a "super-coiled" (Pardon and
Wilkins 1972) or, now widely accepted, "string-of-beads" form
(Kornberg 1974, 1977a,b; Thomas 1978).

I. A·T Specificity and Binding Forces for Polylysine-DNA

Poly-L-lysine (Fig. 1) possesses no chromophoric ring system and, therefore, no light absorption properties which could be used to indicate binding to DNA. The compound is a polycation which carries one positive charge per lysine residue at near-neutral pH at which most interaction studies, reported in the literature, have been performed. As polycation, polylysine neutralizes phosphates of DNA with the consequent formation of a complex in a separate aggregate phase (Leng and Felsenfeld 1966; Shapiro et al. 1969). The interaction is stoiciometric, as one would expect for mixtures of polyions of opposite charges, and removes from solution one nucleotide for each lysine residue added, with all of the added polylysine bound in the aggregate phase and only free DNA in solution (Leng and Felsenfeld 1966). The aggregate or precipitate so formed in the presence of approximately 1 M NaCl, LiCl, or CsCl contains preferentially A·T-rich DNA (Leng and Felsenfeld 1966; Shapiro et al. 1969), freely exchanges DNA with the soluble phase (Leng and Felsenfeld 1966), and is dissociated at higher sodium chloride concentrations (Spitnik et al. 1955; Leng and Felsenfeld 1966), first for G·C-rich DNA and subsequently for DNA rich in A·T

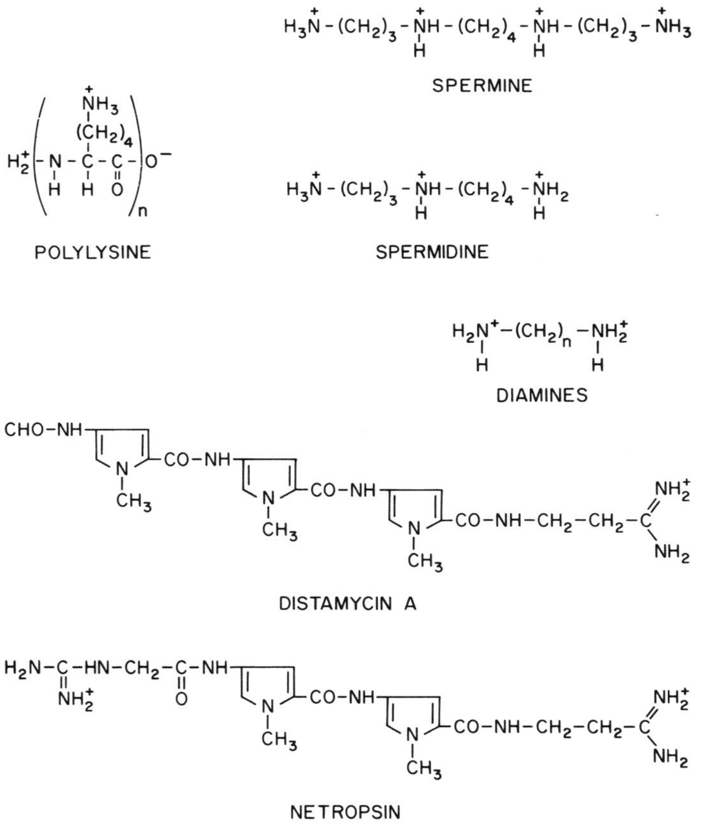

$$H_3\overset{+}{N}-(CH_2)_3-\overset{+}{N}H-(CH_2)_4-\overset{+}{N}H-(CH_2)_3-\overset{+}{N}H_3$$

SPERMINE

POLYLYSINE

$$H_3\overset{+}{N}-(CH_2)_3-\overset{+}{N}H-(CH_2)_4-\overset{+}{N}H_2$$

SPERMIDINE

$$H_2\overset{+}{N}-(CH_2)_n-\overset{+}{N}H_2$$

DIAMINES

DISTAMYCIN A

NETROPSIN

Fig. 1. Chemical structures of the compounds discussed. The protonated forms expected to predominate at near neutral pH are drawn

(Wehling et al. 1975). Preferential interaction with A·T-rich DNA is also observed for the optical isomer, poly-D-lysine, and the copolymer, poly-D,L-lysine (Shapiro et al. 1969), and for shorter lysines with as few as seven residues (Leng and Felsenfeld 1966), but not in the presence of tetramethylammonium chloride in which polylysine precipitates G·C-rich DNA material first (Leng and Felsenfeld 1966; Shapiro et al. 1969).

The A·T specificity of polylysine observed in the presence of 1 M salts is less apparent at lower ionic conditions. At lower ionic strengths, electrostatic attractions become more important and the polypeptide binds irreversibly to DNA (Tsuboi et al. 1966) and forms, either after reconstitution, i.e., after gradient dialysis of the high-salt polylysine-DNA aggregates against decreasing concentrations of salt (Olins et al. 1967), or upon its slow direct addition to DNA in low salt concentrations (Spitnik et al. 1955; Tsuboi et al. 1966, Li et al. 1973), complexes with DNA which are soluble at higher polylysine contents than the polylysine-DNAs which are prepared under high-salt conditions. Solubility of the low-salt complexes persists until about one lysine residue has been added per DNA phosphate; in contrast, the high-salt aggregates precipitate at any input of polylysine to DNA (Spitnik et al. 1955; Olins et al. 1967; Li et al. 1973, 1975).

Contributory to polylysine's binding process is a cooperativity which facilitates occupancy of certain entire DNA molecules by polylysine while other molecules of DNA remain completely vacant. Such cooperativity occurs in the high-salt polylysine-DNA complexes which form the separate aggregate phase. The same attachment cooperativity has been observed for polylysine in its low-salt reconstituted complexes with DNA: in "melting" experiments, the free DNA molecules melt in a denaturation transition separate from the transition of DNA molecules that are stoichiometrically covered by polylysine (Olins et al. 1967, 1968; Shih and Bonner 1970). Polylysine binding cooperativity has also been determined for low-salt directly mixed polylysine-DNAs (Li et al. 1973), but the observed biphasic melting phenomenon is attributed, like that of nucleoproteins in general (Li and Bonner 1971; Li 1972), to a melting of free segments of DNA rather than of molecules, independently from segments occupied by the polypeptide (Li 1973).

The distinction between entire DNA molecules and between segments of DNA in the cooperative binding of polylysine has been derived from the biphasic melting profiles also in the presence of shorter lysines with as few as 20 or less (but not four) residues (Olins et al. 1968; Inoue and Ando 1970; Li and Bonner 1971), it depends on the A·T content of DNA (Ohba 1966; Olins et al. 1967; Li et al. 1974a, 1975), and suggests a distribution of the polypeptide in reconstituted DNA-polylysine complexes which is different from the distribution in the directly mixed complexes of polylysine with DNA. That a difference exists in the distribution of polylysine in the two complexes has been demonstrated in "renaturation" experiments (Li et al. 1974c): in directly mixed complexes polylysine clamps together

individual or successive segments of *all* DNA molecules while
in reconstituted complexes only *some* DNA molecules are held together by the polypeptide (Li et al. 1974c). Indications for
the binding of polylysine other than renaturations also point
to a different polylysine attachment and thus to different
properties of the two complexes in question.

II. DNA Secondary Structural Considerations for the Binding of Polylysine

One indication of a difference between the two complexes is the
solubility characteristic of the two polylysine-DNAs. Reconstituted polylysine-DNA complexes show centrifuged pellets of gel-
like appearance which can easily be brought back into solution,
while the pellets of directly mixed complexes are hard solids
which cannot easily be redissolved. This difference in solubility has been attributed to different hydrational states of the
two complexes (Li et al. 1974b), which may, in turn, indicate
different conformational properties of the two polylysine-bound
DNAs. The conformation of DNA in directly mixed and reconstituted
complexes with polylysine has been investigated in detail by
optical rotatory dispersion (ORD) and circular dichroism (CD).

Direct mixing of DNA and polylysine yields a circular dichroism
spectrum which possesses a positive band at 276 nm, that is
sensitive to the polypeptide content in the complex, and a
negative band near 250 nm and optical anomalies below 240 nm,
which are independent of the amount of polypeptide in the polylysine-DNA (Chang et al. 1973). The positive band of this spectrum is reduced by polylysine; these reductions have been interpreted as effects produced by the polypeptide in the circular
dischroism of DNA. The reductions are proportional to the lysine
input into the complex from which has been inferred the existence
of two fractions of DNA, that is, a fraction of free segments
and a fraction of segments covered by polylysine. With the assumption of only one type of bound segments, the optical anomalies observed for increasing polypeptide contents could be ex-
trapolated to a spectrum of bound DNA in which the positive
band, typical for the DNA B-form, was greatly reduced. A similar
reduction in the positive CD band of B-DNA is also produced by
the addition of ethylene glycol (Green and Mahler 1968; Nelson
and Johnson 1970), or high concentrations of salt (Tunis and
Hearst 1968; Tunis-Schneider and Maestre 1970; Ivanov et al.
1973b), or upon dehydration of the nucleic acid to a low-humi-
dity film (Tunis-Schneider and Maestre 1970). This has been attributed to the formation of the more dehydrated C-form of DNA
so that the DNA spectral changes produced by polylysine have
been ascribed to a conformational transition of free B-DNA towards more dehydrated C-DNA covered by polylysine (Chang et al.
1973).

Formation of the C-form upon binding of polylysine leaves DNA
in the B-family of conformations. Also to the B-conformational
family appears to belong DNA in reconstituted complexes (as well
as high-salt complexes) with polylysine in spite of the fact

that its ORD and CD spectra under such conditions fail to resemble the conservative spectra typical for the canonical (10-
base-pairs-per-helical-turn) B-form or the somewhat altered
spectra of the C-conformation. The rotatory spectra of the
polylysine-reconstituted DNAs differ most prominently from
those of the B- and C-forms (also from those of the A-form)
in their extremely large optical activities of negative sign,
located around 250-270 nm in CD and around 280 nm in ORD (Cohen
and Kidson 1968; Shapiro et al. 1969; Haynes et al. 1970; Olins
and Olins 1971; Carroll 1972). The large activities depend on
the polypeptide content of the complex (Cohen and Kidson 1968;
Carroll 1972) and on the A·T content of the employed DNA (Cohen
and Kidson 1968). They increase with increases in salt content,
used to mix high-salt complexes of polylysine with DNA (Weiskopf and Li 1977), and also increase with decreases in ionic
strengths during the salt-gradient preparation of the reconstituted polylysine-DNAs (Carroll 1972). Large rotations have likewise been observed for DNA in the presence of certain neutral
or anionic polymers plus salt, and have been attributed to a
polymer-and-salt-induced condensed state of DNA, denoted by the
acronym ψ (Lerman 1971a; Jordan et al. 1972).

Causation of a condensed state appears to be a group property
of agents which produce the large negative rotations (ψ spectra)
for DNA. To this DNA-condensing group seems to belong polylysine
(Sponar and Frič 1972; Cheng and Mohr 1975). An explanation for
the polylysine-produced large ψ-like rotations, different from
that of a condensed state, is that of a polypeptide-induced
change in secondary structure of DNA (Cohen and Kidson 1968;
Shapiro et al. 1969; Haynes et al. 1970), also suggested by
other indications of the binding of polylysine (Lees and von
Hippel 1968), but a polylysine-caused tertiary (or higher-order)
arrangement of DNA is probably the more widely accepted interpretation for the unusual ORD and CD observed with the polypeptide.

Support for the formation of a tertiary structure comes from
observations which suggest that DNA undergoes long-range ordering in the presence of polylysine. Such ordering can be inferred
from the huge viscosity decreases produced by polylysine (Zama
and Ichimura 1971) and from the formation of condensed particles
that is suggested by findings of light scattering entities of
spherical shape (Shapiro et al. 1969) and by observations, in
the electron microscope, of hollow doughnut- and stem- or rod-
like particles of polylysine-DNA (Haynes et al. 1970; Olins and
Olins 1971; Laemmli 1975). Formation of defined particles has
similarly been inferred from sedimentation increases for the
condensation of DNA with polymers and salt (Lerman 1971b), and
an intraparticle structure with interhelical order and parallelism
recognized by small-angle X-ray techniques (Maniatis et al. 1974),
by birefringence observations (Lerman 1974) and, most prominently, by the appearance of the nonconservative ψ spectra during
condensation, which resemble the large negative rotations observed with polylysine (Jordan et al. 1972). The pattern of
helix order which this rotatory similarity suggests for the particles with polylysine is supported by an interpretation that

the ψ-like rotations observed in the presence of the polypeptide (Shapiro et al. 1969; Haynes et al. 1970) imply a polylysine-produced liquid crystalline arrangement of DNA (de Vries 1951; Robinson 1961; Holzwarth and Holzwarth 1973). A helix parallelism indicative of some form of tertiary arrangement of DNA has been observed recently with the electron microscope for DNA particles condensed with polylysine (Laemmli 1975). Other properties of these particles, including an intraparticle periodicity (Haynes et al. 1970) and particle hydration (Shapiro et al. 1969), typical of liquid crystals, are likewise indicative of helix ordering during polylysine's production of a tertiary (or higher-order) structure of DNA (von Hippel and McGhee 1972).

The alternative possibility which views a polylysine-produced DNA secondary structural change as reason for the appearance of the unusual ψ-like ORD and CD spectra appears to be less supported. Speaking against a change of DNA, such as from the B- to the A-conformation, are the dissimilarity of A spectra and the nonconservative ψ spectra produced by polylysine; the preference of the polypeptide for A·T pairs (known to impose upon DNA a B- and not an A-conformation); and the following other observations which suggest preservation of DNA in its polylysine complex in a secondary structural B-form. Thus, DNA in its B-form (as typically obtained in aqueous solution) has been observed by X-ray techniques in high-humidity gels of high-salt (ψ exhibiting) DNA-polylysine complexes (Haynes et al. 1970) and a modified B-form, denoted B* (with a slightly different orientation of its phosphate groups compared to B), been suggested by infrared linear dichroism in films of similar complexes of polylysine-DNA (Liquier et al. 1975). DNA in the B-form has also been observed in X-ray diagrams of high-humidity fibers of the polylysine-DNA complex, and a C- but not an A-form was evident in these fibers at lower relative humidities (Suwalsky and Traub 1972). These observations rule out a polylysine conversion of B-DNA to A. They allow the general inference that the ψ-like spectra produced by polylysine are caused by an interhelix interaction among DNA molecules in a B-form to aggregate into the compact and ordered tertiary ψ structure. Which member of the B-conformational family serves in this compact arrangement, whether it is B* (Liquier et al. 1975), C (Li et al. 1974d), or a different B-form, remains to be established.

III. The Non-Intercalative Binding of Polylysine

The structural considerations for the binding of polylysine have up to this point only dealt with the DNA participant of the association. While DNA, when encountered by polylysine as a member of the B family of conformations, remains in that conformational family upon association, polylysine assumes an ordered disposition in the complex which differs considerably from its random-coil solution form. How does the polypeptide bind to B-DNA? Binding by intercalation can be ruled out since the molecule lacks a planar structural prerequisite for an insertion. Intercalation has been ruled out with the X-ray diffraction observations on nucleoproteins and polylysine-DNA complexes which have demonstrated an attachment of the polypeptide to the exterior of DNA's helical structure (Feughelman et al. 1955; Wilkins 1956). Molecular model building studies have placed polylysine into the small groove of the helix structure (Wilkins 1956; Li 1975) or into its large groove (Tsuboi 1967) or were indecisive (Suwalsky and Traub 1972) while Fourier analyses of the X-ray data on polylysine-DNA have decided for a binding of the polypeptide in the minor helical groove (with perhaps some intermolecular DNA crosslinking) (Suwalsky and Traub 1972). Placement of polylysine into the minor DNA groove has also been concluded from the inability of DNA to prevent the digestion of the polypeptide by trypsin (Li 1975) and from the capability of a lysine oligomer to bind to DNA, regardless of major-groove glucosylation (Carroll and Botchan 1972). The accepted model which emerges from these observations for the binding of polylysine has been described (Wilkins 1956; Suwalsky and Traub 1972); according to this model, polylysine is constrained to follow the nucleic acid's secondary structure in the minor groove of the double helix, pointing the side chains of its trans-configuration (Suwalsky and Traub 1972) alternately up and down so that the strongly basic end groups of these side chains lie close to the phosphate groups of the sugar-phosphate backbone in the B-form of DNA.

C. Spermine and Related Polyamines

The oligo-amines spermine and spermidine (Fig. 1) are widely distributed in animal tissues and microorganisms, as are the diamines, 1,3-diaminopropane, 1,4-diaminobutane (putrescine), and 1,5-diaminopentane (cadaverine), which in the literature are customarily included with the oligo-amines in the term "polyamines". The distribution of the polyamines has been detailed in numerous reviews, as well as their biosynthesis and metabolism and their biological effects (Tabor et al. 1961; Tabor and Tabor 1964, 1972, 1976; Bachrach 1970; Stevens 1970). Most outstanding among their biological activities is the link of the polyamines to cellular growth processes. They are necessary for the growth of microorganisms, increase in content in animal tissues during periods of rapid growth, and stimulate in cell-free systems the synthesis of DNA, RNA, and proteins. Furthermore important is the role of the amines in the folding of DNA in viruses into a compact form to permit packaging of the nucleic acid into the limited space afforded inside the viral capsules. In view of the fact that diverse activities of the polyamines appear to be caused by their interactions with DNA or RNA, nucleic acid-polyamine associations have been extensively studied in vitro, stimulated by the early findings of a polyamine neutralization of bacteriophage DNA (Ames et al. 1958; Ames and Dubin 1960).

I. Qualitative and Quantitative Observations on the Polyamine-DNA Complex

The polyamines are, like polylysine, devoid of a chromophore and therefore possess no optical properties that can be used as binding indicators. They exist at physiological pHs as polycations, expected to bind to cellular anions such as DNA. Attachment to DNA has been demonstrated by the formation of insoluble DNA complexes with spermine or spermidine (Razin and Rozansky 1959; Agrell and Heby 1968), that depends on the number of amino groups in the polyamine and is favored by high amine concentrations but is reversed by inorganic ions (Razin and Rozansky 1959; Horáček and Cernohorský 1968). In the presence of higher concentrations of inorganic ions, soluble polyamine-DNA complexes are formed (Leibo and Mazur 1966; Horáček and Cernohorský 1968). Forming these complexes, the doubly charged diamines $^{2+}$, the triamine spermidine $^{3+}$, and the tetraamine spermine $^{4+}$ compete successfully with Na^+, Mg^{2+}, or Ca^{2+} ions for attachment to the phosphates of DNA (Felsenfeld and Huang 1961; Chang and Carr 1968; Rubin 1977). Evidently, the polyamine-DNA interaction is of predominantly electrostatic nature and proceeds until polyamines are bound with an equivalent charge per DNA phosphate (Felsenfeld and Huang 1961; Rubin 1977).

Neutralization of the phosphates of DNA by the polyamines stabilizes double-helical DNA with respect to heat. The largest helix stabilization is produced by spermine and a smaller increase in the thermal denaturation temperature of DNA is produced by spermidine with effective concentrations of both polyamines (10^{-5} to 10^{-4} M) lower than the concentrations of diamines (10^{-2} M), which produce a comparable stabilization of DNA (Tabor 1961, 1962). Among the diamines, those containing three to five central carbon atoms are most effective in raising the T_m of DNA (Mahler et al. 1961; Mahler and Mehrotra 1962, 1963) and a similar trend in raising DNA's T_m has also been observed by varying the length of the hydrocarbon chain in the series of spermine-related tetraamines, in which the member with four carbons between the secondary amino groups (i.e., spermine itself) and its 5-carbon homolog produced the maximum stabilization effect (Stevens 1967). It thus appears that the distance which four or five central methylene groups provide between two charged nitrogens in a polyamine is most favorable for the efficient stabilization of DNA.

Neutralization of DNA phosphates by polyamines produces thermal stabilizations of DNA even when the DNA helix is part of the triple-stranded hybrid polymer, poly (dA·dT)rU (Herbst and Tanguay 1971). Low concentrations of spermine, spermidine, and the diamine putrescine release at low temperatures the rU strand from the triple helix and produce a stabilization at higher temperatures of the DNA-like double helix of poly dA·dT, while higher polyamine concentrations prevent the addition of the rU strand to poly dA·dT and produce an even larger increase in the T_m of poly dA·dT.

The thermal protection of DNA which occurs as a consequence of
the neutralization by polyamines has been reported to depend
on the base composition of the employed DNA. A preferential increase in T_m by spermine (Mandel 1962) and by aliphatic diamines
(Mahler and Mehrotra 1962, 1963) has been observed for A·T-rich
DNAs suggesting a preferred interaction of the amines with A·T-
rich regions of DNA (Mahler and Mehrotra 1963). This interpretation has been challenged in view of the fact that the average
charge on spermine is reduced at higher temperatures so that a
consequently smaller amount of the amine binds to higher melting, G·C-rich DNAs than does to the lower melting, A·T-rich
DNAs (Hirschman et al. 1967). When, instead, the charge on
spermine was kept constant in equilibrium dialysis experiments,
it was found that spermine binds to a single class of strong
binding sites in DNA with a maximum stoichiometry (about 0.23
equivalent of charge per nucleotide) and an association constant which are both independent of base composition (Hirschman et al. 1967).

II. Structure of the Polyamine-DNA Complex

The counterionic action of the polyamines, which leads to the
stabilization of the secondary structure of DNA, also produces
a folding of the nucleic acid into a tertiary compact form.
This folding has been demonstrated through use of a linear
flow dichroism method which suggested a cooperative transition
of extended rod-like native DNA molecules to molecules of lesser degree of flow orientation compacted by spermidine or spermine into a more spherical shape (Gosule and Shellman 1976).
Condensation of DNA by spermidine and spermine has also been
observed in the electron microscope (Gosule and Shellman 1976),
while ultracentrifugation, conventionally used to characterize
the more spherical compact form (Lerman 1971a,b), has been employed to show a protection by spermidine of the condensed form
of DNA under normally unfolding conditions (Flink and Pettijohn
1975).

Folding of the nucleic acid into the compact form and protection
against unfolding conditions could not have come about by intercalation of polyamines. Titration of closed circular DNA with
spermine failed to produce the decreases and subsequent increases in the rate of sedimentation, which are indicative of
intercalation binding (Waring 1970a,b, 1971). The following
melting, molecular model building, X-ray, and electron microscope studies describe the amines' external attachment to DNA.

Three binding models have been considered: (1) Melting experiments have suggested an intrastrand binding of polyamines to
adjacent phosphates on *one* strand of duplex DNA (Glaser and
Gabbay 1968). (2) A molecular model has been derived, based
on X-ray analysis of spermine and spermidine, which has suggested a bridging of the central methylene moieties of polyamines across the minor groove of DNA (Liquori et al. 1967).
(3) X-ray diffraction studies of the spermine-DNA complex and
the formation of the compact form of DNA by amines have indi-

cated an inter- or intramolecular crosslinking by polyamines
of more than one DNA molecule (Suwalsky et al. 1969) or of
non-adjacent segments of individual molecules of DNA (Harrison
and Bode 1975). Binding across the minor DNA groove, for which
X-ray (Suwalsky et al. 1969) and ORD (Maestre and Tinoco 1967)
studies indicate preservation of the B-conformation, explains
best the thermal stabilization of DNA which is maximal with
those polyamines whose molecular dimensions of four or five
central methylene groups provide the most suitable fit across
the small groove. However, X-ray observations of the binding
of spermine rule out a monomodal attachment of amines (Suwalsky
et al. 1969) since only an assumption of bridging of the small
groove as well as of intermolecular crosslinks, or of even
greater variety of binding modes, did lead to a satisfactory
explanation of the diffraction patterns (Suwalsky et al. 1969).
At least two modes of binding are suggested by the polyamine-
produced compactization of DNA, attributed to small-groove
bridging and to inter- or intramolecular crosslinking (Harrison
and Bode 1975; Osland and Kleppe 1977). Observations in the
electron microscope have demonstrated the occurrence of all
three modes of electrostatic outside interaction of the polyamines with B-DNA (Chevaillier 1969).

D. Distamycin A and Netropsin

Distamycin A (DMC) and netropsiñ (Nt) are fermentation products
of *Streptomyces distallicus* (Arcamone et al. 1961) and of *Streptomyces
netropsis* or *Streptomyces* IA 2814 (Finlay et al. 1951; Thrum 1959),
respectively. The chemical structure of distamycin has been determined by chemical analysis and confirmed by total synthesis
(Arcamone et al. 1964; Penco et al. 1967), and the structure of
netropsin has also been determined by organic chemical methods
(Nakamura et al. 1964). The two antibiotics are structurally
related basic oligopeptides with identical N-methylpyrrole ring
systems and with the same propionamidine side chain; their
chemical structures differ in the number of peptidically linked
N-methylpyrrole rings and in the guanidinoacetamidino side
chain of netropsin which in distamycin is replaced by a formyl
group (Fig. 1).

Distamycin exhibits antifungal and antimitotic activities, in-
terferes with bacterial adaptive enzyme syntheses, inhibits RNA
oncogenic viruses whose replication depends on DNA synthesis,
and directs its most important antibiotic effects against viruses
containing DNA (review: Hahn 1975). The anti-DNA viral activity
of distamycin depends on structural features such as chromophores and side chains of the DMC molecule (Chandra et al. 1971,
1972a,b) as does the anti-oncogenic-RNA-virus action of the
antibiotic (review: Chandra et al. 1976). Netropsin also possesses anti-DNA viral activity and inhibits the growth of a variety
of bacteria (review: Hahn 1975).

Some biological effects of distamycin and netropsin have been
attributed to the antibiotics' inhibitions of nucleic acid
biosynthesis reactions directed by DNA. Macromolecular biosyntheses directed by DNA are most sensitive to distamycin
and netropsin, while polymerizations of nucleic acids and proteins mediated by RNA are less affected (Haupt and Thrum 1971;
Müller et al. 1974). The inhibitions of the DNA-programmed
macromolecular biosyntheses by distamycin depend on the same
structural parameters as do the antibiotic's effects against
DNA-containing viruses; they also depend on structural parameters of netropsin and have been attributed to an interaction
of DMC and Nt with the DNA templates (review: Hahn 1975, 1977;
Zimmer 1975), in particular with templates which are rich in
A·T (Wartell et al. 1974; Wähnert et al. 1975) or with A·T-rich
initiation sites in template DNA (Puschendorf et al. 1971, 1974).

I. The Distamycin A-DNA Interaction

Double-helical calf thymus DNA produces bathochromic shifts of
the first absorption maximum of distamycin (Zimmer et al. 1971b;
Zimmer 1975) from 303 to 321 nm but no significant decreases
in the absorption intensity of the antibiotic (Krey and Hahn
1970). Bathochromic shifts generally indicate the binding of
individual ligand molecules rather than of dimers or higher
aggregates (Michaelis 1947). The absence of large hypochromic
effects has been interpreted as an indication of a non-place-
ment of distamycin into the less polar environment between base
pairs of DNA (Krey and Hahn 1970).

Double-stranded DNA furthermore induces Cotton effects in the
ORD and CD spectrum of the antibiotic (Krey and Hahn 1970;
Zimmer and Luck 1970; Zimmer et al. 1971b; Zimmer et al. 1972)
while free distamycin in solution is optically inactive. The
effects induced in DMC by double-helical calf thymus DNA are
of much larger molecular amplitudes than the intrinsic Cotton
effects of DNA; they are probably associated with both electronic
transitions of distamycin A. DNA renders optically active the
transition of bound distamycin at around 321 nm by inducing a
first positive Cotton effect - either with an ORD peak and
through at about 351 and 315 nm, respectively, and a transition
midpoint at approximately 333 nm, or with a CD maximum at 328
nm - and it probably also renders optically active the second
transition - at 237 nm - of the bound antibiotic. The author
refrains from a discussion of effects observed for the distamycin-DNA complex below 300 nm.

Finally, double-helical DNA is stabilized by distamycin with
respect to heat (Chandra et al. 1970; Zimmer et al. 1970a,
1971a, 1972). Distamycin shifts the melting temperature of calf
thymus DNA to higher temperatures, increases the hyperchromicity
the nucleic acid undergoes during the thermal denaturation process, and renders this process more cooperative. The observed
changes increase systematically with the concentration of DMC
and are larger than the stabilizations most substances produce
for DNA.

These three biophysical indications of the binding of distamycin provide information on the A·T specificity in the interaction of the antibiotic with DNA. The bathochromic shifts produced by DNA in the absorption spectrum of the antibiotic increase with the A·T content of the employed DNA (Table 1); the induced Cotton effects of the antibiotic likewise depend on the A·T content of DNA (Zimmer and Luck 1970); and DNAs are stabilized by DMC to heat as a function of their A·T content (Zimmer et al. 1970a, 1971b, 1972; Zimmer and Luck 1970). These observations suggest preferential interaction of distamycin with DNAs of high A·T base composition. To investigate the structural and composi- tional requirements for the binding of distamycin, the studies of its interaction with natural DNAs were extended to synthetic DNA-like duplex polymers (Krey et al. 1973) with the following results:

Double-helical poly d(A-T), poly dA·dT, and poly dI·dC produce bathochromic shifts in the absorption spectrum of distamycin similar to those observed with the A·T-rich DNAs. They induce Cotton effects of large molecular amplitudes in the ORD spec- trum of the antibiotic and are strongly stabilized by the anti- biotic with respect to heat. In contrast, double-helical poly dG·dC causes only a minor redshift for DMC but a decrease in the antibiotic's absorption intensity, induces a small Cotton effect, and is stabilized by distamycin to a lesser extent to heat. These results are in agreement with a preferential dista- mycin binding to duplex polydeoxynucleotides of high A·T base content. The binding of the antibiotic to poly dG·dC is differ- ent and weaker than that to polymers of high A·T base composi- tion and is determined by the presence of guanine.

An A·T preference of distamycin is additionally suggested by electro-optical studies with calf thymus DNA and with poly d(A-T) and poly dG·dC (Krey, to be published). Orientation of calf thymus DNA by an electric field aligns the long axis of the double helix in the direction of the applied field so that the base pairs of the nucleic acid assume a roughly perpendicu- lar direction. This produces an increase in the 260-nm absor- bancy of the bases for incident light polarized perpendicular to the field and an absorbancy decrease for light of parallel

Table 1. Wavelengths of absorption maxima of the distamycin-DNA complex as function of DNA A·T content. (Data from Zimmer et al. 1971b)

Molar ratio of DNA phosphorus over distamycin A	DNA from Strepto- myces chrysomallus (28 mol % A·T)	DNA from calf thymus (58 mol % A·T)
0	303 nm	303 nm
5	305 nm	307 nm
10	306 nm	312 nm
17	–	315 nm
20	312 nm	–

polarization. Absorbancy changes of equal magnitude but opposite sign from those obtained for the base pairs of DNA are observed at 315 nm for bound distamycin. The resulting dichroic effects suggest for a composite transition moment of the three N-methyl-pyrroles of DNA-bound distamycin an orientation (calculated according to established principles) (Yamaoka and Charney 1972; Fredericq and Houssier 1973) of about 39 degrees versus the helical axis when base pairs are assumed perpendicular to the helix axis in the B form of DNA. Distamycin's binding to poly d(A-T) results in dichroic effects similar to those observed with DNA. On the other hand, interaction of distamycin with poly dG·dC, a duplex for which X-ray scattering studies indicate a secondary structure different from that of calf thymus DNA and poly d(A-T) (Bram 1971b) and which, according to X-ray diffraction results, coexists in the B- and A-conformation of DNA (Arnott and Selsing 1974), produces dichroic effects of smaller magnitude and opposite sign. These results indicate that the inclination of the antibiotic's transition moment versus the helical axis is different in double-helical poly dG·dC from that in poly d(A-T) and in calf thymus DNA.

The heterogeneity of distamycin's binding to double-helical polydeoxynucleotides of different base composition is interesting in view of the suggestion that the antibiotic interacts with native duplex DNA in a bimodal manner. It has been found that distamycin reacts strongly in an irreversible process with large molar excesses of DNA, and that it attaches weakly in a reversible manner to smaller amounts of the nucleic acid (Mazza et al. 1973). A combined plot of sedimentation data, separately obtained for the strong and the weak process, showed the presence of two independent attachment processes with different stoichiometries (weak process: 1 DMC per 3 base pairs; strong process: 1 per 8 pairs) and different association constants (weak: 5 to $11.6 \cdot 10^5$ M^{-1}, strong $2.4 \cdot 10^9$ M^{-1}) (Luck et al. 1974). Such bimodal interaction may suggest the existence of base-content-determined *regional differences* in the binding of the antibiotic to natural double-helical DNAs.

The possibility of such regional heterogeneity in the binding of distamycin was investigated with a spectrophotometric titration of DMC with calf thymus DNA (Krey, to be published). Increasing concentrations of calf thymus DNA produce progressive bathochromic shifts in the absorption spectrum of DMC, and cause reductions in the antibiotic's absorption intensity that are followed by absorbancy increases at higher DNA concentrations. The observed spectra are devoid of an isosbestic point. This indicates the existence of more than one mode of binding of distamycin to calf thymus DNA. The spectra further reveal red-shifts in the presence of high concentrations of DNA which resemble the large bathochromic shifts observed with poly d(A-T), poly dA·dT, and poly dI·dC but not the small shift produced by poly dG·dC. The preferred attachment, which these spectral similarities suggest, of distamycin to A·T-rich regions of DNA may constitute the strong binding process postulated by Luck and co-workers (1974). In contrast, distamycin spectral shifts and reductions, at low concentrations of DNA, similar to the

spectral changes observed with poly dG·dC, point to a second
mode of antibiotic attachment which is perhaps the weaker process of Luck (Luck et al. 1974). This weaker process may be an
interaction of DMC with G·C-abundant regions of DNA that is determined by guanine.

The stoichiometry of distamycin's binding to A·T-rich regions
of calf thymus DNA has been estimated from three different
binding indications (Krey, to be published). First, the stoichiometry has been derived from an electro-optical titration of
distamycin with calf thymus DNA in which graded concentrations
of the nucleic acid produce increasing amounts of bound antibiotic and therefore also progressively increasing dichroic effects. The observed effects approach saturation in the presence
of high concentrations of DNA at which distamycin binds to the
A·T-rich regions, suggesting a stoichiometry for this attachment of one DMC molecule bound per approximately six nucleotide
pairs. Secondly, distamycin's binding stoichiometry has been
estimated from the spectral shifts oligomers produce in the absorption spectrum of the antibiotic. Equimolar mixtures of oli-
godeoxynucleotides containing adenine and thymine produce, with
increasing chain length, progressively larger bathochromic
shifts for DMC, yielding again a stoichiometry of one antibiotic molecule bound per six base pairs. Finally, distamycin's
binding stoichiometry to the A·T-rich regions of DNA has been
obtained from thermal denaturations of poly d(A-T) with DMC.
The melting curves of poly d(A-T) are biphasic in the presence
of limiting amounts of DMC, indicating a thermal denaturation
(at lower temperatures) of polydeoxynucleotide segments free
of distamycin, followed by denaturations (at higher temperatures)
of poly d(A-T) regions covered with the antibiotic. Progressive
reductions of the excess of poly d(A-T) in the complex reduces
the hyperchromicity of free polymer segments until zero hyperchromicity is obtained when poly d(A-T) is completely occupied
by distamycin. This extrapolation likewise arrives at the
stoichiometry of one distamycin molecule bound per six base
pairs.

The differential interaction of distamycin with individual regions of double-helical DNAs according to A·T base composition
may be determined by a high affinity of DMC for A or T, by a
direct exclusion by G, or may represent a differential attachment of the antibiotic to individual DNA segments on which
their base composition imposes a characteristic conformation.
In view of the proposal of intramolecular differences in secondary structure of duplex DNA in solution (Bram 1971c) and of the
suggestion that "clusters of G·C in natural DNA" could adopt
the unique secondary structure of poly dG·dC which is different
from the B-form of poly d(A-T) and DNA (Bram 1971b), it appears
that the A·T preference in the binding of distamycin may constitute a preference for the B-form which is regionally imposed on
DNA by a high A·T base composition. On the other hand, the weaker
attachment of the antibiotic to G·C-rich regions may occur to
the different poly dG·dC secondary form. A preference of DMC
for the A·T-determined B-conformation is supported by the observation that a conformational transition of DNA from B to A

produced by ethanol (Ivanov et al. 1973a,b, 1974; Usatyi and
Shlyakhtenko 1974; Malenkov et al. 1975) reduces the magnitude
of the first Cotton effect induced in the CD spectrum of distamycin (Zimmer 1975), and also by the finding that the DNA-RNA
hybrid polymers, poly rA·dT and poly dA·rU, with a secondary
structure in the A-form, produce smaller antibiotic spectral
shifts than do the A·T-rich natural DNAs and DNA-like polymers
which are in the B-conformation (Krey, to be published).

The alternative possibility to the conformational hypothesis,
viz. an actual base specificity of the attachment of distamycin,
has also been investigated (Krey et al. 1976). Single-stranded
poly dA, poly dC, poly dG, and poly dT produce bathochromic
shifts in the absorption spectrum of DMC and reductions in the
absorption intensity of the antibiotic; the absorption redshift
is small with poly dC and increases progressively when produced
by poly dG, poly dT, and poly dA and is even more pronounced
with ØX 174 single-stranded DNA. Single-stranded calf thymus DNA
(Krey and Hahn 1970), poly dA, and to a lesser extent poly dG
induce Cotton effects in the ORD spectrum of distamycin A. In
melting experiments, ØX 174 single-stranded DNA, like heat-
denaturated DNA (Chandra et al. 1970), exhibits a reversible
cooperative transition in the presence of distamycin which has
been attributed (Krey et al. 1976) to the dissociation and re-
formation of an ordered complex of the antibiotic with single-
stranded DNA. The complexes with poly dA, poly dC, and poly dT,
but not with poly dG, do not dissociate with heat. These results indicate a base specificity of distamycin for A or T
when these bases are constituents of single-stranded polymers.
While this base specificity may contribute to the interaction
of DMC with double-stranded DNAs, the antibiotic's attachment
to A·T-rich duplexes produces much stronger binding indications
for distamycin, so that preference can be given to the conformational hypothesis, rather than to a singular base specificity, for the interaction of the antibiotic with DNA.

DNA conformational requirements for the binding of distamycin
imply an alignment of the antibiotic with the nucleic acid's
secondary structure similar to that observed upon orientation
of the DNA-DMC complex by an electric field. Analogous results
are obtained when the complex of the antibiotic with calf thymus
DNA is oriented by flow, that is, dichroic effects are measured
for the base pairs of DNA and for the chromophores of DMC which
are of equal magnitude but opposite sign (Krey and Hahn 1970).
From this can again an orientation be derived of the transition
moment of DMC of 39 degrees versus the helical axis in the
B-form of DNA. Although flow as well as electric dichroism experiments give information about the orientation of the transition moment of bound distamycin, such experiments do not
provide an indication of the orientation of bound DMC itself
because of uncertainly of the direction of its transition moment relative to its molecular sturcture; they do indicate, however, that the chromophores of the bound antibiotic are aligned
in the DNA-DMC complex in a very orderly array relative to the
helix structure.

This alignment with the helix is not the result of an intercalative mode of binding. Distamycin, added at increasing concentrations, progressively decreases the viscosity of calf
thymus DNA until a molar ratio of one DMC per 6 1/4 base pairs
is attained beyond which it again increases the viscosity of
DNA towards the original value (Reinert and Thrum 1970; Zimmer
et al. 1971b). The initial decrease in DNA viscosity produced
by the lower distamycin concentrations has been attributed to
an intramolecular aggregation of DNA and, on the basis of a
more recent analysis of viscosity changes (Reinert 1972), to a
decrease of the persistence length of the nucleic acid (Zimmer
1975), that is, to an increase in its bending (Luck et al. 1977;
Reinert 1978). The viscosity increase at higher antibiotic concentrations is thought to signal intermolecular DNA crosslinks
(Reinert and Thrum 1970; Zimmer et al. 1971b).

The non-intercalative mode of attachment these observations suggest for distamycin is in agreement with the fact that the antibiotic's N-methylpyrrole ring systems do not possess the minimal
planar area of 38 $\mathring{A}^2$, considered prerequisite for a helix insertion (Hahn 1971). The possibility of an insertion of the ring
systems of distamycin between the base pairs of DNA has been
eliminated by the failure of increasing concentrations of the
antibiotic to produce the rise and subsequent fall in the viscosity of closed circular PM2 DNA, which would indicate loss and
reversal of supercoils as consequence of an insertion binding
(Krey et al. 1973). Furthermore, the inability of distamycin
to cause an analogous dip in the circular nucleic acid's rate
of sedimentation also provides evidence against intercalation
binding (Luck et al. 1974). Finally, methylation experiments
of the DMC-DNA complex (Kolchinskii et al. 1975) as well as the
production of redshifts for DMC by major-groove-brominated poly
d(A-BrU), equal to the shifts observed with DNA (Krey, to be
published), suggest an attachment of the antibiotic to the outside of DNA in the minor groove of the helix structure.

The distamycin-DNA complex is held together by strong forces
which probably involve electrostatic as well as non-polar interactions. Indicative of the strength of the binding is the unusually large association constant of the stronger of distamycin's attachment processes (Luck et al. 1974). Furthermore, sodium
chloride (10^{-1} M), magnesium acetate (10^{-2} M) or 6 M urea do not
reverse the bathochromic shift produced by DNA in the absorption
spectrum of distamycin (Krey and Hahn 1970), neither does exhaustive dialysis of the antibiotic-DNA complex against $5 \cdot 10^{-3}$ M
Tris buffer or 1% sodium lauryl sulfate (Hahn and Krey 1971a,b).
The stability of the distamycin-DNA complex this indicates to
ionic competition suggests a contribution of non-polar forces
to distamycin's interaction with DNA.

A non-ionic contribution to the binding of distamycin is made
by the antibiotic's N-methylpyrrole ring systems. DNA produces
for a series of distamycin congeners with from 3 to 5 N-methylpyrrole rings redshifts and hyperchromicities (Zunino and Di
Marco 1972) as well as Cotton effects (Zimmer et al. 1972) which
increase with increasing numbers of the chromophores. Stabili-

zation to heat (Zimmer et al. 1970b, 1972) and viscosity decreases (Reinert and Thrum 1970) of DNA are also functions of the number of N-methylpyrrole rings. Finally, the displacement of methyl green from DNA increases with increasing numbers of the N-methylpyrrole rings (Zunino and Di Marco 1972).

An unambiguous interpretation in regard to strength or type of binding forces cannot be given for decreases in distamycin-DNA interaction indications by dimethyl sulfoxide, methanol (Zimmer and Luck 1972), or ethylene glycol (Zimmer et al. 1972), since in the presence of organic solvents, e.g., ethylene glycol, DNA undergoes a conformational transition from the B- to the C-form (Green and Mahler 1968; Nelson and Johnson 1970). The identical reasoning applies to decreases in antibiotic binding indicators by 7.2 M sodium perchlorate or 2 M lithium chloride (Zimmer and Luck 1972) since high salt concentrations likewise convert B-DNA to the C-form (Tunis-Schneider and Maestre 1970; Zimmer and Luck 1973). This transition diminishes the size of the minor groove of the canonical B-form (Ivanov et al. 1973b) in which distamycin is thought to bind.

The only method which dissociates the distamycin-DNA complex is an extraction with a biphasic phenol-water system in which DNA is found in the phenol-poor phase while distamycin is extracted into the phenol-rich phase (Krey et al. 1973). In this respect, the distamycin-DNA complex is reminiscent of the behaviour of nucleoproteins.

II. The Netropsin-DNA Interaction

Netropsin exhibits binding indications similar to those observed for distamycin. Thus, double-helical DNA produces bathochromic shifts of the first absorption maximum of netropsin from 296 nm to about 310 nm without significant absorbancy decreases (Zimmer 1975). Duplex DNA induces first Cotton effects in the ORD (Zimmer and Luck 1970; Zimmer et al. 1971b) and CD (Zimmer et al. 1972) spectra of the antibiotic at about 315 nm; these effects are, however, of smaller molecular amplitude than those observed for distamycin. Finally, DNA is stabilized by Nt to heat and undergoes in the presence of the antibiotic increases in melting cooperativity, hyperchromicity, and T_m, most of which are even more pronounced than the unusually large effects observed with distamycin (Zimmer et al. 1970a, 1971a,b, 1972).

Like for distamycin, binding indications for netropsin increase with increasing A·T content of DNAs. Natural DNAs cause absorption changes (Wartell et al. 1974) and Cotton effects (Zasedatelev et al. 1974) which yield netropsin attachment parameters that depend on A·T content. An A·T preference of netropsin is also indicated by DNA-induced anomalies in the ORD (Zimmer and Luck 1970) and CD (Luck et al. 1974) spectra of the antibiotic, by DNA melting experiments (Zimmer et al. 1970a, 1971b, 1972), and by density changes in isopycnic centrifugations of netropsin-DNAs (Matthews et al. 1978). Furthermore, spectrophotometric and polarimetric indications of netropsin's binding to the synthetic DNA-like duplexes, poly d(A-T) and poly dA·dT, are of similar magnitude as those with the A·T-rich natural DNAs, while poly dG·dC and poly d(G-C) cause no effects (Wartell et al. 1974; Zasedatelev et al. 1974). Large effects are however also observed with the I·C containing duplexes, poly dI·dC and poly d(I-C) (Wartell et al. 1974), which suggests that the A·T preference of netropsin is, like that of distamycin, determined by the absence of guanine.

The A·T preference of netropsin is not an individual base specificity since the single-stranded homopolymers poly dA, poly dC, poly dG, poly dI, and poly dT produce no binding effects (Wartell et al. 1974). The affinity this rather suggests of netropsin for A·T pairs may constitute the only mode of interaction of the antibiotic with natural DNAs. Absorption difference spectra of netropsin-DNA complexes reveal an isosbestic point. Also, linear dichroism of these complexes yield only *one* orientation of bound netropsin in accord with only one type of antibiotic attachment to DNA (Wartell et al. 1974). The stoichiometry of netropsin's binding to its one class of sites in DNA has been derived from a comparison of DNA-produced spectrophotometric binding data and an allosteric model of DNA-ligand interactions as one antibiotic molecule per three A·T pairs (Wartell et al. 1974), or from netropsin-poly d(A-T) melting experiments as one antibiotic molecule per four or five base pairs (McGhee 1976; Patel and Canuel 1977).

DNA interaction studies carried out at lower ionic strengths yield manifestations of two modes of attachment of netropsin. Natural DNAs produce in polarimetric titrations families of CD spectra for Nt which reveal an isosbestic point only for high concentrations of DNA while at lower DNA concentrations the spectra are displaced from this point, pointing to more than one form of association of netropsin with DNA (Luck et al. 1974). The circular dichroic effects induced by high concentrations of DNA resemble the anomalies produced by the A·T-containing DNA-like duplexes, so that at high DNA concentrations the antibiotic binds strongly to a class of sites in DNA which is abundant in A·T pairs. Weaker types of attachment at lower DNA concentrations involve antibiotic interactions with sites containing a smaller number of A·T pairs (Luck et al. 1974). Evidently, the second class of sites with the lesser number of A·T pairs manifests itself more clearly at lower ionic strengths and involves a type of Nt attachment which accounts for the indications of the reduced binding of the antibiotic to natural DNAs, observed under high-salt conditions (Zasedatelev et al. 1974; Zimmer 1975).

The strong preference of netropsin for A·T pairs of DNA may be interpreted, like that of distamycin, as an affinity for regions of DNA on which the A·T pairs impose the B-form. Additional support for the importance of the B-conformation in the binding of netropsin comes from the fact that A-form double-stranded f2 phage RNA (Luck et al. 1974) and RNAs or hybrids of DNA and RNA (Wartell et al. 1974) produce nonspecific or no-attachment indications for the antibiotic, and from observations of a release of Nt from DNA during the B to A conformational transition of the nucleic acid produced by ethanol (Luck and

Zimmer 1973). This release is most prominent from A·T-poor
DNAs, but practically nonexistent for netropsin bound to A·T-
rich DNAs. Netropsin prevents the A·T-rich DNAs from adopting
the A-form (Luck and Zimmer 1973) and reverses, with increasing
concentrations, the B to A transition of (58% A·T) calf thymus
DNA and returns the nucleic acid to the B-conformation (Ivanov
et al. 1974; Zasedatelev et al. 1974).

Netropsin binds, like distamycin, to the outside of DNA in the
minor groove. However, while distamycin produces viscosity decreases for DNA, typical of an outside attachment, netropsin
causes viscosity increases (Reinert and Thrum 1970; Zimmer et
al. 1971b) which have been attributed to an A·T-specific elongation of the contour length of DNA and stiffening of the DNA
helix (Reinert 1972). Increases in the contour length of DNA
and stiffening are conventionally attributed to intercalation
(Lerman 1961); such binding can, however, be ruled out for
netropsin: the antibiotic hardly distorts adjacent regions of
DNA but allows the vicinal insertion of actinomycin D (Wartell
et al. 1974, 1975), and it also does not unwind and rewind superhelical M13 RFI (Wartell et al. 1974) or PM2 (Luck et al.
1974) DNAs sufficient to indicate intercalation binding. Spectrophotometric shifts of netropsin by the duplex polymers, poly
dI·dBrC and poly d(A-BrU) (which carry bromine in the major
groove), are equal to the shifts produced by A·T-rich DNAs and
suggest the placement of the antibiotic into the minor helical
groove (Wartell et al. 1974).

Netropsin binds to DNA with ionic and non-polar forces. These
forces are weaker than those which bind distamycin, since they
permit some release of netropsin at higher ionic conditions
(Zasedatelev et al. 1974; Zimmer 1975). This release is most
prominent for netropsin from its secondary sites in DNA with
the smaller number of A·T pairs and allows an attachment of the
antibiotic to poly dG·dC only at very low ionic strengths in
contrast to its interaction with poly dA·dT at higher salt concentrations. Almost complete binding to poly dA·dT occurs in
5 M lithium chloride in which the interaction with poly d(A-T)
is, however, decreased and that with calf thymus DNA reversed
(Luck et al. 1974), as it is in the presence of 7.2 M sodium
perchlorate or in ethylene glycol (Zimmer et al. 1971b; Zimmer
and Luck 1972). The almost complete interaction of netropsin
with poly dA·dT in 5 M lithium chloride may, in the absence of
polymer conformational changes (Zimmer and Luck 1974), be caused
by strong, predominantly non-ionic attachment forces (Luck et
al. 1974). Elimination of the basic side chains of netropsin,
bearing a number of hydrogen donor sites (Zimmer 1975), sharply
reduces the antibiotic's interaction with DNA (Reinert and Thrum
1970; Zimmer et al. 1970b, 1972), and suggests for this attachment the involvement of ionic as well as non-polar binding forces.

The optical and hydrodynamic observations of the binding of
netropsin form the basis of two almost identical, tentative
models that have been proposed for the netropsin-DNA complex
(Wartell et al. 1974; Zimmer 1975). Both models emphasize the
requirement of the B-form of DNA and both propose a non-intercalative attachment of netropsin. They also agree on a placement

of the antibiotic into the minor helical groove, and emphasize
the importance of electrostatic interactions of the two positively charged side chains of netropsin with the phosphates of
DNA. In one model (Wartell et al. 1974), the predominantly
planar netropsin molecule attaches to three base pairs by span-
ning alongside one A·T pair at an angle of 30 degrees, and by
forming hydrogen bonds, involving its side chains. In the second
model (Zimmer 1975), netropsin is placed into the minor groove
so that it contacts five base pairs, with additional possibilities of hydrogen bond formation. The important fact, in spite
of the uncertainty of netropsin's disposition in the small
groove, is that the antibiotic produces lengthening and stiffening effects of DNA which have not been observed with distamycin. There is one significant difference between the bound
forms of the antibiotics, viz. the direction of the two substances
in the small groove. Planar netropsin spans this groove, according to the first model, obliquely with an orientation almost
perpendicular to the helical axis, while longer and perhaps
more flexible distamycin (Krey, electric dichroism studies to
be published) adapts itself instead more lengthwise into that
groove (Zasedatelev et al. 1976, 1978) to wrap itself around
the helix at a more-nearly-parallel-to-the-helical-axis inclination (Poletaev et al. 1977).

E. Methyl Green

Methyl green (MG) belongs to the class of triphenylmethane dyes.
Its chemical structure is shown in Fig. 2. The dye is important
as a histochemical stain for DNA (Kurnick 1950a) and serves,
when complexed with the nucleic acid, as substrate for the determination of deoxyribonuclease activity (Kurnick 1950b, 1962)
and as indicator of the binding of chemotherapeutic drugs whose
bioreceptor is DNA (Krey and Hahn 1975). The underlying reason
for the histochemical and biochemical usefulness of methyl green
is that its colored quinonoid form (Fig. 2) is stabilized by
complexation with DNA, while the free dye in solution spontaneously rearranges into a colorless benzenoid form. This conversion appears to be a group property of triphenylmethane dyes
(Goldacre and Phillips 1949).

To understand this selective stabilization of methyl green, it
has been necessary to study its binding to DNA. The interaction
of other triphenylmethane dyes with the nucleic acid has also
been investigated, though not as extensively as that of MG.
Rosaniline, the methylated parent compound, binds to DNA (Cavalieri and Angelos 1950; Cavalieri et al. 1951) with shifts in its
absorption maximum, like that of the unmethylated parent substance, para-rosaniline, to longer wavelengths (Lawley 1956;
Yamaoka 1972). DNA produces bathochromic shifts for malachite
green and crystal violet (Adams 1968; Yamaoka 1972) and induces
Cotton effects in the ORD spectra of these dyes (Yamaoka 1972).
The attachment of the triphenylmethane dyes is specific for A·T
pairs and therefore for the B-form of DNA (Müller and Gautier 1975).

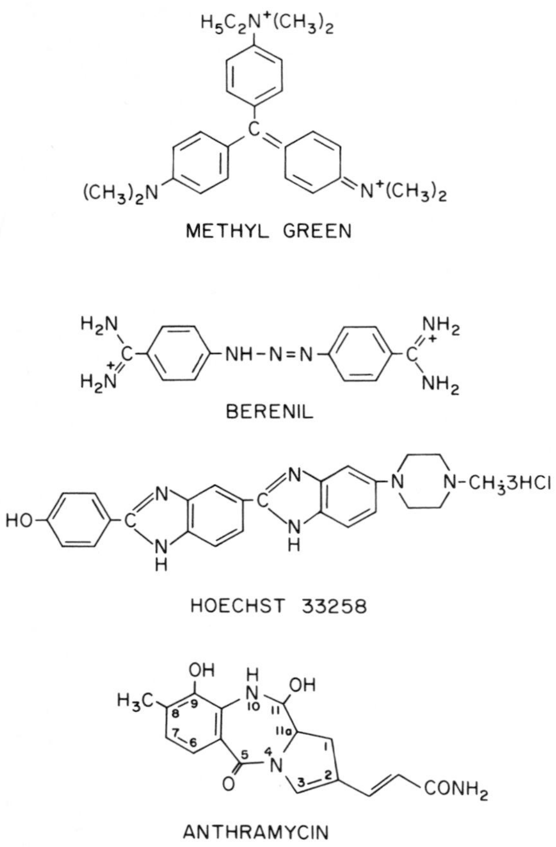

$H_5C_2N^+(CH_3)_2$

$(CH_3)_2N$ $N^+(CH_3)_2$

METHYL GREEN

H_2N
     C $NH-N=N$ C
H_2N NH_2
                                             NH_2

BERENIL

HO C C N $N-CH_3 \cdot 3HCl$

HOECHST 33258

H_3C OH   H   OH
              N
              10 11
      8   9   11a
   7 6   5 4
        O   N   3 2
                        CONH_2

ANTHRAMYCIN

Fig. 2. Chemical structures of non-intercalators

I. Binding of Methyl Green to the B-Form of DNA

Methyl green absorbs prominently in the visible region of the spectrum with an absorption maximum at 632 nm and a shoulder (due to a 587-nm absorption band) at shorter wavelengths and possesses less intense absorption maxima at 421 nm, 312 nm, and around 260 nm (Rottier 1953; Krey, unpublished observations). The visible and near-UV absorption bands of methyl green in solution decrease in intensity as a consequence of the dye's intramolecular rearrangement to the benzenoid carbinol (Kurnick and Foster 1950). The conversion is favored by pHs outside the range of 3.5 to 5.0 (Kurnick and Foster 1950) and proceeds, at pH 7.5, in 19 h to an equilibrium in which 98% of MG is in the colorless state (Kurnick and Foster 1950; Krey and Hahn 1975).

Methyl green's rearrangement can be prevented by double-stranded DNA. Duplex DNA forms a stable complex with methyl green (Kurnick and Foster 1950) and shifts its absorption maximum to longer wavelengths (Kurnick and Foster 1950; Kurnick and Mirsky 1950; Scott 1967; Krey and Hahn 1975) with increases in absorption intensities (Kurnick and Mirsky 1950) to values about 40% above those of the free dye (Krey and Hahn 1975). These bathochromic shifts and intensity increases of methyl green can be ascribed to a binding of single dye molecules by a non-intercalative process. The absorption intensity of bound methyl green remains constant over a period of hours or days (Krey and Hahn 1975), indicative of the stabilization of the colored form of the dye, and is even regained upon addition of native calf thymus DNA to completely decolorized MG (Kurnick and Foster 1950). Methyl green binds to DNA with a stoichiometry of one dye molecule per 13 DNA phosphates (Kurnick and Foster 1950).

The colored form of methyl green is also stabilized by binding to DNA which has been transversely fragmented into shorter double-helical rods (Rosenkranz and Bendich 1958) and upon re-naturation of DNA (Pellicciari and Fraschini 1978), but fails to be preserved by heat-denaturated DNA and by DNA which under-goes enzymatic hydrolysis (Kurnick 1950c,d, 1954; Kurnick and Foster 1950; Kurnick and Mirsky 1950; Rosenkranz and Bendich 1958; Scott and Willett 1966; Scott 1967). The selectivity this suggests of methyl green for duplex DNA is supported by obser-vations that single-stranded ØX 174 DNA and transfer RNA fail to conserve the absorption intensity of the dye. Instead, the dye's absorbancies decrease gradually in the presence of the polymers over several days to final values which are below those of freshly dissolved MG (Krey and Hahn 1975). Evidently, only a fraction of methyl green becomes stably bound to single-stranded DNA and tRNA, as does to heat- and enzymatically-altered DNA.

The preference for the double-helical secondary structure of DNA is, however, not a sufficient condition for the stable binding of methyl green. While double-helical poly d(A-T), like native calf thymus DNA, increases the absorption inten-sity of methyl green and restores the visible and near-UV ab-sorbancies of degraded MG, duplex poly dG·dC, which in contrast to DNA and to poly d(A-T) prefers the A-form, fails to protect the dye. This suggests that the preference of methyl green for DNA's double-helical structure may be a preference for DNAs and DNA-like polymers which are in the B-form (Krey and Hahn 1975).

The preference for the B-conformation may be imposed by two ad-jacent A·T pairs, necessary for the efficient binding of methyl green (Müller and Gautier 1975), but the release of methyl green during the B to A conformational transition of DNA produced by ethanol suggests that the determining factor in the stable attach-ment of the dye is the B-form. Thus, ethanol, added to an aqueous solution of the methyl green-DNA complex, releases methyl green in two distinct reactions: A slow zero-order kinetic process and a rapid initial release after which MG decays according to first-order kinetics. The amount of initially released MG is moderate at low ethanol concentrations which produce for DNA only small structural changes within the B-family of conformations (Girod et al. 1973). At higher ethanol concentrations above 65 vol%, however, large structural changes occur which transform DNA from the B- to the A-forms (Ivanov et al. 1973a,b, 1974; Usatyi and Shlyakhtenko 1974; Malenkov et al. 1975). This conformational transition to A-DNA releases the bulk of the dye which is pre-ferentially bound to DNA in the B-form (Krey and Hahn 1975).

II. The Non-Intercalative Binding of Methyl Green

The mode of binding of methyl green has remained a continued
question as it has for other triphenylmethane dyes. Intercalation has been suggested for para-rosaniline, neofuchsin, and
Doebner's violet with speculations that two of the three phenyl
rings of these dyes are inserted with enforced coplanarity between the base pairs of DNA (Lerman 1964a). Extending the arguments of an enforced coplanarity of two phenyl rings to methyl
green, some authors (Scott and Willet 1966; Scott 1967) have
speculated on the intercalation of MG.

Intercalation by enforced coplanarity of two phenyl rings appears however to be an unlikely mode of binding of the triphenylmethane dyes. These dyes assume - in order to alleviate over-
crowding of hydrogens in the 2- and 6-positions (Lewis et al.
1942) - a non-planar, propeller-like conformation of their
phenyl rings (Sharp and Sheppard 1957; Gust and Mislow 1973).
Accordingly, no evidence of intercalation has been obtained for
the parent compound para-rosaniline in X-ray fiber analyses of
its complex with DNA (Neville and Davies 1966) and for crystal
violet in viscometric titrations of DNA (Müller and Gautier 1975).

An intercalative mode of binding for methyl green is contraindi-
cated by observations that in methyl green-DNA fibers the dye's
"light absorbing groups" are lying approximately parallel to the
fiber axis (White and Elmes 1952) and that in flow-oriented
methyl green-DNAs MG's 642-nm electronic transition is directed
towards the flow direction (Hahn and Krey 1969) at an angle,
derived from dichroic values (Krey and Hahn 1971; Nordén and
Tjerneld 1977) according to established principles (Yamaoka and
Charney 1972; Fredericq and Houssier 1973), of less than 50 degrees versus the helix axis. To study the mode of methyl green
binding conclusively, the influence of the dye on the viscosity
of DNA was investigated in a viscometer equipped with an ultra-
sensitive timing device, necessary to detect the relatively
small hydrodynamic effects produced by MG (Krey and Hahn 1975).
The results showed the viscosity decreases of DNA which provide
evidence against intercalation of methyl green and suggest a
binding of the dye to the outside of the double helix.

III. Electrostatic Binding of Methyl Green

The presence of two positive charges in the methyl green molecule suggests an electrostatic binding to DNA. Sodium ions interfere at concentrations above 3 M with the attachment of methyl
green (Scott and Willet 1966) or produce at 1 to 1.5 M a dissociation from DNA (Inagaki and Kageyama 1970). Magnesium ions
also release methyl green from its complex with DNA. Moreover,
graded concentrations of these ions produce a systematic correlation between the first-order rate constants and the end
points of the release reactions which is bimodal, pointing to
differences in the kinetics of the methyl green displacement
mechanism by ionic competition and thus perhaps to more than
one mode of electrostatic attachment of the dye (Krey and Hahn
1975).

In contrast to the low concentrations of magnesium ions ($5 \cdot 10^{-3}$ to $5 \cdot 10^{-1}$ M) which almost completely release methyl green, higher amounts of urea (6 M) cause only a 40% displacement of MG (Krey and Hahn 1975). Urea has the ability to form hydrogen bonds (Mirsky and Pauling 1936) but fails, at 6 M, to produce a denaturation of DNA. Evidently, some non-ionic forces contribute to the attachment of methyl green which are, however, of lesser importance. Binding of the dye is thus predominantly of electrostatic nature.

A counterionic effect of methyl green has been observed in melting experiments with DNA (Krey and Hahn 1975). In its stoichiometric complex with DNA, methyl green shifts DNA's melting profile by $+12^\circ$C, while equimolar crystal violet fails to produce this effect. This indicates the importance of the quarternary ammonium group of MG in the binding of methyl green. Furthermore, the release of MG during the progressive melting of the secondary structure of DNA suggests an attachment of the two cationic charges of the dye to phosphates of both strands of DNA. Methyl green fits - according to model building experiments (Krey, unpublished observation) - with its two positive charges to phosphates across the minor groove of the DNA helix. The fact that MG binds to a lesser extent also to single-stranded DNA and to DNA which has been heat-denaturated, suggests a second mode of attachment of the dye to phosphates along the same DNA chain.

IV. Bimodal Binding of Methyl Green from MG Displacements by DNA-Complexing Drugs

Methyl green is liberated from its complex with DNA by drugs and dyes which form an association with DNA. Such release of MG has been first observed with the antimalarials quinacrine and chloroquine (Kurnick and Radcliffe 1962) while subsequently displacements of the dye have been used to study the binding of antibiotics (Rauen et al. 1965; Zeleznick and Sweeney 1967; Bates et al. 1969; Di Marco et al. 1971; Zunino and Di Marco 1972) and of other substances to DNA (Zeleznick et al. 1969). These investigations report only arbitrary end points of the displacement reactions, while, in fact, their kinetics and absolute end points can be determined systematically.

The progress of the methyl green displacement reactions can be followed spectrophotometrically as decrease in the 642-nm absorbancy of the triphenylmethane dye. Since free methyl green decays at a rate faster than the release reactions, the displacements themselves rather than the intramolecular rearrangement of free MG constitute the rate-limiting factor. Evaluation of the displacements so obtained for a series of DNA-binding drugs and dyes (Krey and Hahn 1970, 1971, 1975) shows that the release of methyl green by some substances follows a kinetic course which cannot unambiguously be interpreted as either first- or second-order, while the displacements by other compounds, particularly by quinacrine and other substances which displace methyl green rapidly and almost completely, are initially of first-order and change with decreasing concentrations of free displacing compound to a subsequent second-order reaction. An entirely first-order reaction is obtained with 10^{-4} M quinacrine while $2.5 \cdot 10^{-5}$ M quinacrine produces a second-order displacement course (Krey and Hahn 1974).

The end points of the methyl green displacements by various DNA-complexing substances have a systematic bimodal relation to the first- as well as second-order rates of the displacement reactions (Krey and Hahn 1975). It has been pointed out (Krey and Hahn 1975) that this relation, which is obtained for different DNA-binding compounds, can virtually be superimposed on the similar correlation that is observed for a series of magnesium ion concentrations. These correlations are interpreted as an indication of a bimodal binding of methyl green (Krey and Hahn 1975).

Methyl green is displaced rapidly and almost completely by the phenanthridines propidium and ethidium and by the aminoacridines quinacrine, Nitroakridin 3582, Nitroakridin 2, and proflavine, i.e., by substances known to be typical strong intercalators. It is also displaced efficiently by the antibiotic daunomycin which belongs into this group of DNA-binding compounds (review: Di Marco et al. 1975) and by the antiviral substance tilorone, considered to be an intercalator (Chandra et al. 1972c). In contrast, the isolated dicationic side chain of quinacrine and chloroquine, which is devoid of a chromophore necessary for intercalation, and the 8-aminoquinoline primaquine, which binds weakly to DNA (Morris et al. 1970), displace only small amounts of MG. Between these two extremes is a heterogeneous group of substances with intermediate methyl green displacing capabilities: the antibiotic distamycin A which binds strongly to DNA but which does not intercalate and the benzimidazole "Hoechst 33258" which also binds strongly but which likewise is a non-intercalative compound (Section F of this review); 6-chloro-2-methoxy-9-methylaminoacridine, which as methylated acridine ring system of quinacrine can be assumed to intercalate but perhaps with a lesser affinity for the DNA helix; the thiaxanthenone miracil D, which does intercalate (Weinstein and Hirschberg 1971; Hirschberg 1975) but probably also with lesser affinity than quinacrine (Weinstein and Hirschberg 1971); chloroquine and quinine which intercalate partly, since their quinoline ring system occupies only the minimal planar area for the insertion (Hahn et al. 1966; Estensen et al. 1969); actinomycin D which intercalates with low stoichiometry (Wells and Larson 1970); acridine orange which predominantly stacks to the outside of the double helix (Bradley and Wolf 1959); and finally, berberine for which intercalation is indicated but whose conformation in aqueous solution is probably only partly planar (Krey and Hahn 1969; Hahn and Krey 1971c).

The efficient displacement of methyl green from DNA by intercalating substances probably results from the change in helical parameters which releases the triphenylmethane dye that is attached to phosphates of both companion strands of DNA across the minor groove of the helix structure.

The methyl green displacements further suggest a liberation of dye from its attachment to phosphates of one strand of DNA. Such a release may be accomplished without a change in helical parameters by substances which attach themselves to the exterior of DNA along one strand of the DNA helix. Non-intercalative anthramycin (Section F) fulfills this condition (Glaubiger et al. 1974) and displaces, with its one-strand binding, methyl green (Bates et al. 1969) which may be bound to individual DNA chains. One can thus conclude that the methyl green displacements provide an indication of a bimodal binding of methyl green: across the minor groove of B-DNA and along individual strands of the DNA helix.

F. Other Non-Intercalators

I. Berenil

The aromatic diamidine berenil, B (Fig. 2), possesses trypanocidal and babesicidal properties and is used for the treatment of bovine trypanosomiasis and, in limited clinical trials, against early stages of sleeping sickness in man (review: Newton 1975). B also exerts antibacterial and antifungal activities (review: Newton 1975) and produces respiratory deficient mitochondrial mutants in yeast (Mahler and Perlman 1973). The trypanocidal and perhaps also the mutagenic activity of berenil may be attributed to a preferential inhibition by B of the replication of kinetoplast or respectively mitochondrial DNA through an interaction of the drug with the extranuclear (circular) DNA template (Newton and Le Page 1967; Perlman and Mahler 1973; Nagley et al. 1975; Newton 1975). Action of berenil on A·T-richer extranuclear kinetoplast (Riou and Paoletti 1967) or mitochondrial (Tewari et al. 1966) DNA rather than on the A·T-poorer chromosomal nucleic acid may have as one explanation the base specificity found for the binding of B. Structurally related hydroxystilbamidine resembles berenil with its non-intercalative (Festy and Daune 1973), bimodal (Festy et al. 1975; Sturm et al. 1975), and A·T-specific (Festy et al. 1975) binding to DNA.

Upon binding, DNA produces a bathochromic shift in the 370-nm absorption maximum of berenil. Spectrophotometric titration of B yields a maximum attachment of one drug molecule per four to five nucleotides of calf thymus DNA (Newton 1967; Newton and Le Page 1968) as does a gel filtration technique (Newton 1972). Berenil also binds to heat denatured DNA, RNA, synthetic homopolymers, and a poly AU copolymer with no evidence of an attachment to mononucleotides (Newton 1967). In its complex with DNA, B raises the melting temperature of the nucleic acid and decreases its buoyant density (Newton 1967; Newton and Le Page 1968; Newton 1972). The decrease in the buoyant density depends on the A·T content of DNA and constitutes evidence for an A·T specificity in the binding of B (Newton and Le Page 1968). Such specificity has likewise been derived from the progressive effects of increasing concentrations of berenil on the sedimentation of DNA which demonstrated a bimodal attachment of B: the stronger mode of attachment is characterized by an association constant which depends on A·T base composition (Festy et al. 1970).

Berenil causes no increase in the viscosity of DNA (Newton 1967; Newton and Le Page 1968) and fails to produce, upon progressive binding, sedimentation decreases and subsequent increases indicative of changes in the superhelical density of closed circular DNA (Waring 1970a,b, 1971). These hydrodynamic results are evidence against intercalation binding.

Berenil binds to the outside of DNA. Its two amidine groups are electrostatically attached to phosphates of DNA. This adds to the non-ionic DNA binding process observed at higher ionic strengths (Newton 1975). Important for the binding of B is the spacing of its two terminal amidine groups: the amino-azo derivative of the drug with two nitrogens instead of the central triazine bridge, obtained after rearrangement in neutral solution, fails to cause a decrease of the buoyant density of DNA (Newton 1967, 1975) as does "Compound 1" which likewise lacks one central N (Newton 1975), while "Compound 2", devoid of a second N, affects, like B, the hydrodynamic property of DNA (Newton 1975). It thus appears that an 11- to 13-Å distance between the terminal amidine groups in berenil or in Compound 2 (rather than the 5-Å distance in Compound 1) (Newton 1975), is most suitable for the binding to DNA and allows probably a bimodal attachment of B to the double helix which resembles the across-the-minor-groove interaction proposed for hydroxystilbamidine (Festy et al. 1975) and the additional intrastrand binding described (in previous sections) for polyamines and methyl green.

II. Hoechst 33258

The bis-benzimidazole derivative Hoechst 33258 (Fig. 2) has been introduced as fluorescent probe for the structure (Hilwig and Gropp 1972) and function (Latt 1973) of chromosomes. It produces chromosomal banding patterns which differ from those observed with the intercalator quinacrine (review: Latt 1976). One of the reasons for this difference lies in the fact that 33258 binds to DNA with a specificity for a high A·T base composition.

This A·T specificity provides for the more prominent of two attachment processes of Hoechst 33258 and has been investigated by several nonfluorescent techniques: First, increasing concentrations of DNA progressively shift 33258's absorption maximum to longer wavelengths as well as decrease and subsequently increase the absorption intensity of the dye. No isosbestic point is observed pointing to the existence of more than one form of bound dye. The absorption redshift produced for 33258 depends on the A·T content of the employed DNA and furnishes one indication for the base specificity of the Hoechst compound (Bontemps et al. 1975; Latt and Wohlleb 1975; Latt and Stetten 1976). Second, Hoechst stabilizes DNA to heat with increases

in the melting temperature which likewise suggest an A·T specificity of the dye (Comings 1975). Third, DNA renders Hoechst
optically active which as free dye does not exhibit CD (Latt
and Wohlleb 1975). High concentrations of DNA induce a large
near-UV Cotton effect for 33258 at 360 nm, which resembles the
strong optical activities produced by poly d(A-T). On the other
hand, low DNA concentrations produce only weak anomalies with a
shoulder at shorter wavelengths (335 nm) where poly d(G-C) induces a relatively small effect (Bontemps et al. 1975; Latt and
Wohlleb 1975; Latt and Stetten 1976). These three optical techniques demonstrate the A·T preference in the binding of Hoechst
33258 to natural DNAs which prevails at high DNA concentrations
and requires three adjacent A·T pairs in the binding site
(Müller and Gautier 1975). They also suggest a weaker type of
dye attachment to secondary sites at low concentrations of DNA,
that is determined by the presence of G·C pairs.

The preferential binding of Hoechst to A·T-rich regions of DNA
is not reversed at higher ionic strengths (Latt and Wohlleb
1975; Latt and Stetten 1976) and may be the stronger of two attachment processes, derived from equilibrium dialysis titrations
for 33258's association with DNA (Bontemps et al. 1975). Hoechst's
preference for A·T-rich regions at high concentrations of DNA is
explained by the requirement for three adjacent A·T pairs (Müller
and Gautier 1975) whose frequency in natural DNAs may be as low
as the stoichiometry of the stronger binding process (Bontemps
et al. 1975). If one equates A·T specificity and strong binding
of 33258 at high concentrations of DNA, one can consider that
its G·C-determined attachment at low DNA concentrations represents a relatively weak interaction with DNA. Indeed, it has
been found that the G·C-determined attachment process of 33258
is easily reversed by 0.4 M NaCl with no effect on the A·T-
specific binding of the dye (Latt and Wohlleb 1975; Latt and
Stetten 1976).

The A·T-specific binding of the benzimidazole plays an important
role in the fluorescence properties of Hoechst. Fluorescence of
free Hoechst is weak (quantum yield 0.01) and exhibits an excitation maximum at 345 and an emission maximum at 505 nm, but is
increased upon addition of DNA and shifted to longer excitation
or shorter emission wavelengths (Latt 1973; Latt and Wohlleb
1975). The increases in the fluorescence intensity of Hoechst
depend exponentially on the amount of added DNA (Weisblum and
Haenssler 1974; Comings 1975) and resemble at higher DNA concentrations an enhancement observed with A·T-rich regions of
DNA (Bontemps et al. 1975) or with A·T-rich DNAs (Latt and Wohlleb 1975). Fluorescence intensity increases of Hoechst are
furthermore proportional to the A·T content of added natural
DNAs or synthetic DNA-like polymers (Comings 1975) and constitute increases not only by A·T-rich regions of DNAs but also,
though less efficiently, by G·C-rich regions (Weisblum and
Haenssler 1974). Hoechst's fluorescence increases are in contrast to the fluorescence quenching produced for quinacrine by
(regularly interspersed) G·C pairs (Weisblum 1973, 1974) and
have been attributed to changes in the quantum yield of the
benzimidazole (Bontemps et al. 1975); however, the dye's A·T

binding specificity also contributes to the fluorescence properties of the Hoechst compound.

Increases in the fluorescence of Hoechst can, at higher ionic
strengths, serve as measure for A·T-rich regions of natural DNAs
(Latt and Stetten 1976). The fluorescence intensity of Hoechst
bound to A·T-rich regions or to A·T-containing synthetic polydeoxynucleotides is, however, reduced upon introduction of bromine into the major helical groove (Latt 1973; Latt et al. 1975).
This suggests binding of the dye in the large groove (Làtt 1973).

Attachment of 33258 in one of the grooves instead of by intercalation appears likely since the compound produces viscosity
decreases of DNA (Comings 1975). External attachment is also
suggested by electro-optical determinations of the orientation
of 33258 in its complex with DNA. The benzimidazole binds to
DNA with a long-axis orientation of 45 degrees versus the helix
axis of A·T-rich regions of DNA, i.e., of those regions which
produce strong attachment with high fluorescence. It furthermore interacts weakly with G·C-rich regions of the nucleic acid
at an inclination which, though somewhat closer to that of the
base pairs, suggests a second external attachment of the dye
(Bontemps et al. 1975).

III. Anthramycin

The antibiotic anthramycin (Fig. 2) possesses antitumor, anti-
microbial and chemosterilant properties (review: Horwitz 1971;
Kohn 1975), selectively inhibits DNA and RNA (but not protein)
synthesis in vivo, and inhibits the DNA-dependent DNA and RNA
polymerases and DNAse I in vitro (review: Kohn 1975). Although
the binding of anthramycin to DNA templates has been studied
by several groups, evidence of an unequivocal base specificity
in the attachment process has not been obtained yet. In that
respect anthramycin differs from other non-intercalators in
this review.

For binding studies with DNA, anthramycin has frequently been
used in the form of its methyl ether. The use of the methyl
ether was necessitated by the instability of the antibiotic
in aqueous solution. The compound yielded, however, the same
biological, biochemical, and interaction responses as anthramycin itself (Horwitz 1971; Kohn 1975). For this reason, both
anthramycin proper as well as its methyl ether will be referred
to in the review here as "anthramycin" (abbreviated: A).

Binding stabilizes anthramycin in aqueous solution and protects
the antibiotic against degradation by heat or high pH (Kohn and
Spears 1970). It alters A's absorption spectrum with bathochromic shifts and hypochromicities which depend on the concentrations of DNA added (Stefanović 1968) and on the time allowed
for the binding reaction; binding has been reported completed
in several seconds (Stefanović 1968) or in 60 to 95 min (Kohn
et al. 1968; Horwitz 1971; Hurley et al. 1977). Binding of
anthramycin stabilizes the secondary structure of DNA to heat

and decreases the buoyant density of the nucleic acid (Kohn et al. 1968; Bates et al. 1969; Kohn and Spears 1970). The binding has also been investigated by equilibrium dialysis and an adsorption isotherm derived which suggests a bimodal attachment of A to a small set of stronger sites, observed at high DNA concentrations, and a large set of weaker sites at low concentrations of DNA (Bates et al. 1969).

A's absorption spectrum is not altered either by deoxyguanosine (Adamson et al. 1968), by the natural purines and pyrimidines and their oxy- and deoxynucleoside and -nucleotide derivatives (Stefanović 1968), or by the four deoxynucleotides, individually (Kohn and Spears 1970) or in mixture (Kohn et al. 1968). On the other hand, the antibiotic shows large spectrophotometric interaction indications with poly dG·dC and poly dG but not with poly dC or with DNA-like duplexes which contain I and C or A and T (Kohn et al. 1974). The preference this would suggest anthramycin possesses for guanine manifests itself, however, only after the antibiotic has reacted with the polymers for a period of approximately 40 min. In contrast, A attaches itself during the earlier stages of the binding process faster to A·T-rich DNA, suggesting the existence of a class of fast reacting A·T-rich sites in addition to a set of slower binding G·C-rich sites for anthramycin in natural DNAs (Kohn and Spears 1970). If the class of fast A·T-rich sites represents the small set of stronger anthramycin attachment sites, then interaction of A with the slower, but more abundant G·C-rich sites could disguise an A·T binding specificity of the antibiotic.

Anthramycin produces hydrodynamic and electric dichroism changes of DNA which suggest its attachment to the outside of the double helix. A causes viscosity increases of DNA which are, however, unlike those produced by intercalating agents, not accompanied by reductions in sedimentation rates. The viscosity increases are not proportional to the extent of antibiotic binding and no increases are observed with low-molecular-weight, rod-like DNA molecules. The conclusion from these observations (Glaubiger et al. 1974) that anthramycin stiffens but does not lengthen the DNA helix is supported by the findings that increases in DNA's radius of gyration are, like the viscosity increases, abolished when the molecular weight of the polymer is reduced to about 10^6 daltons. The possibility of intercalation is further diminished by electric dichroism measurements which have shown that A's transition moment deviates from the plane of the DNA base pairs by at least 50° (Glaubiger et al. 1974).

One of the modes of outside attachment of anthramycin (i.e., the weaker mode, observed at low DNA concentrations) is easily reversed at higher ionic strenghts (Stefanović 1968). In contrast, the stronger mode of binding (observed at higher concentrations of DNA) cannot be abolished: neither exhaustive dialysis of the anthramycin-DNA complex for 14 days removes the antibiotic from its association with DNA, nor does alcohol extract anthramycin from its DNA complex or do 1% sodium lauryl sulfate or 10^{-3} M silver ions restore the spectrum of the free antibiotic (Kohn and Spears 1970). The unusually strong binding of anthramycin

may be mediated through positions 9, 10 and/or 11 in the antibiotic molecule since certain substituted anthramycins fail to
raise the melting temperature of DNA (Horwitz 1971) and since
the CD spectrum of bound anthramycin resembles that of free
anthramycin deprotonated in position 9 (Glaubiger et al. 1974).
The strength of anthramycin's binding resembles that of distamycin A (Sect. D); however, unlike distamycin which can be extracted from its DNA complex with phenol, anthramycin has been
suggested to form a labile covalent bond with DNA (Kohn and
Spears 1970).

G. Discussion and Conclusions

To the knowledge of this author, this is the first categorical
review of the peripheral binding of ligands to DNA. For this
reason, references to publications have been cited to document
the review adequately even if their dates would not characterize
them as "progress" in molecular biology.

The non-intercalative compounds reviewed here share certain
trends of their interactions with DNA:

1. Preference for A·T pairs. This preference is for some substances a specificity for DNA in its B-conformation, since
   A·T pairs predispose against the A-form.
2. Placement into one of the grooves of the double helix, most
   often into the minor groove.
3. Extraordinary strength of the stronger of two binding processes (observed for most, but not all, substances). The stronger
   process represents the specificity for A·T pairs and/or the
   B-form.

Compounds were selected for review of the peripheral DNA binding
phenomenon for which there existed a sufficient amount of biophysical evidence that they do bind in this manner. Inasmuch as
DNA binding studies continue, it is reasonable to assume that
additional instances of external binding will be investigated
and described.

References

Adams, E.: Binding of crystal violet by nucleic acids of *Escherichia coli*.
   J. Pharm. Pharmacol. Suppl. 20, 18S-22S (1968)
Adamson, R.H., Hart, L.G., DeVita, V.T., Oliverio, V.T.: Antitumor activity
   and some pharmacologic properties of anthramycin methyl ether. Cancer Res.
   28, 343-347 (1968)
Agrell, I.P.S., Heby, O.: Interactions of polyamines and acid macromolecules
   observed in double diffusion in gel experiments. Exp. Cell Res. 50, 668-
   671 (1968)

Ames, B.N., Dubin, D.T.: The role of polyamines in the neutralization of bacteriophage deoxyribonucleic acid. J. Biol. Chem. 235, 769-775 (1960)

Ames, B.N., Dubin, D.T., Rosenthal, S.M.: Presence of polyamines in certain bacterial viruses. Science 127, 814-816 (1958)

Arcamone, F., Bizioli, F., Canevazzi, G., Grein, A.: Distamycin and distacin. German Pat. 1,039,198 (1958), thru Chem. Abstr. 55, 2012f (1961)

Arcamone, F., Penco, S., Orezzi, P., Nicolella, V., Pirelli, A.: Structure and synthesis of distamycin A. Nature (London) 203, 1064-1065 (1964)

Arnott, S.: Secondary structures of polynucleotides. In: Proceedings of the First Cleveland Symposium on Macromolecules (ed. A.G. Walton), pp. 87-104. Amsterdam-Oxford-New York: Elsevier 1977

Arnott, S., Selsing, E.: The structure of polydeoxyguanylic acid-polydeoxy-cytidylic acid. J. Mol. Biol. 88, 551-552 (1974)

Bachrach, U.: Metabolism and function of spermine and related polyamines. Annu. Rev. Microbiol. 24, 109-134 (1970)

Bates, H.M., Kuenzig, W., Watson, W.B.: Studies on the mechanism of action of anthramycin methyl ether, a new antitumor antibiotic. Cancer Res. 29, 2195-2205 (1969)

Blake, A., Peacocke, A.R.: Perspectives report: The interaction of amino-acridines with nucleic acids. Biopolymers 6, 1225-1253 (1968)

Bontemps, J., Houssier, C., Fredericq, E.: Physico-chemical study of the complexes of '33258 Hoechst' with DNA and nucleohistone. Nucleic Acids Res. 2, 971-984 (1975)

Bradley, D.F., Wolf, M.K.: Aggregation of dyes bound to polyanions. Proc. Natl. Acad. Sci. (USA) 45, 944-952 (1959)

Bram, S.: The secondary structure of DNA in solution and in nucleohistone. J. Mol. Biol. 58, 277-288 (1971a)

Bram, S.: Polynucleotide polymorphism in solution. Nature New Biol. 233, 161-164 (1971b)

Bram, S.: Secondary structure of DNA depends on base composition. Nature New Biol. 232, 174-176 (1971c)

Bram, S.: The polymorphism of DNA. Prog. Mol. Subcell. Biol. 4, 1-15 (1976)

Cairns, J.: The application of autoradiography to the study of DNA viruses. Cold Spring Harbor Symp. Quant. Biol. 27, 311-318 (1962)

Carroll, D.: Optical properties of deoxyribonucleic acid-polylysine com-plexes. Biochemistry 11, 421-426 (1972)

Carroll, D., Botchan, M.R.: Competition between pentalysine and actino-mycin D for binding to DNA. Biochem. Biophys. Res. Commun. 46, 1681-1687 (1972)

Cavalieri, L.F., Angelos, A.: Studies on the structure of nucleic acids. I. Interaction of rosaniline with desoxypentose nucleic acid. J. Am. Chem. Soc. 72, 4686-4690 (1950)

Cavalieri, L.F., Angelos, A., Balis, M.E.: Studies on the structure of nucleic acids. IV. Investigation of dye interactions by partition analysis. J. Am. Chem. Soc. 73, 4902-4906 (1951)

Chandra, P., Zimmer, Ch., Thrum, H.: Effect of distamycin A on the structure and template activity of DNA in RNA-polymerase system. FEBS Lett. 7, 90-94 (1970)

Chandra, P., Götz, A., Wacker, A., Verini, M.A., Casazza, A.M., Fioretti, A., Arcamone, F., Ghione, M.: Some structural requirements for the antibiotic action of distamycins. FEBS Lett. 16, 249-252 (1971)

Chandra, P., Götz, A., Wacker, A., Verini, M.A., Casazza, A.M., Fioretti, A., Arcamone, F., Ghione, M.: Some structural requirements for the antibiotic action of distamycins. II. Structural modification of the side chains in distamycin A molecule. FEBS Lett. 19, 327-330 (1972a)

Chandra, P., Götz, A., Wacker, A., Zunino, F., Di Marco, A., Verini, M.A., Casazza, A.M., Fioretti, A., Arcamone, F., Ghione, M.: Some structural requirements for the antibiotic action of distamycins. III. Possible interaction of formyl group of distamycin side chain with adenine. Hoppe-Seyler's Z. Physiol. Chem. 353, 393-398 (1972b)

Chandra, P., Zunino, F., Gaur, V.P., Zaccara, A., Woltersdorf, M., Luoni, G., Götz, A.: Mode of tilorone hydrochloride interaction to DNA and poly-deoxyribonucleotides. FEBS Lett. 28, 5-9 (1972c)

Chandra, P., Steel, L.K., Ebener, U., Woltersdorf, M., Laube, H., Will, G.: Inhibitors of DNA synthesis in RNA tumor viruses: Biological implications and their mode of action. Prog. Mol. Subcell. Biol. 4, 167-226 (1976)

Chang, C., Weiskopf, M., Li, H.J.: Conformational studies of nucleoprotein. Circular dichroism of deoxyribonucleic acid base pairs bound by polylysine. Biochemistry 12, 3028-3032 (1973)

Chang, K.Y., Carr, C.W.: The binding of calcium with deoxyribonucleic acid and deoxyribonucleic acid-protein complexes. Biochim. Biophys. Acta 157, 127-139 (1968)

Cheng, S.M., Mohr, S.C.: Condensed states of nucleic acids. II. Effects of molecular size, base composition, and presence of intercalating agents on the ψ transition of DNA. Biopolymers 14, 663-674 (1975)

Chevaillier, P.: Etude au microscope électronique de l'interaction de l'acide désoxyribonucléique et des polyamines. Exp. Cell Res. 58, 213-224 (1969)

Cohen, G., Eisenberg, H.: Viscosity and sedimentation study of sonicated DNA-proflavine complexes. Biopolymers 8, 45-55 (1969)

Cohen, P., Kidson, C.: Conformational analysis of DNA-poly-L-lysine complexes by optical rotatory dispersion. J. Mol. Biol. 35, 241-245 (1968)

Comings, D.E.: Mechanisms of chromosome banding. VIII. Hoechst 33258-DNA interaction. Chromosoma 52, 229-243 (1975)

De Vries, H.: Rotatory power and other optical properties of certain liquid crystals. Acta Crystallogr. 4, 219-226 (1951)

Di Marco, A., Zunino, F., Silvestrini, R., Gambarucci, C., Gambetta, R.A.: Interaction of some daunomycin derivatives with deoxyribonucleic acid and their biological activity. Biochem. Pharmacol. 20, 1323-1328 (1971)

Di Marco, A., Arcamone, F., Zunino, F.: Daunomycin (Daunorubicin) and adriamycin and structural analogs: Biological activity and mechanism of action. In: Antibiotics (eds. J.W. Corcoran, F.E. Hahn), Vol. III, pp. 101-128. Berlin, Heidelberg, New York: Springer 1975

Estensen, R.D., Krey, A.K., Hahn, F.E.: Studies on a deoxyribonucleic acid-quinine complex. Mol. Pharmacol. 5, 532-541 (1969)

Felsenfeld, G., Huang, S.L.: Some effects of charge and structure upon ionic interactions of nucleic acids. Biochim. Biophys. Acta 51, 19-32 (1961)

Festy, B., Daune, M.: Hydroxystilbamidine. A nonintercalating drug as a probe of nucleic acid conformation. Biochemistry 12, 4827-4834 (1973)

Festy, B., Lallemant, A.M., Riou, G., Brack, C., Delain, E.: Mécanisme d'action des diamidines trypanocides. Importance de la composition en bases dans l'association berenil-polynucléotides. C.R. Acad. Sci. Paris Ser. D 271, 684-687 (1970)

Festy, B., Sturm, J., Daune, M.: Interaction between hydroxystilbamidine and DNA. I. Binding isotherms and thermodynamics of the association. Biochim. Biophys. Acta 407, 24-42 (1975)

Feughelman, M., Langridge, R., Seeds, W.E., Stokes, A.R., Wilson, H.R., Hooper, C.W., Wilkins, M.H.F., Barclay, R.K., Hamilton, L.D.: Molecular structure of deoxyribose nucleic acid and nucleoprotein. Nature (London) 175, 834-838 (1955)

Finlay, A.C., Hochstein, F.A., Sobin, B.A., Murphy, F.X.: Netropsin, a new antibiotic produced by a streptomyces. J. Am. Chem. Soc. 73, 341-343 (1951)

Flink, I., Pettijohn, D.E.: Polyamines stabilize DNA folds. Nature (London) 253, 62-63 (1975)

Fredericq, E., Houssier, C.: Electric dichroism and electric birefringence. Oxford: Clarendon Press 1973

Freifelder, D.: Electron microscopic study of the ethidium bromide-DNA complex. J. Mol. Biol. 60, 401-403 (1971)

Girod, J.C., Johnson, W.C., Jr., Huntington, S.K., Maestre, M.F.: Conformation of deoxyribonucleic acid in alcohol solutions. Biochemistry 12, 5092-5096 (1973)

Glaser, R., Gabbay, E.J.: Topography of nucleic acid helices in solutions. III. Interactions of spermine and spermidine derivatives with polyadenylic-polyuridylic and polyinosinic-polycytidylic acid helices. Biopolymers 6, 243-254 (1968)

Glaubiger, D., Kohn, K.W., Charney, E.: The reaction of anthramycin with DNA. III. Properties of the complex. Biochim. Biophys. Acta 361, 303-311 (1974)

Goldacre, R.J., Phillips, J.N.: The ionization of basic triphenylmethane dyes. J. Chem. Soc. 1724-1732 (1949)

Gosule, L.C., Schellman, J.A.: Compact form of DNA induced by spermidine. Nature (London) 259, 333-335 (1976)

Green, G., Mahler, H.R.: Optical properties of DNA in ethylene glycol. Biopolymers 6, 1509-1514 (1968)

Gust, D., Mislow, K.: Analysis of isomerization in compounds displaying restricted rotation of aryl groups. J. Am. Chem. Soc. 95, 1535-1547 (1973)

Hahn, F.E.: Complexes of biologically active substances with nucleic acids - Yesterday, today, tomorrow. Prog. Mol. Subcell. Biol. 2, 1-9 (1971)

Hahn, F.E.: Distamycin A and netropsin. In: Antibiotics (eds. J.W. Corcoran, F.E. Hahn), Vol. III, pp. 79-100. Berlin, Heidelberg, New York: Springer 1975

Hahn, F.E.: Distamycins and netropsin as inhibitors of RNA and DNA polymerases. Pharmacol. Ther. A 1, 475-485 (1977)

Hahn, F.E., Krey, A.K.: Deoxyribonucleic acid-induced anomalous optical rotatory dispersion of antimalarial drugs and dyes. Antimicrob. Agents Chemother. 1968, 15-20 (1969)

Hahn, F.E., Krey, A.K.: Complex of DNA with the antibiotic, distamycin A. Fed. Proc. 30, 1095 (1971a)

Hahn, F.E., Krey, A.K.: Complex of DNA with the antibiotic distamycin A. Abstr. VIIth Int. Congr. Chemother. 1, A-11/11 (1971b)

Hahn, F.E., Krey, A.K.: Interactions of alkaloids with DNA. In: Progress in Molecular and Subcellular Biology (ed. F.E. Hahn), Vol. II, pp. 134-151. Berlin, Heidelberg, New York: Springer 1971c

Hahn, F.E., O'Brien, R.L., Ciak, J., Allison, J.L., Olenick, J.G.: Studies on modes of action of chloroquine, quinacrine, and quinine and on chloroquine resistance. Mil. Med. Suppl. 131, No. 9, 1071-1089 (1966)

Harrison, D.P., Bode, V.C.: Putrescine and certain polyamines can inhibit DNA injection from bacteriophage lambda. J. Mol. Biol. 96, 461-470 (1975)

Haupt, I., Thrum, H.: Wirkung von Netropsin auf das Wachstum, die Nuclein-säure- und Proteinbiosynthese von Escherichia coli. Z. Allg. Mikrobiol. 11, 457-460 (1971)

Haynes, M., Garrett, R.A., Gratzer, W.B.: Structure of nucleic acid-poly base complexes. Biochemistry 9, 4410-4416 (1970)

Herbst, E.J., Tanguay, R.B.: The interaction of polyamines with nucleic acids. Prog. Mol. Subcell. Biol. 2, 166-180 (1971)

Hilwig, I., Gropp, A.: Staining of constitutive heterochromatin in mammalian chromosomes with a new fluorochrome. Exp. Cell Res. 75, 122-126 (1972)

Hirschberg, E.: Thiaxanthenones: Miracil D and hycanthone. In: Antibiotics (eds. J.W. Corcoran, F.E. Hahn), Vol. III, pp. 274-303. Berlin, Heidelberg, New York: Springer 1975

Hirschman, S.Z., Leng, M., Felsenfeld, G.: Interaction of spermine and DNA. Biopolymers 5, 227-233 (1967)

Holzwarth, G., Holzwarth, N.A.W.: Circular dichroism and rotatory dispersion near absorption bands of cholesteric liquid crystals. J. Opt. Soc. Am. 63, 324-331 (1973)

Horáček, P., Cernohorský, I.J.: Dependence of T_m of DNA complexes with putrescine, hexandiamine and spermidine on the ionic strength. Biochem. Biophys. Res. Commun. 32, 956-962 (1968)

Horwitz, S.B.: Anthramycin. Prog. Mol. Subcell. Biol. 2, 40-47 (1971)

Hurley, L.H., Gairola, Ch., Zmijewski, M.: Pyrrolo(1,4)benzodiazepine antitumor antibiotics. In vitro interaction of anthramycin, sibiromycin and tomaymycin with DNA using specifically radiolabelled molecules. Biochim. Biophys. Acta 475, 521-535 (1977)

Inagaki, A., Kageyama, M.: Interaction of antibiotics with deoxyribonucleic acid. II. DNA-cellulose chromatography of antibiotics and related compounds. J. Biochem. (Tokyo) 68, 187-192 (1970)

Inoue, S., Ando, T.: Interaction of clupeine with deoxyribonucleic acid. I. Thermal melting and sedimentation studies. Biochemistry 9, 388-394 (1970)

Ivanov, V.I., Malenkov, G.G., Minchenkova, L.E., Minyat, E.E., Frank-Kamenetskii, M.D., Schyolkina, A.K.: B-A transition of DNA in solution. Stud. Biophys. 40, 1-5 (1973a)

Ivanov, V.I., Minchenkova, L.E., Schyolkina, A.K., Poletayev, A.I.: Different conformations of double-stranded nucleic acid in solution as revealed by circular dichroism. Biopolymers 12, 89-110 (1973b)

Ivanov, V.I., Michenkova, L.E., Minyat, E.E., Frank-Kamenetskii, M.D., Schyolkina, A.K.: The B to A transition of DNA in solution. J. Mol. Biol. 87, 817-833 (1974)

Jordan, C.F., Lerman, L.S., Venable, J.H. Jr.: Structure and circular dichroism of DNA in concentrated polymer solutions. Nature New Biol. 236, 67-70 (1972)

Kohn, K.W.: Anthramycin. In: Antibiotics (eds. J.W. Corcoran, F.E. Hahn), Vol. III, pp. 3-11. Berlin, Heidelberg, New York: Springer 1975

Kohn, K.W., Spears, C.L.: Reaction of anthramycin with deoxyribonucleic acid. J. Mol. Biol. 51, 551-572 (1970)

Kohn, K.W., Bono, V.H., Kann, H.E.: Anthramycin, a new type of DNA-inhibiting antibiotic: Reaction with DNA and effect on nucleic acid synthesis in mouse leukemia cells. Biochim. Biophys. Acta 155, 121-129 (1968)

Kohn, K.W., Glaubiger, D., Spears, C.L.: The reaction of anthramycin with DNA. II. Studies of kinetics and mechanism. Biochim. Biophys. Acta 361, 288-302 (1974)

Kolchinskii, A.M., Mirazabekov, A.D., Zasedatelev, A.S., Gurskii, G.V., Grokhovskii, S.L., Zhuze, A.L., Gottikh, B.P.: Structure of complexes of antibiotics of the distamycin type and actinomycin D with DNA: New experimental data on the location of the antibiotics within the narrow groove of DNA. Mol. Biol. (USSR) 9, 14-20 (1975)

Kornberg, R.D.: Chromatin structure: A repeating unit of histones and DNA. Science 184, 868-871 (1974)

Kornberg, R.D.: Structure of chromatin. In: The Molecular Biology of the Mammalian Genetic Apparatus (ed. P.O.P. Ts'o), Vol. I, pp. 195-198. Amsterdam, New York, Oxford: North-Holland Publishing Company 1977a

Kornberg, R.D.: Structure of chromatin. Annu. Rev. Biochem. 46, 931-954 (1977b)

Krey, A.K., Hahn, F.E.: Berberine: Complex with DNA. Science 166, 757-758 (1969)

Krey, A.K., Hahn, F.E.: Studies on the complex of distamycin A with calf thymus DNA. FEBS Lett. 10, 175-178 (1970)

Krey, A.K., Hahn, F.E.: Methyl green-DNA complex: Displacement of dye by DNA-binding substances. Proc. 1st Eur. Biophys. Congr. 1, 223-227 (1971)

Krey, A.K., Hahn, F.E.: Optical studies on the interaction of DL-quinacrine with double- and single-stranded calf thymus deoxyribonucleic acid. Mol. Pharmacol. 10, 686-695 (1974)

Krey, A.K., Hahn, F.E.: Studies on the methyl green-DNA complex and its dissociation by drugs. Biochemistry 14, 5061-5067 (1975)

Krey, A.K., Allison, R.G., Hahn, F.E.: Interactions of the antibiotic, distamycin A, with native DNA and with synthetic duplex polydeoxyribo-nucleotides. FEBS Lett. 29, 58-62 (1973)

Krey, A.K., Olenick, J.G., Hahn, F.E.: Interactions of the antibiotic distamycin A with homopolymeric single-stranded polydeoxyribonucleotides and with ØX 174 deoxyribonucleic acid. Mol. Pharmacol. 12, 185-190 (1976)

Kurnick, N.B.: The quantitative estimation of desoxyribosenucleic acid based on methyl green staining. Exp. Cell Res. 1, 151-158 (1950a)

Kurnick, N.B.: The determination of desoxyribonuclease activity by methyl green: Application to serum. Arch. Biochem. 29, 41-53 (1950b)

Kurnick, N.B.: Methyl green-pyronin. I. Basis of selective staining of nucleic acids. J. Gen. Physiol. 33, 243-264 (1950c)

Kurnick, N.B.: Methyl green-pyronin: Basis of selective staining and histo-chemical application. J. Natl. Cancer Inst. 10, 1345 (1950d)

Kurnick, N.B.: Mechanism of desoxyribonuclease depolymerization: Effect of physical and enzymatic depolymerization on the affinity of methyl green and of desoxyribonuclease for desoxyribonucleic acid. J. Am. Chem. Soc. 76, 417-424 (1954)

Kurnick, N.B.: Assay of deoxyribonuclease activity. In: Methods of Bio-chemical Analysis (ed. D. Glick), Vol. IX, pp. 1-38. New York, London: Interscience 1962

Kurnick, N.B., Foster, M.: Methyl green. III. Reaction with desoxyribonucle-ic acid, stoichiometry, and behavior of the reaction product. J. Gen. Physiol. 34, 147-159 (1950)

Kurnick, N.B., Mirsky, A.E.: Methyl green-pyronin. II. Stoichiometry of re-action with nucleic acids. J. Gen. Physiol. 33, 265-274 (1950)

Kurnick, N.B., Radcliffe, I.E.: Reaction between DNA and quinacrine and other antimalarials. J. Lab. Clin. Med. 60, 669-688 (1962)

Laemmli, U.K.: Characterization of DNA condensates induced by poly(ethylene oxide) and polylysine. Proc. Natl. Acad. Sci. (USA) 72, 4288-4292 (1975)

Latt, S.A.: Microfluorometric detection of deoxyribonucleic acid replication in human metaphase chromosomes. Proc. Natl. Acad. Sci. USA 70, 3395-3399 (1973)

Latt, S.A.: Optical studies of metaphase chromosome organization. Annu. Rev. Biophys. Bioeng. 5, 1-37 (1976)

Latt, S.A., Stetten, G.: Spectral studies on 33258 Hoechst and related bisbenzimidazole dyes useful for fluorescent detection of deoxyribonucleic acid synthesis. J. Histochem. Cytochem. 24, 24-33 (1976)

Latt, S.A., Wohlleb, J.C.: Optical studies of the interaction of 33258 Hoechst with DNA, chromatin, and metaphase chromosomes. Chromosoma 52, 297-316 (1975)

Latt, S.A., Stetten, G., Juergens, L.A., Willard, H.F., Scher, C.D.: Recent developments in the detection of deoxyribonucleic acid synthesis by 33258 Hoechst fluorescence. J. Histochem. Cytochem. 23, 493-505 (1975)

Lawley, P.D.: Interaction studies with DNA. I. The binding of rosaniline at low ratio of concentrations rosaniline : DNA, and competitive effect of sodium and other metal cations. Biochim. Biophys. Acta 19, 160-167 (1956)

Lees, C.W., Von Hippel, P.H.: Hydrogen-exchange studies of deoxyribonucleic acid-protein complexes. Development of a filtration method and application to the deoxyribonucleic acid-polylysine system. Biochemistry 7, 2480-2488 (1968)

Leibo, S.P., Mazur, P.: Effect of osmotic shock and low salt concentration on survival and density of bacteriophages T4B and T4BΘ_1. Biophys. J. 6, 747-772 (1966)

Leng, M., Felsenfeld, G.: The preferential interactions of polylysine and polyarginine with specific base sequences in DNA. Proc. Natl. Acad. Sci. USA 56, 1325-1332 (1966)

Lerman, L.S.: Structural considerations in the interaction of DNA and acridines. J. Mol. Biol. 3, 18-30 (1961)

Lerman, L.S.: The structure of the DNA-acridine complex. Proc. Natl. Acad. Sci. USA 49, 94-102 (1963)

Lerman, L.S.: Amino group reactivity in DNA-aminoacridine complexes. J. Mol. Biol. 10, 367-380 (1964a)

Lerman, L.S.: Acridine mutagens and DNA structure. J. Cell. Comp. Physiol. Suppl. 64, 1-18 (1964b)

Lerman, L.S.: Intercalability, the ψ transition, and the state of DNA in nature. Prog. Mol. Subcell. Biol. 2, 382-391 (1971a)

Lerman, L.S.: A transition to a compact form of DNA in polymer solutions. Proc. Natl. Acad. Sci. USA 68, 1886-1890 (1971b)

Lerman, L.S.: Chromosomal analogues: Long-range order in ψ-condensed DNA. Cold Spring Harbor Symp. Quant. Biol. 38, 59-73 (1974)

Lewis, G.N., Magel, T.T., Lipkin, D.: Isomers of crystal violet ion. Their absorption and re-emission of light. J. Am. Chem. Soc. 64, 1774-1782 (1942)

Li, H.J.: Thermal denaturation of nucleohistones - Effects of formaldehyde reaction. Biopolymers 11, 835-847 (1972)

Li, H.J.: Helix-coil transition in nucleoprotein - Theory and applications. Biopolymers 12, 287-296 (1973)

Li, H.J.: A model for chromatin structure. Nucleic Acids Res. 2, 1275-1289 (1975)

Li, H.J., Bonner, J.: Interaction of histone half-molecules with deoxyribonucleic acid. Biochemistry 10, 1461-1470 (1971)

Li, H.J., Chang, C., Weiskopf, M.: Helix-coil transition in nucleoprotein-chromatin structure. Biochemistry 12, 1763-1772 (1973)

Li, H.J., Brand, B., Rotter, A.: Thermal denaturation of calf thymus DNA: Existence of a G·C-richer fraction. Nucleic Acids Res. 1, 257-265 (1974a)

Li, H.J., Brand, B., Rotter, A., Chang, C., Weiskopf, M.: Helix-coil transition in nucleoprotein. Effect of ionic strength on thermal denaturation of polylysine-DNA complexes. Biopolymers 13, 1681-1697 (1974b)

Li, H.J., Chang, C., Weiskopf, M., Brand, B., Rotter A.: Helix-coil transition in nucleoprotein: Renaturation of polylysine-DNA and polylysine-nucleohistone complexes. Biopolymers 13, 649-667 (1974c)

Li, H.J., Epstein, P., Yu, S.S., Brand, B.: Investigation of huge negative circular dichroism spectra of some nucleoproteins. Nucleic Acids Res. 1, 1371-1383 (1974d)

Li, H.J., Herlands, L., Santella, R., Epstein, P.: Studies on interaction between poly(L-lysine) and DNA of varied G+C contents. Biopolymers 14, 2401-2415 (1975)

Liquier, J., Pinot-Lafaix, M., Taillandier, E., Brahms, J.: Infrared linear dichroism investigations of deoxyribonucleic acid complexes with poly(L-arginine) and poly(L-lysine). Biochemistry 14, 4191-4197 (1975)

Liquori, A.M., Constantino, L., Crescenzi, V., Elia, V., Giglio, E., Puliti, R., De Santis Savino, M., Vitagliano, V.: Complexes between DNA and poly-amines: A molecular model. J. Mol. Biol. 24, 113-122 (1967)

Luck, G., Zimmer, Ch.: Interaction of netropsin with DNA in the course of the B-A transition. Stud. Biophys. 40, 9-12 (1973)

Luck, G., Triebel, H., Waring, M., Zimmer, Ch.: Conformation dependent binding of netropsin and distamycin to DNA and DNA model polymers. Nucleic Acids Res. 1, 503-530 (1974)

Luck, G., Zimmer, Ch., Reinert, K.E., Arcamone, F.: Specific interactions of distamycin A and its analogs with (A·T)-rich and (G·C)-rich duplex regions of DNA and deoxypolynucleotides. Nucleic Acids Res. 4, 2655-2670 (1977)

Luzzati, V., Masson, F., Lerman, L.S.: Interaction of DNA and proflavine: A small-angle X-ray scattering study. J. Mol. Biol. 3, 634-639 (1961)

Maestre, M.F., Tinoco, I., Jr.: Optical rotatory dispersion of viruses. J. Mol. Biol. 23, 323-335 (1967)

Mahler, H.R., Mehrotra, B.D.: Dependence of deoxyribonucleic acid-amine interactions on deoxyribonucleic acid composition. Biochim. Biophys. Acta 55, 252-254 (1962)

Mahler, H.R., Mehrotra, B.D.: The interaction of nucleic acids with diamines. Biochim. Biophys. Acta 68, 211-233 (1963)

Mahler, H.R., Perlman, P.S.: Induction of respiration deficient mutants in *Saccharomyces cerevisiae* by berenil. I. Berenil, a novel, non-intercala-ting mutagen. Mol. Gen. Genet. 121, 285-294 (1973)

Mahler, H.R., Mehrotra, B.D., Sharp, C.W.: Effects of diamines on the thermal transition of DNA. Biochem. Biophys. Res. Commun. 4, 79-82 (1961)

Malenkov, G., Minchenkova, L., Minyat, E., Schyolkina, A., Ivanov, V.: The nature of the B-A transition of DNA in solution. FEBS Lett. 51, 38-42 (1975)

Mandel, M.: The interaction of spermine and native deoxyribonucleic acid. J. Mol. Biol. 5, 435-441 (1962)

Maniatis, T., Venable, J.H., Jr., Lerman, L.S.: The structure of ψ DNA. J. Mol. Biol. 84, 37-64 (1974)

Matthews, H.R., Johnson, E.M., Steer, W.M., Bradbury, E.M., Allfrey, V.G.: The use of netropsin with CsCl gradients for the analysis of DNA and its application to restriction nuclease fragments of ribosomal DNA from *Physarum polycephalum*. Eur. J. Biochem. 82, 569-576 (1978)

Mauss, Y., Chambron, J., Daune, M., Benoit, H.: Etude morphologique par diffusion de la lumière du complexe formé par le DNA et la proflavine. J. Mol. Biol. 27, 579-589 (1967)

Mazza, G., Galizzi, A., Minghetti, A., Siccardi, A.: Interaction between deoxyribonucleic acid and distamycin A studied by transformation in *bacillus subtilis*. Antimicrob. Agents Chemother. 3, 384-391 (1973)

McGhee, J.D.: Theoretical calculations of the helix-coil transition of DNA in the presence of large, cooperatively binding ligands. (Appendix. The melting of poly[d(A-T)] in the presence of netropsin). Biopolymers 15, 1345-1375 (1976)

Michaelis, L.: The nature of the interaction of nucleic acids and nuclei with basic dyestuffs. Cold Spring Harbor Symp. Quant. Biol. 12, 131-142 (1947)

Mirsky, A.E., Pauling, L.: On the structure of native, denatured, and coagulated proteins. Proc. Natl. Acad. Sci. USA 22, 439-447 (1936)

Morris, C.R., Andrew, L.V., Whichard, L.P., Holbrook, D.J.: The binding of antimalarial aminoquinolines to nucleic acids and polynucleotides. Mol. Pharmacol. 6, 240-250 (1970)

Müller, W., Crothers, D.M.: Studies on the binding of actinomycin and related compounds to DNA. J. Mol. Biol. 35, 251-290 (1968)

Müller, W., Crothers, D.M.: Interactions of heteroaromatic compounds with
nucleic acids. 1. The influence of hereoatoms and polarizability on the
base specificity of intercalating ligands. Eur. J. Biochem. 54, 267-277
(1975)

Müller, W., Gautier, F.: Interactions of heteroaromatic compounds with
nucleic acids. A·T-specific non-intercalating DNA ligands. Eur. J. Biochem. 54, 385-394 (1975)

Müller, W.E.G., Obermeier, J., Maidhof, A., Zahn, R.K.: Distamycin: An
inhibitor of DNA-dependent nucleic acid synthesis. Chem. Biol. Interact.
8, 183-192 (1974)

Müller, W., Bünemann, H., Dattagupta, N.: Interactions of heteroaromatic
compounds with nucleic acids. 2. Influence of substituents on the base
and sequence specificity of intercalating ligands. Eur. J. Biochem. 54,
279-291 (1975)

Nagley, P., Mattick, J.S., Hall, R.M., Linnane, A.W.: Biogenesis of mitochondria. 43. A comparative study of petite induction and inhibition of
mitochondrial DNA replication in yeast by ethidium bromide and berenil.
Mol. Gen. Genet. 141, 291-304 (1975)

Nakamura, S., Yonehara, H., Umezawa, H.: On the structure of netropsin.
J. Antibiot. (Tokyo) Ser. A 17, 220-221 (1964)

Nelson, R.G., Johnson, W.C.: Conformation of DNA in ethylene glycol. Biochem.
Biophys. Res. Commun. 41, 211-216 (1970)

Neville, D.M., Davies, D.R.: The interaction of acridine dyes with DNA:
An X-ray diffraction and optical investigation. J. Mol. Biol. 17, 57-74
(1966)

Newton, B.A.: Interaction of berenil with deoxyribonucleic acid and some
characteristics of the berenil-deoxyribonucleic acid complex. Biochem. J.
105, 50P-51P (1967)

Newton, B.A.: Recent studies on the mechanism of action of berenil (Diminazene)
and related compounds. In: Comparative Biochemistry of Parasites (ed.
H. Van den Bossche), pp. 127-138. New York, London: Academic Press 1972

Newton, B.A.: Berenil: A trypanocide with selective activity against extranuclear DNA. In: Antibiotics (eds. J.W. Corcoran, F.E. Hahn), Vol. III,
pp. 34-47 Berlin, Heidelberg, New York: Springer 1975

Newton, B.A., Le Page, R.W.F.: Preferential inhibition of extranuclear deoxyribonucleic acid synthesis by the trypanocide berenil. Biochem. J. 105,
50P (1967)

Newton, B.A., Le Page, R.W.F.: Interaction of berenil with trypanosome DNA.
Trans. R. Soc. Trop. Med. Hyg. 62, 131-132 (1968)

Nordén, B., Tjerneld, F.: Binding of methyl green to deoxyribonucleic acid
analyzed by linear dichroism. Chem. Phys. Lett. 50, 508-512 (1977)

Ohba, Y.: Structure of nucleohistone. II. Thermal denaturation. Biochim.
Biophys. Acta 123, 84-90 (1966)

Olins, D.E., Olins, A.L.: Model nucleohistones: The interaction of F1 and
F2a1 histones with native T7 DNA. J. Mol. Biol. 57, 437-455 (1971)

Olins, D.E., Olins, A.L., Von Hippel, P.H.: Model nucleoprotein complexes:
Studies on the interaction of cationic homopolypeptides with DNA. J. Mol.
Biol. 24, 157-176 (1967)

Olins, D.E., Olins, A.L., Von Hippel, P.H.: On the structure and stability
of DNA-protamine and DNA-polypeptide complexes. J. Mol. Biol. 33, 265-
281 (1968)

Osland, A., Kleppe, K.: Polyamine induced aggregation of DNA. Nucleic Acids
Res. 4, 685-695 (1977)

Pardon, J.F., Wilkins, M.H.F.: A super-coil model for nucleohistone. J. Mol. Biol. 68, 115-124 (1972)

Patel, D.J., Canuel, L.L.: Netropsin-poly(dA-dT) complex in solution: Structure and dynamics of antibiotic-free base pair regions and those centered on bound netropsin. Proc. Natl. Acad. Sci. USA 74, 5207-5211 (1977)

Peacocke, A.R., Skerrett, J.N.H.: The interaction of aminoacridines with nucleic acids. Trans. Faraday Soc. 52, 261-279 (1956)

Pellicciari, C., Fraschini, A.: Methods of denaturation and renaturation of DNA in interphase chromatin: Cytochemical quantitative analysis by methyl green staining. Histochem. J. 10, 213-222 (1978)

Penco, S., Redaelli, S., Arcamone, F.: Distamicina A. II. Sintesi totale. Gazz. Chim. Ital. 97, 1110-1115 (1967)

Perlman, P.S., Mahler, H.R.: Induction of respiration deficient mutants in *Saccharomyces cerevisiae* by berenil. II. Characteristics of the process. Mol. Gen. Genet. 121, 295-306 (1973)

Poletaev, A.I., Makarov, V.L., Sveshnikov, P.G., Kondrat'eva, N.O., Vol'Kenshtein, M.V.: Circular dichroism of DNA-dye complexes. II. Anisotropy of circular dichroism long-wave effect and structure of complex. Mol. Biol. (USSR) 11, 710-720 (1977)

Puschendorf, B., Petersen, E., Wolf, H., Werchau, H., Grunicke, H.: Studies on the effect of distamycin A on the DNA dependent RNA polymerase system. Biochem. Biophys. Res. Commun. 43, 617-624 (1971)

Puschendorf, B., Becher, H., Böhlandt, D., Grunicke, H.: Effect of distamycin A on T4-DNA-directed RNA synthesis. Eur. J. Biochem. 49, 531-537 (1974)

Rauen, H.M., Norpoth, K., Unterberg, W., Haar, H.: Coplanare Heterooligobasen (Phthalanilide), hochaktive Cytostatica. Experientia 21, 300-304 (1965)

Razin, S., Rozansky, R.: Mechanism of the antibacterial action of spermine. Arch. Biochem. Biophys. 81, 36-54 (1959)

Reinert, K.E.: Adenosine-thymidine cluster-specific elongation and stiffening of DNA induced by the oligopeptide antibiotic netropsin. J. Mol Biol. 72, 593-607 (1972)

Reinert, K.E.: Interaction of the oligopeptides distamycin (Dst) and actinomycin (Acm) with DNA; Viscometric criteria for binding and bending. Stud. Biophys. 67, 49-50 (1978)

Reinert, K.E., Thrum, H.: Conformational changes of DNA by interactions with oligopeptide antibiotics as studied by viscometric investigations. Stud. Biophys. 24/25, 319-325 (1970)

Riou, G., Paoletti, C.: Preparation and properties of nuclear and satellite deoxyribonucleic acid of *Trypanosoma cruzi*. J. Mol. Biol. 28, 377-382 (1967)

Robinson, C.: Liquid-crystalline structures in polypeptide solutions. Tetrahedron 13, 219-234 (1961)

Rosenkranz, H.S., Bendich, A.: On the nature of the deoxyribonucleic acid-methyl green reaction. J. Biophys. Biochem. Cytol. 4, 663-664 (1958)

Rottier, P.B.: Spectrophotometry of dyes: 1. Methyl green. 2. Pyronin. Stain Technol. 28, 265-273 (1953)

Rubin, R.L.: Spermidine-deoxyribonucleic acid interaction in vitro and in *Escherichia coli*. J. Bacteriol. 129, 916-925 (1977)

Scatchard, G.: The attractions of proteins for small molecules and ions. Ann. N. Y. Acad. Sci. 51, 660-672 (1949)

Scott, J.E.: On the mechanism of the methyl green-pyronin stain for nucleic acids. Histochemie 9, 30-47 (1967)

Scott, J.E., Willett, I.H.: Binding of cationic dyes to nucleic acids and other biological polyanions. Nature (London) 209, 985-987 (1966)

Shapiro, J.T., Leng, M., Felsenfeld, G.: Deoxyribonucleic acid-polylysine complexes. Structure and nucleotide specificity. Biochemistry 8, 3219-3232 (1969)

Sharp, D.W.A., Sheppard, N.: Complex fluorides. Part VIII. The preparation and properties of salts of the triphenylmethyl cation: The infrared spectrum and configuration of the ion. J. Chem. Soc. 674-682 (1957)

Shih, T.Y., Bonner, J.: Template properties of DNA-polypeptide complexes. J. Mol. Biol. 50, 333-344 (1970)

Spitnik, P., Lipshitz, R., Chargaff, E.: Studies on nucleoproteins. III. Deoxyribonucleic acid complexes with basic polyelectrolytes and their fractional extraction. J. Biol. Chem. 215, 765-775 (1955)

Sponar, J., Frič, I.: Complexes of histone F1 with DNA in O.15 M NaCl. Circular dichroism and structure of the complexes. Biopolymers 11, 2317-2330 (1972)

Stefanović, V.: Spectrophotometric studies on the interaction of anthramycin with deoxyribonucleic acid. Biochem. Pharmacol. 17, 315-323 (1968)

Stevens, L.: Studies on the interaction of homologues of spermine with deoxyribonucleic acid and with bacterial protoplasts. Biochem. J. 103, 811-815 (1967)

Stevens, L.: The biochemical role of naturally occuring polyamines in nucleic acid synthesis. Biol. Rev. 45, 1-27 (1970)

Stone, A.L., Bradley, D.F.: Aggregation of acridine orange bound to polyanions: The stacking tendency of deoxyribonucleic acids. J. Am. Chem. Soc. 83, 3627-3634 (1961)

Sturm, J., Zana, R., Daune, M.: Interaction between hydroxystilbamidine and DNA. II. Temperature jump relaxation study. Dynamics of nucleic acids and polynucleotides. Biochim. Biophys. Acta 407, 43-6O (1975)

Suwalsky, M., Traub, W.: A comparative X-ray study of a nucleoprotamine and DNA complexes with polylysine and polyarginine. Biopolymers 11, 2223-2231 (1972)

Suwalsky, M., Traub, W., Shmueli, U., Subirana, J.A.: An X-ray study of the interaction of DNA with spermine. J. Mol. Biol. 42, 363-373 (1969)

Tabor, C.W., Tabor, H.: 1,4-diaminobutane (putrescine), spermidine, and spermine. Annu. Rev. Biochem. 45, 285-3O6 (1976)

Tabor, H.: The stabilization of Bacillus subtilis transforming principle by spermine. Biochem. Biophys. Res. Commun. 4, 228-231 (1961)

Tabor, H.: The protective effect of spermine and other polyamines against heat denaturation of deoxyribonucleic acid. Biochemistry 1, 496-5O1 (1962)

Tabor, H., Tabor, C.W.: Spermidine, spermine, and related amines. Pharmacol. Rev. 16, 245-3OO (1964)

Tabor, H., Tabor, C.W.: Biosynthesis and metabolism of 1,4-diaminobutane, spermidine, spermine, and related amines. Adv. Enzymol. 36, 2O3-268 (1972)

Tabor, H., Tabor, C.W., Rosenthal, S.M.: The biochemistry of the polyamines: Spermidine and spermine. Annu. Rev. Biochem. 3O, 579-6O4 (1961)

Tewari, K.K., Vötsch, W., Mahler, H.R., Mackler, B.: Biochemical correlates of respiratory deficiency. VI. Mitochondrial DNA. J. Mol. Biol. 2O, 453-481 (1966)

Thomas, J.O.: Chromatin structure. Int. Rev. Biochem. 17, 181-232 (1978)

Thrum, H.: Eine neue, von einer Spezies der Streptomyces-reticuli-Gruppe gebildete Antibiotikakombination. Naturwissenschaften 46, 87 (1959)

Tsuboi, M.: Helical complexes of poly-L-lysine and nucleic acids. In: Conformation of Biopolymers (ed. G.N. Ramachandran), Vol. II, pp. 689-7O2. New York: Academic Press 1967

Tsuboi, M., Matsuo, K., Ts'o, P.O.P.: Interaction of poly-L-lysine and nucleic acids. J. Mol. Biol. 15, 256-267 (1966)

Tunis, M.J.B., Hearst, J.E.: Optical rotatory dispersion of DNA in concentrated salt solutions. Biopolymers 6, 1218-1223 (1968)

Tunis-Schneider, M.J.B., Maestre, M.F.: Circular dichroism spectra of oriented and unoriented deoxyribonucleic acid films - A preliminary study. J. Mol. Biol. 52, 521-541 (1970)

Usatyi, A.F., Shlyakhtenko, L.S.: Melting of DNA in ethanol-water solutions. Biopolymers 13, 2435-2446 (1974)

Von Hippel, P.H., McGhee, J.D.: DNA-protein interactions. Annu. Rev. Biochem. 41, 231-300 (1972)

Wähnert, U., Zimmer, Ch., Luck, G., Pitra, Ch.: (dA·dT)-dependent inactivation of the DNA template properties by interaction with netropsin and distamycin A. Nucleic Acids Res. 2, 391-404 (1975)

Waring, M.: Drugs and DNA: Uncoiling of the DNA double helix as evidence of intercalation. Humangenetik 9, 234-236 (1970a)

Waring, M.: Variation of the supercoils in closed circular DNA by binding of antibiotics and drugs: Evidence for molecular models involving intercalation. J. Mol. Biol. 54, 247-279 (1970b)

Waring, M.: Binding of drugs to supercoiled circular DNA: Evidence for and against intercalation. Prog. Mol. Subcell. Biol. 2, 216-231 (1971)

Wartell, R.M., Larson, J.E., Wells, R.D.: Netropsin: A specific probe for A-T regions of duplex deoxyribonucleic acid. J. Biol. Chem. 249, 6719-6731 (1974)

Wartell, R.M., Larson, J.E., Wells, R.D.: The compatibility of netropsin and actinomycin binding to natural deoxyribonucleic acid. J. Biol. Chem. 250, 2698-2702 (1975)

Watson, J.D., Crick, F.H.C.: The structure of DNA. Cold Spring Harbor Symp. Quant. Biol. 18, 123-131 (1953)

Wehling, K., Arfmann, H.A., Standke, K.H.C., Wagner, K.G.: Specificity of DNA-basic polypeptide interactions. Influence of neutral residues incorporated into polylysine and polyarginine. Nucleic Acids Res. 2, 799-807 (1975)

Weinstein, I.B., Hirschberg, E.: Mode of action of miracil D. Prog. Mol. Subcell. Biol. 2, 232-246 (1971)

Weisblum, B.: Why centric regions of quinacrine-treated mouse chromosomes show diminished fluorescence. Nature (London) 246, 150-151 (1973)

Weisblum, B.: Fluorescent probes of chromosomal DNA structure: Three classes of acridines. Cold Spring Harbor Symp. Quant. Biol. 38, 441-449 (1974)

Weisblum, B., Haenssler, E.: Fluorometric properties of the bibenzimidazole derivative Hoechst 33258, a fluorescent probe specific for AT concentration in chromosomal DNA. Chromosoma 46, 255-260 (1974)

Weiskopf, M., Li, H.J.: Poly(L-lysine)-DNA interactions in NaCl solutions: B → C and B → ψ transitions. Biopolymers 16, 669-684 (1977)

Wells, R.D., Larson, J.E.: Studies on the binding of actinomycin D to DNA and DNA model polymers. J. Mol. Biol. 49, 319-342 (1970)

White, J.C., Elmes, P.C.: Fibres of human sodium deoxyribonucleate and nucleoprotein studied in polarized light by a simple method. Nature (London) 169, 151-152 (1952)

Wilkins, M.H.F.: Physical studies of the molecular structure of deoxyribose nucleic acid and nucleoprotein. Cold Spring Harbor Symp. Quant. Biol. 21, 75-90 (1956)

Yamaoka, K.: Effect of chemical structures of acridine and triphenylmethane dyes on the induced optical activity of DNA-dye complexes. Biopolymers 11, 2537-2561 (1972)

Yamaoka, K., Charney, E.: Electric dichroism studies of macromolecules in solutions. I. Theoretical considerations of electric dichroism and electrochromism. J. Am. Chem. Soc. 94, 8963-8974 (1972)

Zama, M., Ichimura, S.: Difference between polylysine and polyarginine in changing DNA structure upon complex formation. Biochem. Biophys. Res. Commun. 44, 936-942 (1971)

Zasedatelev, A.S., Gursky, G.V., Zimmer, Ch., Thrum, H.: Binding of netropsin to DNA and synthetic polynucleotides. Mol. Biol. Rep. 1, 337-342 (1974)

Zasedatelev, A.S., Zhuze, A.L., Zimmer, Ch., Grokhovskiy, S.L., Tumanyan, V.G., Gurskiy, G.V., Gottikh, B.P.: Stereochemical model of the molecular mechanism of "recognition" of AT-pairs during the binding of the antibiotics distamycin A and netropsin to DNA. Dokl. Akad. Nauk SSSR 231, 1006-1009 (1976)

Zasedatelev, A., Zhuze, A., Zimmer, Ch., Grokhovsky, S., Tumanyan, V., Gursky, G., Gottikh, B.: A stereochemical model for molecular mechanism of A·T-pair recognition exhibited by binding of distamycin A and netropsin to DNA. Stud. Biophys. 67, 47-48 (1978)

Zeleznick, L.D., Sweeney, C.M.: Inhibition of deoxyribonuclease action by nogalamycin and U-12241 by their interaction with DNA. Arch. Biochem. Biophys. 120, 292-295 (1967)

Zeleznick, L.D., Crim, J.A., Gray, G.D.: Immunosuppression by compounds which complex with deoxyribonucleic acid. Biochem. Pharmacol. 18, 1823-1827 (1969)

Zimmer, Ch.: Effects of the antibiotics netropsin and distamycin A on the structure and function of nucleic acids. Prog. Nucleic Acid Res. Mol. Biol. 15, 285-318 (1975)

Zimmer, Ch., Luck, G.: Optical rotatory dispersion properties of nucleic acid complexes with the oligopeptide antibiotics distamycin A and netropsin. FEBS Lett. 10, 339-342 (1970)

Zimmer, Ch., Luck, G.: Stability and dissociation of the DNA complexes with distamycin A and netropsin in the presence of organic solvents, urea and high salt concentration. Biochim. Biophys. Acta 287, 376-385 (1972)

Zimmer, Ch., Luck, G.: Conformation and reactivity of DNA. III. Circular dichroism studies of the effects of aqueous concentrated univalent salt solutions upon helix conformation. Biochim. Biophys. Acta 312, 215-227 (1973)

Zimmer, Ch., Luck, G.: Conformation and reactivity of DNA. VI. Circular dichroism studies of salt-induced conformational changes of DNAs of different base composition. Biochim. Biophys. Acta 361, 11-32 (1974)

Zimmer, Ch., Haupt, I., Thrum, H.: Einfluß von Netropsin und Distamycin A auf die zellfreie Proteinsynthese und die Struktur der Nukleinsäuren. In: Int. Symp. Wirkungsmech. Fungizid. Antibiot. Cytostat. (eds. H. Lyr, W. Rawald), pp. 61-71. Berlin: Akademie Verlag 1970a

Zimmer, Ch., Luck, G., Thrum, H.: Changes in the DNA secondary structure by interaction with oligopeptide antibiotics: Thermal melting, ORD and CD of DNA complexes with netropsin and distamycin A derivatives. Stud. Biophys. 24/25, 311-317 (1970b)

Zimmer, Ch., Puschendorf, B., Grunicke, H., Chandra, P., Venner, H.: Influence of netropsin and distamycin A on the secondary structure and template activity of DNA. Eur. J. Biochem. 21, 269-278 (1971a)

Zimmer, Ch., Reinert, K.E., Luck, G., Wähnert, U., Löber, G., Thrum, H.: Interaction of the oligopeptide antibiotics netropsin and distamycin A with nucleic acids. J. Mol. Biol. 58, 329-348 (1971b)

Zimmer, Ch., Luck, G., Thrum, H., Pitra, Ch.: Binding of analogues of the antibiotics distamycin A and netropsin to native DNA. Effect of chromophore systems and basic residues of the oligopeptides on thermal stability, conformation and template activity of the DNA complexes. Eur. J. Biochem. 26, 81-89 (1972)

Zunino, F., Di Marco, A.: Studies on the interaction of distamycin A and its derivatives with DNA. Biochem. Pharmacol. 21, 867-873 (1972)

The Amatoxins[1]

H. Faulstich

A. Introduction

The green death cup *Amanita phalloides* is one of the most toxic plants found in Europe and Northern America and the deleterious activities of extracts of this toadstool have been studied for more than a century. However, investigation of the biological activities became possible only 20 years ago when Th. Wieland and collaborators had isolated in crystalline form the various toxins present in the extract. Two groups of low molecular weight toxins have since been characterized, the phallotoxins and the amatoxins (for reviews see Wieland and Wieland 1959; Wieland 1968; Wieland and Wieland 1972; Wieland and Faulstich 1978).

After ingestion of mushrooms, the phallotoxins are not suffi- ciently resorbed by the gastrointestinal tract to contribute to poisoning. Besides, they are about ten times less toxic than the amatoxins. Only recently, the molecular mechanism of its toxicity has been revealed by Wieland and co-workers. They have found that the toxin binds to actin, preferably to actin of liver cells causing a thorough polymerization of this contrac- tile protein. Our current knowledge on chemistry and toxicity of the phallotoxins was reviewed by Wieland and Faulstich (1978) and Wieland and Faulstich (1979).

The amatoxins are by far the more potent poisons of the green death cup. All symptoms described for poisoning with this mush- room can likewise be produced by pure amatoxins. It is, there- fore, concluded that amatoxins are the sole cause of fatal human poisoning. Fiume and Stirpe (1966) detected in liver nuclei of amanitin-poisoned mice a decreased RNA concentration, which could be attributed to an inhibition of RNA polymerase activity by the amatoxins (Stirpe and Fiume 1967). This was a first step into a new field of toxicological enzymology. Since then several reviews have been published by Fiume and Wieland (1970), Wieland (1972), Chambon et al. (1972), Chambon (1975), Roeder (1976), Wieland and Faulstich (1978).

With further development in this field it became evident that α-amanitin provided a specific tool for molecular biologists. It was found that amatoxins inhibit eukaryotic RNA polymerases only, and therefore can be used to distinguish them from pro-

---

1 Dedicated to Prof. Th. Wieland on the occasion of his 65th birthday

karyotic, viral or mitochondrial RNA polymerases. Furthermore, amatoxins specifically inhibit only two of the three RNA polymerases present in eukaryotic cells. One kind of RNA polymerases, nucleoplasmic enzymes, are very sensitive to the toxin. A second type of nucleoplasmic and sometimes cytoplasmic RNA polymerases is only inhibited by high concentrations of the toxin, while the third group of enzymes, the nucleolar RNA polymerases, are completely resistant to α-amanitin. This distinct pattern of sensitivity has permitted the differentiation by α-amanitin of the enzymic activities as well as of the transcription products. Finally, in the last few years, amatoxins have been applied to a large variety of in vivo systems, in order to study complex processes such as virus replication, hormonal induction, development of eggs, or behavior of animals. In most cases the use of α-amanitin allowed to ascertain whether the various biological processes were under the control of RNA synthesis. In this review the biochemistry of the amatoxins as well as their biological applications will be discussed.

B. Occurrence and Structure

In the fresh *Amanita phalloides* mushroom the amatoxin concentration is 0.2 mg per g tissue (Faulstich et al. 1973, 1974). Comparable levels of amatoxins were found only in the white species *Amanita virosa* and *Amanita verna*, and in some other Amanitas existent exclusively in North America (Yocum and Simons 1977). A considerably smaller amount of amatoxin, 0.03 mg/g fresh tissue, has been determined in *Galerina marginata* (Faulstich et al. 1974).

Within the last 30 years the laboratory of Th. Wieland has isolated and characterized the amatoxin components (Wieland 1972). The toxins represent a family of nine cyclic peptides, which differ only slightly in structure and biological activity. The naturally occurring amatoxins known so far are depicted in Fig. 1.

Six of the cyclic peptides are toxic in vivo and inhibit RNA synthesis in vitro at concentrations of 10^{-9} to 10^{-8} M. Two others, amanullin and amanullinic acid, are non-toxic in vivo, but still inhibit RNA synthesis in vitro at somewhat higher concentrations. The remaining peptide of this family, proamanullin, is possibly a biological precursor and lacks toxicity as well as any inhibitory capacity. Only recently, a description of the crystal structure of β-amanitin has been published by Kostansek et al. (1977). For further details see Wieland and Faulstich (1978).

The compound most widely used as an inhibitor in biological systems is α-amanitin. It crystallizes with 4 molecules of water: $C_{39}H_{54}N_{10}O_{14}S \cdot 4H_2O$ (m.wt. = 990). In solution, the amatoxins can be detected by their characteristic UV spectrum distinguishable even in the presence of proteins by their 310 nm absorption band (ε_{310} = 13.500; Wieland and Faulstich 1978). On

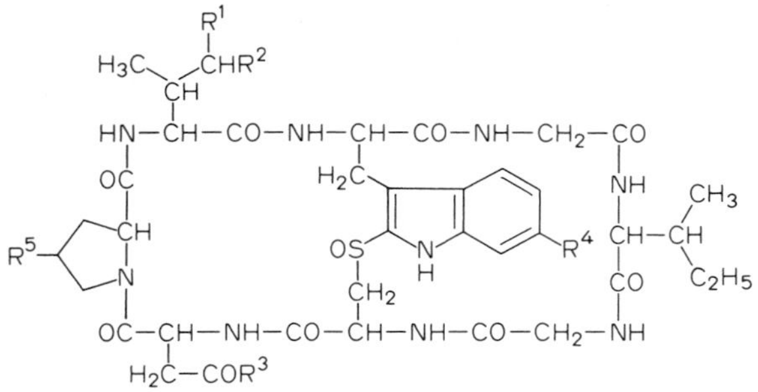

The structural formula with substituents R1, R2, R3, R4, R5.

Name	R^1	R^2	R^3	R^4	R^5	LD_{50} (mg/kg white mouse)
α-amanitin	CH_2OH	OH	NH_2	OH	OH	0.3
β-amanitin	CH_2OH	OH	OH	OH	OH	0.5
γ-amanitin	CH_3	OH	NH_2	OH	OH	0.2
ε-amanitin	CH_3	OH	OH	OH	OH	0.3
amanin	CH_2OH	OH	OH	H	OH	0.5
amanin amide[a]	CH_2OH	OH	NH_2	H	OH	0.3
amanullin	CH_3	H	NH_2	OH	OH	>20
amanullinic acid	CH_3	H	OH	OH	OH	>20
proamanullin	CH_3	H	NH_2	OH	H	>20

[a]in *A. virosa* only

Fig. 1. Naturally occuring amatoxins

thin layer plates, the amatoxins can best be characterized by
their typical violet color reaction with cinnamic aldehyde in
hydrochlorid acid, the limit of detection being 0.1 µg per spot.
For the determination of minute amounts of amatoxins, two kinds
of bioassays have been introduced. One assay makes use of the
inhibition of RNA polymerase II and allows to detect 1 ng ama-
toxins/ml (Preston et al. 1975; Johnson and Preston 1976). Further, radioimmunoassays were established using sera raised
against amanitin-conjugates in rabbits (Faulstich et al. 1975)
or rats (Fiume et al. 1975); the limit of detection in these
cases was 0.5 ng amatoxins/ml.

In the last years extensive chemical studies have been performed
on the amatoxin peptides. For a comprehensive review see Wieland
and Faulstich (1978). In the present article only those amatoxin
derivatives will be considered which were prepared to either introduce radioactive labels (Fig. 2) or to conjugate amatoxins
to macromolecules (Fig. 3). By the latter derivatization the
penetration rates into cells and, concomitantly, the specificity of amatoxin activities could be modified.

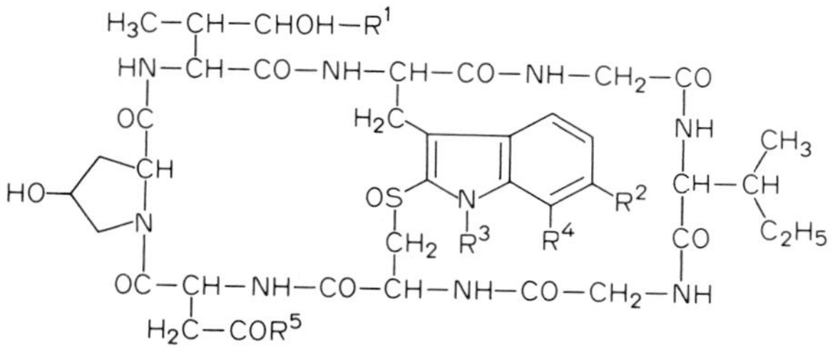

Fig. 2. Amatoxin derivatives with radioactive labels

Name	R^1	R^2	R^3	R^4	R^5	spec. activity of preparation [Ci/mmol]	Notes	References
[^{3}H]O-Methylde-hydroxy-methyl-α-amanitin (old name: O-methyl-demethyl-γ-amanitin)	[^{3}H]	OCH_3	H	H	NH_2	3.8	K_i 3 times higher than K_i of α-amanitin	Wieland and Fahrmeir (1970)
[^{14}C]O-Methyl-α-amanitin	CH_3	$O[^{14}C]H_3$	H	H	NH_2	0.07	lable partly removable by microsomal oxidation	Govindan (1969)
[^{3}H]O-Methyl-α-amanitin	CH_2OH	$OC[^3H]_3$	H	H	NH_2	4.5		Faulstich et al. (in prep.b)
[^{3}H]Dimethyl-α-amanitin	CH_2OH	$OC[^3H]_3$	$C[^3H]_3$	H	NH_2	8.3	K_i ~2 times higher than K_i of α-amanitin	Faulstich et al. (in prep.b)
[^{3}H]-6'-amanin	CH_2OH	[^{3}H]	H	H	OH	1.4		Wieland and Brodner (1976)
[^{125}J]-7'-α-amanitin	CH_2OH	OH	H	[^{125}J]	NH_2	0.3	K_i ~ 5 times lower than that of α-amanitin	Morris et al. (1978)

C. Toxicity

Susceptibility to amatoxins differs among various animal species. For the white mouse the parenteral LD_{50} of α-amanitin is 0.3 mg/kg body weight; the other naturally occuring amatoxins show only slight differences from this value (see Fig. 1). Dogs and guinea pigs are more sensitive than mice (LD_{50} = 0.1 mg/kg), whereas rats are rather resistant (LD_{50} > 3 mg/kg). Amphibians are significantly less sensitive than mammals, as shown by the resistance of *Rana esculenta* (LD_{50} = 15 mg/kg body weight) (Wieland and Dose 1954).

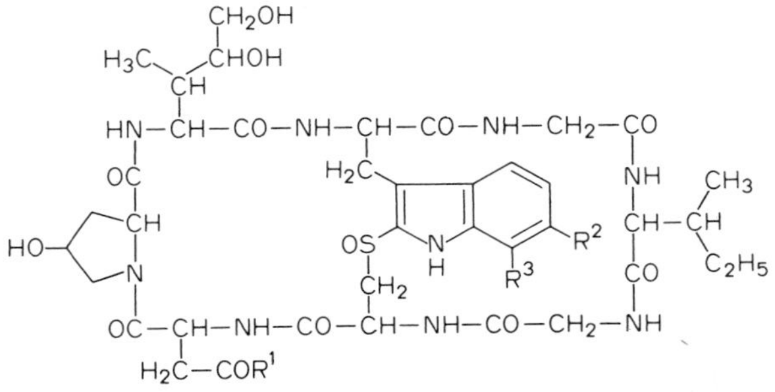

Protein linkage (derivative of)		R^1	R^2	R^3	Reference
R^1 = ■HN-protein R^2 = -O-■CO-protein } (mixture),	(β)	- OH	OH -	H H	Cessi and Fiume (1969)
R^1 = ■HN-protein,	(β)	-	OH	H	Faulstich and Trischmann (in preparation)
R^3 = -N=N- -CONH-(CH$_2$)$_6$-NH■CO-protein (tentatively),	(α)	NH$_2$	OH	-	Faulstich and Trischmann (1973)
R^2 = -O-(CH$_2$)$_5$-CONH-(CH$_2$)$_2$-NHCO-(CH$_2$)$_2$-CO ■NH-protein,	(α)	NH$_2$		H	Faulstich and Trischmann (in preparation)

■ = bond between the toxin derivative and the protein

Fig. 3. Protein conjugates of α- and β-amanitin

Death occurs in mammals 2-7 days after administration of the toxin and is normally caused by severe liver damage. On the other hand, cases were described in the dog (Fauser et al., in preparation) and the mouse (Fiume et al. 1969) where fatality was due exclusively to kidney failure. This was true especially with low doses of α-amanitin. In these cases, lesions were found exlusively in the cells of the proximal convoluted tubules. It therefore appears that penetration of the toxin into these cells is the toxic effector, probably by reabsorption of the toxin from glomerular filtrate. Such lesions were never found in rat kidneys and, most probably, rats lack the capacity to reabsorb amatoxins (Fiume et al. 1969).

After oral administration amatoxin absorption must be very slow in mice and rats because they can withstand 10 mg/kg without any symptoms of intoxication. Dogs and cats exhibit more efficient resorption from the gastro-intestinal tract; their oral LD$_{50}$ values are 1-5 mg/kg. Humans obviously absorb the toxins rather quickly, since the ingestion of a single 30 g mushroom (equal to 5-6 mg of amatoxins) is sufficient to kill an adult. No oral absorption is reported for the larvae of *Chironomus* which remain unaffected after being fed with 10^{-6} M α-amanitin in tap water (Beermann 1971). After intracerebral-ventricular (i.c.v.) administration, α-amanitin is extremely toxic; the LD$_{50}$ values in the mouse and rat are 0.002 mg/kg, and 0.008 mg/kg, respectively (Habermann and Cheng-Raude 1975).

The complex symptoms of amanitin intoxication have been analyzed in Beagle dogs (Faulstich and Fauser 1979; Fauser and Faulstich 1975). Predominant symptoms are: hypoglycemia and severe hemorrhagia, followed by acute liver dystrophy (Faulstich and Fauser, in preparation; Fauser and Faulstich, in preparation). From these experiments and other pharmacokinetic studies (Faulstich et al., in preparation a) it became evident that amatoxins are rapidly excreted into the bile. One important treatment in the therapy of human *Amanita* poisoning, therefore, must be the interruption of the enterohepatic circulation. Also, amatoxins are dialyzable and hence quickly excreted into urine (Fauser and Faulstich 1973; Faulstich et al., in preparation c).

Among the amatoxin derivatives the β-amanitin serum albumin conjugates possess a six to seven times greater toxicity than β- or α-amanitin (Cessi and Fiume 1969; Fiume et al. 1969). This has been explained by its preferential uptake by protein-consuming cells of the liver and kidney. Another reason for the extremely high toxicity may be that the conjugates are not dialyzable in the kidney and hence are excreted very slowly. In fact, toxicity of amatoxins is widely determined by the pharmacodynamic behaviour. One example for this is the non-toxicity of amanullin. Cochet-Meilhac and Chambon (1974) determined a high dissociation rate of the amanullin complex with RNA polymerases II. As a consequence, administration of even high doses of amanullin cannot maintain the block of RNA polymerases II long enough to kill the animal.

D. Inhibition of RNA Polymerases in Vitro

In 1967, Stirpe and Fiume observed that α-amanitin inhibited RNA synthesis in isolated mouse liver nuclei, with preference for the synthesis catalyzed at high ionic strength (Stirpe and Fiume 1967). Through this, they found the molecular mechanism of amanitin toxicity and, simultaneously, early evidence for the existence of multiple RNA polymerases. The events following this observation have been described in an historical review by Fiume and Wieland (1970).

Currently, there are three classes of RNA polymerases: I, II, and III. They are defined by their order of elution from DEAE-Sephadex (Roeder and Rutter 1969); further elucidation comes from the dependance of enzymic activity on salt concentration and on the ratio of Mn^{2+}/Mg^{2+} ions, and from the type of DNA preferred as template. In addition, these enzymes differ in their sensitivity to α-amanitin, (Kedinger et al. 1970; Chambon et al. 1974). The strongest evidence for the existence of three types of RNA polymerases was obtained after their resolution by chromatography. This permitted their identification by molecular weight of their subunit polypeptides.

Mammalian cell RNA polymerases I are totally insensitive to α-amanitin ($K_i > 10^{-3}$ M). They are eluted as the first enzyme activity from DEAE-Sephadex and have been further resolved into two forms, I_A and I_B.

Rat liver enzyme I_A has subunits with molecular weights 195 000; 130 000; 65 000; 40 000; and 19 000. Form I_B is identical except for the absence of the 65 000 subunit.

RNA polymerases II of mammalian cells are highly sensitive to α-amanitin and inhibited by ~10^{-8}M of the toxin. These RNA polymerases show their greatest activity at high salt concentrations and represent, therefore, probably the enzymes inhibited in situ in the experiments of Stirpe and Fiume (1967). In most tissues they exist in two forms, II_A and II_B, both equally sensitive to α-amanitin (Kedinger et al. 1971). In rat livers and mouse plasma cytoma cells, one finds an additional third form II_0. The three forms differ only in the molecular weight of their largest polypeptide chain: 220,000 or 240,000 in II_0, 214,000 in II_A, and 180,000 in II_B. The other subunits are polypeptides with molecular weights 140,000, 34,000, 25,000, 21,000, and 16,500 for the rat liver and calf thymus enzymes. II_B has proved to be heterogenous in charge, and could be further resolved electrophoretically into two species (Kedinger and Chambon 1972).

RNA polymerases III of mammalian cells are inhibited only by high concentrations of α-amanitin ($K_i = 10^{-5}$M to 10^{-4}M). These enzymes are probably identical to RNA polymerase C, as described by Seifart et al. (1972), which is partially located in the cytoplasm of rat liver. Two forms also exist, III_A and III_B. The subunit composition of III_A in mouse plasmacytoma cells is: 155,000, 138,000, 89,000, 70,000, 53,000, 49,000, 41,000, 32,000, 29,000, and 19,000. Form III_B has a subunit of 33,000 rather than 32,000. For further details see the reviews of Roeder (1976) and Chambon (1975).

A second nomenclature has been suggested by Chambon et al. (1974) different from the nomenclature originally derived from the chromatographic behavior of the enzymes I, II and III (Roeder and Rutter 1969). This new system is based upon the distinct sensitivities of the enzymes to α-amanitin. Both nomenclature systems are equally useful classifications for mammalian enzymes, where I = A, II = B, and III = C. Both, however, must be regarded as tentative classifications with respect to the RNA · polymerases of other eukaryotes, especially those of lower genera. With them it has been reported that enzymes I, II, and III have been eluted in a sequence different from the mammalian enzymes. Furthermore, it has been clearly proved that enzyme I of yeast is sensitive to α-amanitin, while enzyme III not. Recent publications have used the nomenclature of Roder and Rutter based on subunit structure and other criteria rather than on chromatographic behavior.

The three classes of RNA polymerases are further characterized by their transcription products. Various species of RNA could be attributed to the synthetic activities of enzymes I, II, and III, using the distinct sensitivities of these enzymes to α-amanitin. RNA polymerase I transcribes 45 S RNA, the precursor of 18 S and 28 S rRNA. This has been concluded from the observation that the synthesis of rRNA in isolated mouse plasmacytoma nucleoli, and in nuclei isolated from *Xenopus laevis* oocytes, was fully resistant to high concentrations of α-amanitin ($>10^{-4}$M). Such concentrations of the toxin completely inhibited the activities of the corresponding RNA polymerases II and III (Roeder 1976). RNA polymerase II transcribes heterogenous nuclear RNA, the precursor of mRNA. Synthesis of this class of RNA shows the same sensitivity to α-amanitin as RNA polymerases II (10^{-8}M αamanitin); for references see Roeder 1976. Direct evidence for this phenomenon has been recently provided by Suzuki and Giza (1976). They inhibited the synthesis of fibroin mRNA synthesis in nuclei isolated from *Bombyx mori* with $2 \cdot 10^{-7}$M α-amanitin. Finally, RNA polymerase III transcribes tRNA (4.5 S RNA) and 5 S RNA. The sensitivity of the synthesis of these RNAs also paralleled exactly that of the purified class III enzymes (10^{-4}M α-amanitin) (see Roeder 1976).

RNA polymerases type II, which are highly sensitive to α-amanitin, have been isolated and characterized from many mammalian tissues and cells. Some of them are listed in Table 1. For further details and references see the reviews of Chambon (1975), and Wieland and Faulstich (1978).

In non-mammalian cells, class II RNA polymerases show a lesser sensitivity to α-amanitin as compared to mammalian cells. For example, in most plant cells these enzymes are about 5 times less sensitive than in mammalian cells. In fungi, low species as well as carpophores, the RNA polymerases II were determined to be even 50 times less susceptible to the toxin. Most recently, also two amatoxin accumulating Amanita species were investigated, which proved to be the most resistant eukaryotes so far even more resistant than cells grown in cultures in the presence of α-amanitin. Inhibition of their RNA polymerases II required amatoxin concentrations being more than 10^5 times higher than those blocking mammalian enzymes.

Independent from this peculiar case the sensitivity to α-amanitin seems to increase with the evolutionary hierarchy. The RNA polymerases of prokaryotic species conform to this rule and are insensitive to α-amanitin. This has been ascertained for the RNA polymerase of *E. coli* and other bacteria (Seifart and Sekeris 1969; Dezelee et al. 1970).

The sensitivity to α-amanitin of mitochondrial RNA polymerases depends on the source of the mitochondria. RNA polymerases of mammalian mitochondria are inhibited by high concentrations of α-amanitin, while the corresponding enzymes of lower eukaryotes are completely resistant. The mitochondrial RNA polymerases of rat liver exhibit no inhibition with 0.15 µg/ml (Gallerani et al. 1972); a higher concentration, 10 µg/ml, caused 10-30% inhibition (Menon 1971), or even 50% (Saccone et al. 1971). Rat liver mitochondrial RNA polymerase thus has a drug sensitivity similar to that of nuclear RNA polymerase III. In *Neurospora crassa*, 0.5 µg/ml did not affect the mitochondrial RNA polymerase; however, an inhibition is not to be expected with the α-amanitin

Table 1. Concentrations of α-amanitin [μg/ml] causing 50% inhibition of RNA polymerases II and III in various eukaryotes

Species	RNA polymerase II	III	References
Various mammalian tissue cells	0.01	10 - 40	Seifart and Benicke (1975)
HeLa cells	0.003 - 0.01	15 - 30	Weil and Blatti (1976) Hossenlopp et al. (1975) Zylber and Penman (1971)
Ehrlich Ascites cells	3 (75%)	3 (84%)[c]	Blair and Dommasch (1972)
Xenopus laevis	0.05	20	Roeder (1974) Wilhelm et al. (1974)
Calliphora erythrocephala	0.08	unknown	Doenecke et al. (1972)
Drosophila melanogaster	0.03		Greenleaf et al. (1976)
Drosophila	-	~15[a]	Phillips and Sumner-Smith (1977)
Bombyx mori	0.01 - 0.05	not inhibited at 10	Sklar and Roeder (1975)
Dictyostelium discoideum	0.03	not inhibited at 33	Pong and Loomis (1973) Yagura et al. (1976)
Physarum polycephalum	5 (95%)	not inhibited at 5	Hildebrandt and Sauer (1973)
	1	not inhibited at 30	Grant (1972)
Mucor rouxii	4.5 (90%)	-	Young and Whiteley (1975)
Blastocladiella emersonii	0.3 (100%)	-	Horgen and Griffin (1971)
Zea mais	<0.1		Strain et al. (1971)
	0.05		Jendrisak and Guilfoyle (1978)
Wheat germ	0.3		Hodo III and Blatti (1977)
	0.05		Jendrisak and Guilfoyle (1978)

Species			Reference
Brassica oleracea	0.05		Jendrisak and Guilfoyle (1978)
Agaricus bisporus	6.5		Vaisius and Horgen (1979)
Amanita brunnescens	~ 9		Johnson and Preston (1979)
Amanita hygroscopica	~ 1800		Johnson and Preston (1979)
Amanita suballiacea	~ 3000		Johnson and Preston (1979)
Saccharomyces cerevisiae	0.8	not inhibited at 2000^b	Hager et al. (1976)
	3		Dezelee et al. (1970)
	1	> 1000	Schultz and Hall (1976)
Tetrahymena pyriformis	3	unknown	Higashinakagawa and Mita (1973)

[a] In intact nuclei; after purification of the enzyme 50% inhibition requires 280 µg/ml.
[b] Enzyme I inhibited 50% at 300 µg/ml.
[c] Possibly an enzyme of class II.

concentration used which was as low as $5 \cdot 10^{-7}$M. In yeast, mitochondria 16 µg/ml applied to a "petite mutant" had no inhibitory
activity (Wintersberger 1970); also high concentration of 40 µg/
ml had no significant effect (Tsai et al. 1971). Comparably,
the RNA polymerases of chloroplasts have been reported to resist
concentrations up to 100 µg/ml of α-amanitin (Bottomley et al.
1971a, Polya and Jagendorf 1971).

Within the last few years RNA polymerases of class III have also
been purified from many mammalian sources. For reviews and references see Austoker et al. (1974) and Chambon (1975). Comparable to the 10^3 to 10^4 sensitivity difference found between
enzymes II and III in mammalian cells, the RNA polymerase III
of lower eukaryotes are in most cases completely resistant to
α-amanitin. Table 1 lists the type II and III RNA polymerases
of various eukaryotes. Unexpectedly, yeast RNA polymerase I is
affected at high concentrations of the toxin (Hager et al. 1976).
During purification this enzyme sheds two polypeptide chains,
which increases its sensitivity to α-amanitin (Huet et al. 1975a).
In contrast, the sensitivity of *Drosophila* RNA polymerase II to
α-amanitin decreases 20-fold during purification (Phillips and
Sumner-Smith 1977).

RNA polymerases appear to be localized in the nucleus. Only RNA
polymerase III has been occasionally found in the cytoplasm. It
is unknown whether this is caused by a leakage of the enzyme
from the nucleus, or by a real function of this enzyme in the
cytoplasm. Within the nucleus, RNA polymerase I is located in
the nucleolus, while RNA polymerases II and III are extranucleolar enzymes in the nucleoplasm. Evidence was obtained by sub-
fractionation of isolated nuclei (see Roeder 1976) and the use
of α-amanitin. In isolated nuclei it selectively inhibits the
incorporation of [^{3}H]-uridine into the nucleoplasmic part of
the nucleus while allowing it into the nucleoli. These results
have been obtained by incubating the nuclei of ethanol/acetone
fixed human fibroblasts with NTP, including [^{3}H] UTP (Moore and
Ringertz 1973). The 0.4 M ammonium sulfate induced incorporation
of [^{3}H] UMP into the nucleoplasm is totally suppressed by $5 \cdot 10^{-6}$M
α-amanitin, while the dense grain distribution over the 2-3 nucleoli of the cells is unaffected. Similar results were reported
from experiments with the isolated nuclei of *Acetabularia* (Brändle
and Zetsche 1973). Analogous localization studies were performed
using intact cells like oocytes of *Triturus* or, preferably, the
salivary glands of insects containing polytene chromosomes. After
8 h treatment of *Triturus* oocytes with α-amanitin [^{3}H]-uridine was
incorporated only into the many nucleoli of the oocytes and no
longer into the lampbrush chromosomes (Bucci et al. 1971). Corresponding results were reported for *Chironomus* (Beermann 1971;
Wobus et al. 1971; Egyhazi et al. 1972); for *Drosophila* (Holt and
Kujpers 1972b); and for *Diptera* (Santelli et al. 1976). Other
interesting observations were reported besides the typical effect of inhibited incorporation of [^{3}H]-uridine into the chromo-
sal part of these nuclei. In *Chironomus*, distinct puffs in the
chromosomes continued to synthesize RNA in the presence of α-
amanitin. In all chromosomes incubated with the toxins, a diffuse, though significant, labeling was observed in the presence

of $5 \cdot 10^{-6}$M α-amanitin. This may be due to the continued activity
of the RNA polymerase III at this concentration of the toxin.
Using the same species, Beermann noticed that approximately 15%
of all cells in a gland resisted the drug. This may be a consequence of the slow penetration of amanitin into distinct cells
of the gland. For the giant chromosomes of *Drosophila* it has
been observed that the nucleoli of α-amanitin-treated glands
appear more densely labeled, suggesting a defect in processes
involved in the release of RNA from the nucleolus.

Autoradiographic results have in many cases been substantiated
by the isolation and chracterization of the different types of
RNA from these systems. These data will be discussed with the
in vivo effects in one of the following sections.

E. Molecular Interaction with RNA Polymerase

Early evidence provided by Seifart and Sekeris (1969), Kedinger
et al. (1970), and Lindell et al. (1970) showed that amatoxins
inhibit RNA synthesis not by binding to the template, but rather
to the enzyme. In fact, RNA polymerase II forms a tight complex
with amatoxins. An example of this stability is shown when a
labeled amatoxin, [^{14}C] methyl-γ-amanitin, cosediments through
a glycerol gradient with the enzyme with no tailing (Meilhac et
al. 1970). The labeled toxin is also eluted from DEAE cellulose
together with RNA polymerase II, as reported by Sekeris and
Schmidt (1972); and in experiments of Mandel and Chambon (1971),
the toxin comigrates with the native enzyme in polyacrylamide
electrophoresis.

From stoichiometric estimations at maximum inhibition of RNA
polymerase II by α-amanitin Seifart and Sekeris (1969) proposed
a 1:1 complex. By means of an appropriate filtering technique
and radioactively labeled toxins at equilibrium Cochet-Meilhac
and Chambon (1974) produced the experimental evidence for this
ratio. Knowledge of the stoichiometry of the complex enabled
Cochet-Meilhac et al. (1974) to estimate the number of RNA polymerase II molecules in various tissues. Most tissues, including
rat liver, have an average content of 14-23 µg RNA polymerase
II protein/g tissue, rat brain has less (7 µg/g), while calf
thymus has significantly more (33-55 µg/g). However, these determinations have been performed under the assumption that RNA
polymerase II is the only molecule which binds tightly to amatoxins. With the detection of another nuclear protein, which
binds amatoxins with a comparable affinity (Brodner and Wieland
1976a), these values have probably to be corrected. This protein
consists of 100,000 and 10,000 dalton polypeptides; its function
is totally unknown.

The strong interaction of amatoxins with RNA polymerases II was
first quantitated by Sperti et al. (1973). At 4°C the dissociation constant for [^{14}C] methyl-γ-amanitin and rat liver RNA polymerase II is $K_D = 3.6 \cdot 10^{-9}$M as determined by equilibrium dialysis.

Because the enzyme is labile, it was advantageous to study the equilibrium, especially at higher temperatures, with the more rapid filter technique introduced by Cochet-Meilhac and Chambon (1974). Using calf thymus RNA polymerase II, they found for the same toxin at $20°$, $K_D = 7.2 \cdot 10^{-10} M$. They also determined the dissociation constant of another labeled amatoxin, $[^3H]$-methyl-dehydroxymethyl-α-amanitin, together with the K_i values at $37°C$ for the two labeled toxins. From the closeness of the values for K_D and K_i, one can conclude that there exists a direct relationship between binding of amatoxins to RNA polymerase II and inhibition of RNA synthesis. Table 2 shows part of the data collected by Cochet-Meilhac and Chambon. Upon examination of the data in Table 2, one can see that the stability of the toxin-protein complex increases with decreasing temperature. Likewise, high ionic strength stabilizes the complex, while stability decreases with the addition of 3% dimethylsulfoxide (DMSO).

The stability of the toxin-protein complex also depends on structural features in the side chains of amatoxins, especially in the hydroxylated isoleucine residue. The δ-C-atom and the two hydroxyl groups in γ- and δ-position are necessary for a strong interaction with the protein. The loss of both hydroxyl groups, as in amanullin, increases the dissociation rate of the complex, the k_2-value. This may occur to such an extent that the component becomes non-toxic by the rapid excretion of the free amatoxin.

Since RNA polymerase II of calf thymus contains at least six different polypeptides, it was of interest which subunit represented the target for amatoxins. Brodner and Wieland (1976b) solved the question by covalently coupling a radioactively labeled amatoxin with a functional carboxylic group, $[^3H]$-6'-amanin (see chap. B), to the enzyme, by means of a water-soluble carbodiimide. SDS-gel electrophoresis showed that the polypeptide with 140,000 daltons had incorporated the label. This subunit is present in all types of RNA polymerase II, i.e., in II_A, II_B, and II_O.

A subunit of similar molecular weight (138,000 daltons) is also present in all enzymes of type III, which bind amatoxins, although with lesser affinity. Hence, it is tempting to attribute the toxin-binding capacity of RNA polymerases III to this subunit. However, no binding studies or affinity labeling experiments have been performed so far with enzymes III, and it may be important to recall that yeast RNA polymerase III is not inhibited by amatoxins. Since this is the only RNA polymerase III so far which lacks the 98,000 polypeptide chain, this subunit has become another possible candidate for the amaxotin-binding subunit of RNA polymerases III.

Cochet-Meilhac and Chambon (1974) also investigated which step in the complex process of transcription might be impeded by the toxin. Amatoxins do not inhibit the binding of DNA to the enzyme, nor do they compete with any of the four nucleoside triphosphates. Likewise, there is no influence of amatoxins on the release of RNA. From PPi exchange experiments it was concluded that ama-

Table 2. Dissociation constants (K_D), inhibition constants (K_i) and rate dissociation constants (k_2) of some amatoxins with calf thymus RNA polymerase II (according to Cochet-Meilhac and Chambon 1974)

Toxin	Conditions	K_D (M)	K_i (M)	k_2 (s^{-1})
α-amanitin	37°		$3.1 \cdot 10^{-9}$	$1.2 \cdot 10^{-4}$
O-Methyl-γ-amanitin	37°	$2.6 \cdot 10^{-9}$	$4.6 \cdot 10^{-9}$	$4.1 \cdot 10^{-4}$
	20°	$7.2 \cdot 10^{-10}$		
O-Methyl-dehydroxy-methyl-α-amanitin	37°	$6.4 \cdot 10^{-9}$	$1.0 \cdot 10^{-8}$	
	30°	$2.8 \cdot 10^{-9}$		
	20°	$6.7 \cdot 10^{-10}$		
	5°	$1.7 \cdot 10^{-10}$		
	37°, 1M (NH$_4$)$_2$SO$_4$	$2.2 \cdot 10^{-9}$		
	37°, 3% DMSO	$1.2 \cdot 10^{-8}$		
Amanullin (nontoxic)	37°		$9.1 \cdot 10^{-9}$	$1.4 \cdot 10^{-3}$

toxins most probably block the formation of phosphodiester bonds in the initiation step, as well as in the following elongation.

F. Structural Lesions in Nuclei

Fiume and Laschi (1965) traced the molecular mechanism of amanitin toxicity through the detection of structural lesions in nuclei caused by amatoxins. They found alterations, especially fragmented nucleoli in mouse liver parenchymal cells poisoned by α-amanitin. A similar change was also observed in a KB cell line (Fiume et al. 1966), and, later, in chick fibroblasts (Paweletz and Hoffmann 1972). This indicates that the lesions caused by the toxin are not specific for liver cells. In a follow-up study using total extracts of the mushrooms, Villa et al. (1968) found that such nucleolar lesions could be observed as early as 30 min after application of the toxin. They also found that the granular and fibrillar components of the nucleolus had segregated.

Morphological studies in depth were done in mouse kidney cells by Fiume et al. (1969) and, in rat liver cells, by Marinozzi and Fiume (1971). The condensation of nuclear and nucleolar-associated chromatin was detected. In kidney and liver cells the sequence of events, which occurs 15-30 min after administration of the toxin, is very similar. Concurrently with the fragmentation of the nucleoli, the fibrillar and granular components separate, while the chromatin condenses at the rim of the nucleus (margination). The nucleoplasm appears bright as a consequence. There is, however, an accumulation of interchromatin granules in the middle of the nucleus. Other electron-dense particles, the so-called perichromatin granules, can also be observed. Later, the granular type nucleolar fragments disappear, followed by the fibrillar component. By comparison with the alterations in the nuclei, the organelles of the cytoplasm change very late, i.e., after 48 h. In the cytoplasm mitochondria swell, the number of dense bodies increases, and autophagic vacuoles appear. These changes are, however, regarded to be secondary in nature.

The nuclear and nucleolar lesions described above are considered as consequences of transcription inhibition by α-amanitin. An inhibition of RNA polymerase II, also by other agents, causes euchromatin to condensate to heterochromatin. Involvement of intranucleolar chromatin in the process of condensation could consequently cause the fragmentation of the nucleolus. Evidence for this mechanism was furnished by Derenzini et al. (1976), who treated regenerating rat liver with α-amanitin: unlike normal liver cells, the chromatin in the regenerating liver cells is totally decondensed. Thirty min after toxin treatment, the chromatin around the nucleolus started to condense anew together with fragmentation of the nucleolus. Structural changes similar to those in the nuclei of mammalian cells were observed in nuclei of chicken liver after amanitin treatment. Sinclair and Brasch (1978) describe an extensive compaction of euchromatin, a disaggregation of nucleoli and a segregation of ribonucleoprotein components. All these changes were transient; after 48 h the nuclear structure was found recuperated. In parallel with the structural analysis, the concentrations of some nuclear components were determined: while the content of RNA and residual non-histone-proteins decreased significantly in concert with the morphological changes, DNA and histones remained unchanged.

Recently, Romen et al. (1977) tried to elucidate the mechanism of nucleolar segregation by observing under an electron microscope the development of lesions in nuclei induced by α-amanitin. Unlike the action of actinomycin D, the segregation (and degranulation) of the nucleolus occurs under α-amanitin influence later than the condensation of chromatin and fragmentation of the nucleolus. This is in accordance with the sequence of events observed by Emanuilov et al. (1974). Using mouse liver nuclei they found that 30 to 60 min after α-amanitin, shearing (a prestage of fragmentation) and fragmentation prevail, while segregation was observed only after 60 to 120 min. It is noteworthy that these authors discuss an alternative mechanism of nucleolus segregation. They suggest that the inhibition of maturation of 45 S 5-RNA by α-amanitin, an event occuring earlier than 30 min, may cause the structural lesion of the nucleolus by an inhibited feedback mechanism (see also Hadjilov et al. 1974). An even more detailed insight into the early alterations caused by α-amanitin is furnished by a most recent study of Barsotti and Marinozzi (1979). These authors used sublethal doses of the toxin which caused a delayed development of nucleolar lesions. They observed the appearance of half-moon-shaped electron dense particles at the rim of the nucleoli which may develop into "multigranular perichromatin bodies".

Similar structures were induced also by other agents and the authors conclude that the nucleolar lesions may in fact be caused by disturbances of maturation or transportation of ribosomal RNA. A similar mechanism for α-amanitin-caused alterations of the nucleoli has been discussed for plant cells by Moreno Diaz de la Espina and Risueno (1976). These authors found a temporary, reversible segregation in the nucleoli of *Allium cepa* and suggested that it was due to an impairment in the processing of ribosomal precursors.

By using a special stain, Petrov and Sekeris (1971) were able to recognize a ribonucleoprotein component, perichromatin fibrils (PCF), which is affected by α-amanitin. This structural component diminishes 30 min after administration, of the toxin; at the same time condensation of chromatin occurs. The perichromatin fibrils are the morphological equivalent of newly synthesized h nRNA, which as a ribonucleoprotein possibly maintains the decondensed state of chromatin.

The nucleoli of CHO cells (Kedinger and Simard 1974) were less sensitive than those in liver cells. These nucleoli fragmented only 8 h after treatment with the toxin. Accordingly, in the first 8 h the synthesis of rRNA is not impaired by α-amanitin in CHO cells. These different results when compared to those in liver or kidney cells may be due to the slow penetration of the toxin into CHO cells. Alternatively considered are factors, protein or polynucleotide in nature, present in both cell species which may regulate the synthesis of rRNA and which depend on mRNA. If such factors exist, they are probably long-lived in CHO-cells and short-lived in hepatocytes.

Structural lesions similar to those found in liver or kidney cells were also observed in the nuclei of transformed lymphocytes (Faulstich et al., in press) and in the nuclei of mouse embryos (Golbus et al. 1973). Derenzini et al. (1973) used a serum albumin-conjugated β-amanitin, instead of α-amanitin, in morphological studies of mouse liver. The high molecular weight conjugate preferentially penetrates the sinusoidal cells; as a result, the first nuclear lesions were observed in sinusoidal cells rather than in hepatocytes. The condensation of the chromatin and fragmentation of the nucleoli, however, looked very similar to the alterations induced in parenchymal cells treated by low molecular weight amatoxins.

G. Lesions Induced in In-Vivo-Systems

This section will deal with biochemical alterations, such as impairments in the synthesis and processing of hnRNA, rRNA, tRNA, DNA, and proteins in in vivo systems.

As expected after results from the in vitro assays of isolated RNA polymerase, many investigators report the inhibition of synthesis of hnRNA and mRNA also in in-vivo systems. For example, as early as 15 min after administration of α-amanitin to rats (Sekeris and Schmid 1972) or mice (Hadjiolov et al. 1974) the synthesis of hnRNA was found randomly suppressed in the isolated nuclei of liver. Inhibition was complete after 30 min.

Unexpectedly, α-amanitin also inhibited the synthesis of rRNA in rat liver (Jacob et al. 1970a; Niessing et al. 1970), mouse liver (Hadjiolov et al. 1974), and some insect larvae (Shaaya and Sekeris 1971; Holt and Kuijpers 1972b; Shaaya and Clever 1972). In the rat liver, inhibition of nucleolar RNA synthesis occurs simultaneously with that of nuclear RNA, and is likewise complete after 30 min. However, unlike the activity of the nucleoplasmic enzymes, the inhibition of nucleolar RNA synthesis is transient and may recover as soon as 2 h after administration of the toxin (Jacob et al. 1970a).

Significant depression of the precursor 45 S and the mature 28 S and 18 S RNAs were observed. These inhibitions were confirmed by several other laboratories (Sekeris and Schmid 1972; Tata et al. 1972) reporting an inhibition of pre-rRNA of 70% 4 h after in vivo intoxication. The authors excluded that the impairment of rRNA is simply based on a decrease of RNA polymerase I molecules by an inhibited transcription. An accurate analysis in the mouse liver of all fractions of synthesized RNAs and the timing involved (Hadjiolov et al. 1974) revealed that the earlier event is not the inhibited synthesis but the maturation of 45 S RNA. Similar impairment by α-amanitin of the processing of ribosomal precursor RNA has been also reported for *Drosophila* larvae by Holt and Kuijpers (1972) and for *Dysdercus intermedius* by Duspiva et al. (1973). In these studies $5 \cdot 10^{-6}$ M α-amanitin decreased the 18 S and 28 S RNA concentrations along with an accumulation of 36 S RNA, which seems to be a direct precursor of 18 S RNA. It may be worthwhile to recall that the impeded maturation of pre-rRNA was proposed by Hadjiolov et al. (1974) to be responsible for the morphological alterations in nucleoli of rat and mouse liver after α-amanitin treatment.

All investigators agree that the enzyme responsible for the synthesis of rRNA precursors, RNA polymerase I, is not directly affected by α-amanitin. There is no significant inhibition of this enzymatic activity in the isolated liver nuclei of poisoned rats and mice (Stirpe and Fiume 1967; Tata et al. 1972; Sekeris and Schmid 1972). The most probable mechanism then for the in vivo inhibition of ribosomal RNA synthesis is that the synthesis of rRNA-precursors is under the control of some nucleoplasmic gene product. Such a regulator molecule is characterized by an exceedingly fast turnover, shown by fully developed inhibitory effects within 30 min. The factor could possibly be a special RNA or a translation product. Such protein factors have been postulated by several laboratories (Raynaud-Jammet et al. 1972; Lampert and Feigelson 1974). The actual existence of such factors has been substantiated by the isolation of a protein factor by Higashinakagawa and Muramatsu (1972) and by the observation of Sekeris and Schmid that in rats the synthesis of liver nucleolar proteins decreased significantly after poisoning. Additionally, as reported by several investigators, the early effects of cycloheximide on maturation and on synthesis of 45 S pre-rRNA may be evidence for the protein nature of factors regulating the activity of RNA polymerase I. In analogy to this, similar protein factors regulating RNA polymerases II have been recently described by several laboratories; for a review see Chambon et al. (1974).

However, when the timing of the α-amanitin and cycloheximide effects on the inhibition of rRNA synthesis were compared, differences were found; (Schmid and Sekeris 1973). Furthermore, Lindell (1976) found similarities in the inhibition of rRNA synthesis by α-amanitin and actinomycin D. These arguments favor a regulator molecule represented by a polyribonucleotide. One possible candidate may be the U1 RNA whose synthesis is strongly inhibited by α-aminitin (Ro-Choi et al. 1976).

Further possible mechanisms to explain the dependance of pre-rRNA synthesis on a synchronous transcription of hnRNA are mechanical in nature. For example, the in situ activity of RNA polymerase I may require either a continuous transcription of hnRNA or certain structural organizers, which may have been disturbed when the RNA polymerase II was inhibited. Impairment of such organizers was also discussed with the morphological lesions observed in the nucleoli after α-amanitin treatment. Also, the existence of an amatoxin metabolite formed by the liver, which possibly affects RNA polymerase I, has not been completely disproved. Such a possibility, however, is unlikely since in a perfusion experiment with rat livers most of [^{3}H]O-methyldehydromethyl-α-amanitin was recovered from the bile and found to be completely unmetabolized (Faulstich, unpublished results).

Finally, one possibility not yet considered is that the protein isolated as a by-product of calf thymus preparations of RNA polymerase II (Brodner and Wieland 1976a), which binds α-amanitin with a similar affinity as RNA polymerase II, may be of nucleolar origin and involved in the synthesis of pre-rRNA. Provided that such a protein does exist also in the liver, its binding by α-amanitin could represent an alternative mechanism of inhibition of pre-rRNA synthesis that is completely independent from the activity of RNA polymerase II.

Contrary to the results found in liver tissue, no impairment of rRNA synthesis by α-amanitin has been detected in tissue culture cells. Hastie and Mahy (1973) reported that treatment with $2 \cdot 10^{-5}$M α-amanitin decreased the RNA polymerase II activity as early as 30 min after addition of the toxin. RNA polymerase I and incorporation of labeled orotate into ribosomal RNA were, however, unaffected for at least 2 h. Similar conditions were found in Chinese hamster ovary (CHO) cells by Kedinger and Simard (1974). Although extensive fragmentation of the nucleoli was observed, synthesis of pre-rRNA remained unaffected for several hours. The pre-rRNA synthesis in salivary glands of *Chironomus* (Egyhazi et al. 1972; Serfling et al. 1972) and of adult female *Aedes Aegypti* (Fong and Fuchs 1976a) were also uneffected by α-amanitin in vivo, while the hnRNA was significantly decreased. I should like to propose that the insensitivity of in vivo pre-rRNA synthesis to α-amanitin is the normal case, and that liver tissue sensitivity is caused possibly by a different organization or regulation in the liver nuclei. Perhaps such differences are the reason why the liver - besides having an enhanced penetration - is the target organ in most animal species during α-amanitin poisoning.

α-Amanitin also impairs the synthesis of low molecular weight RNA, although to a lesser extent. This has been established for rat liver by Niessing et al. (1970). Hadjiolov et al. (1974) report that mouse liver maturation of 4.6 S pre-tRNA is sensistive to α-amanitin in vivo and hence may also be under nucleoplasmic control. Recently, Ro-Choi et al. (1976) analyzed in more detail the impairment by α-amanitin of low molecular weight RNA. By a fine resolution of RNA on polyacrylamide gelelectrophoresis, they found the 4 S- and 5 S RNA significantly decreased, while 4.5 S RNA was nearly unaffected. In addition, the synthesis of U1-RNA was strongly inhibited. Most recently, Frederiksen et al. (1978) confirmed that α-amanitin inhibits the synthesis of low molecular weight RNA of all types: U-1 (D), U-2 (C), and U-3 (A). Since in baby hamster kidney cells such inhibition was caused with amanitin concentrations too low to inhibit the activity of RNA polymerase III, these authors concluded that U-1, U-2, and

U-3 RNAs are products of RNA polymerase II activity. This opinion was substantiated by using amanitin-resistant cells in which the synthesis of that low molecular weight RNA was found to be insensitive to α-amanitin (Jensen et al. 1979). In larvae of *Drosophila* (Holt and Kuijpers 1972b) and *Calliphora* (Shaaya and Clever 1972) tRNA synthesis was found, however, to be unaffected by the drug.

DNA synthesis is almost absent in normal liver cells. In order to study the in vivo influence of α-amanitin on replication in liver cells, Montecuccoli et al. (1973) had to use regenerating rat liver after partial hepatectomy. They administered sublethal doses of α-amanitin 20 h after partial hepatectomy, the time of peak DNA synthesis. It decreased within 2-3 h. This is an indirect effect of the toxin as it most probably acts via the decrease in protein synthesis, which is impaired coincidently.

Impairment of protein synthesis by α-amanitin was observed as early as 1954 (Wieland and Dose 1954), at a time when the molecular mechanism of toxicity was still unknown. The authors measured a 50% decrease in serum albumin concentration and an unaffected globulin level 96 h after administration of α-amanitin to the mouse. Direct information on protein synthesis was obtained by the pulse experiments of Montecuccoli et al. (1973), who found maximum inhibition of protein synthesis (∿50%) about 3 h after administration of α-amanitin. Because of the differing sensitivities of the mouse and rat to α-amanitin, protein synthesis showed a lesser decrease in rat organs than in the liver and kidney of the mouse. Even disaggregation of polysomes was observed after α-amanitin administration (Montanaro et al. 1973) in the mouse. Decreases in protein synthesis as a consequence of impeded RNA synthesis have also been reported for cultures of rat fibroblasts by Kuwano and Ikehara (1973).

Finally, α-amanitin has also been employed to study the mechanisms of other toxins. Prasanna et al. (1975), demonstrated that 10^{-5}M aflatoxin B_1 inhibits RNA synthesis by 40% in rat liver slices. This inhibition is similar to one obtained by a 10^{-8} α-amanitin concentration. Since a combination of the two toxins does not significantly increase the rate of inhibition, the authors conclude that aflatoxins impair the same RNA polymerases as α-amanitin does. Since in the rat liver the nucleolar enzymes are also effected by α-amanitin, the enzymes inhibited by aflatoxin are not necessarily the nucleoplasmic RNA polymerases. The bacterial exotoxin, an ATP-analog, likewise inhibits RNA polymerases (Smuckler and Hadjiolov 1972). By use of α-amanitin it could be stated that the nucleoplasmic RNA polymerase is 50 to 100 times more sensitive to the ATP-analog than the nucleolar enzyme, and hence is preferentially inhibited by the exotoxin. The toxicity of bacterial endotoxin has been studied by Seyberth et al. (1972). Administration of a half LD_{50} of α-amanitin to mice enhances the sensitivity of these animals to endotoxin by about 200-fold. It is proposed that this effect is caused by a decreased synthesis of detoxifying enzymes in the reticulo-endothelial system as a consequence of α-amanitin treatment.

H. Penetration into Various Cells

While isolated nuclei are permeable to amatoxins, the penetration of the toxins into most cells appears to be limited. Correspondingly, RNA polymerase activity in isolated nuclei is inhibited by 10^{-8}M α-amanitin, a concentration that inhibits the solubilized enzyme, while in most cell populations inhibitory effects are observed only with concentrations greater than 10^{-6}M.

An easy penetration of amatoxins was observed into liver and kidney cells in situ. These cells are equipped with membrane structures or organelles, which apparently facilitate the uptake of certain low molecular weight compounds. Possibly the high content of endoplasmic reticulum in those cells is involved. It seems reasonable to assume that one reason for the high susceptibility of liver and kidney cells to amatoxins is the enhanced uptake of the toxin into these cells. After 30 min perfusion of a rat liver with [^{3}H]O-methyl-dehydroxymethyl-α-amanitin, Faulstich et al. (in preparation a) estimated that half the concentration present in the perfusion medium was inside the cells. For example, with a toxin concentration in the perfusion medium as low as 10^{-7}M, the liver takes up 20 p mol of toxin per g liver in 30 min. This is sufficient to inhibit two-thirds of the RNA polymerase II molecules present in this organ as determined by Cochet-Meilhac et al. (1974) provided that dissociation of the complex and bile excretion of the toxin are not accounted for. As a consequence, after administration of the lethal dose to a mouse, corresponding to about $2-5 \cdot 10^{-6}$M α-amanitin in the extracellular fluid, the liver cells exhibit nuclear lesions 15-30 min after the administration of the toxin. In comparison, CHO cells treated with α-amanitin in similar concentration show such lesions not before 6-8 h treatment.

The first experiments with α-amanitin in cell cultures were performed by Fiume et al. (1966) with KB-cells and human amnionic cells. They detected no cytopathic effects with a concentration of ∿1 μg/ml (∿10^{-6}M), while with 2 μg/ml lesions became obvious after 2 days and were complete after 6 days. Other cell lines, HeLa, KB, or HEp-2-cells, were more resistant and developed lesions only with toxin concentrations as high as 3.5 to 10 μg/ml (Fiume and Barbanti-Brodano 1974). Human embryonic fibroblasts were inhibited by 5 μg/ml α-amanitin. These levels are in agreement with the observation by Hastie and Mahy (1973) that $2-3 \cdot 10^{-6}$M α-amanitin inhibits RNA polymerase II by 50% for several hours in the same cell line. Hastie et al. (1972) also found that $2 \cdot 10^{-5}$M α-amanitin caused a 50% decline of DNA after 8 h, while protein synthesis remained unaltered.

Several laboratories have reported that the basal activities of RNA polymerases in tumor cells can be several times higher than in normal cells. In rat hepatoma cells the RNA polymerase II activity is twofold, and the RNA polymerase I_B is even ninefold higher than those for normal cells (Chesterton et al. 1972).

Nevertheless, transformed cells, as a rule, are less sensitive
to α-amanitin than the corresponding normal cells. For example,
Boctor and Grossman (1973) report that hepatoma cells are resistant to a concentration as high as 10^{-5}M α-amanitin. Although
Damle and Novack (1972) reported on a RNA polymerase II isolated
from L 1210 leukemic cells, which is resistant to α-amanitin,
the resistance of tumor cells in most other cases is explained
by the poor penetration of the drug into the cells rather than
by the presence of an enzyme insensitive to α-amanitin. In some
cases, however, the elution pattern from DEAE cellulose of RNA
polymerases in tumor cells was changed as compared to normal
cells. Capobianco et al. (1977) found in extracts of normal
cells that the RNA polymerase I was eluted in one peak, while
two peaks of amanitin-resistant RNA polymerase were isolated
from tumor cells. The RNA polymerase II of these tumor cells
was as sensitive to α-amanitin as that of normal cells. A different elution pattern of RNA polymerases from Ehrlich ascites
cells was reported by Blair and Dommasch (1972). These cells
had an unusual "enzyme III" that was inhibited by 50% with 5 μg/
ml α-amanitin. Amalric et al. (1972) report that RNA polymerase
II isolated from hepatoma ascites cells had lost part of its
amanitin sensitivity. This enzyme showed a 10% inhibition with
1 μg/ml α-amanitin, while the corresponding enzyme of calf thymus
was blocked by 50%.

A direct comparison between the sensitivities of tumor and normal cells was possible in cell lines where tumors could be induced by viruses. Fiume and Barbanti-Brodano (1974) found SV
40-transformed embryonic fibroblasts to be two times more resistant to α-amanitin than the corresponding normal cells.
Dinowitz et al. (1977) compared normal and RSV-transformed
chick cells and found with 1 μg/ml α-amanitin 5% RNA synthesis
activity in normal cells, whereas the transformed cells retained
70% of their RNA synthesis activity. Resistance was developed
14 h after virus infection and is explained by changes in metabolic factors or by an alteration of the plasma membrane concomitant with the transformation. A change that causes resistance
to the toxin may also occur in normal cells during certain phases
of the cell cycle.

It has been reported by Ledinko (1971) that α-amanitin can inhibit the process of transformation by viruses. In hamster cells,
the number of transformation foci was reduced to 50% by 0.01 μg/
ml, a concentration which did not impair the viability of the
infected cells. A similar process in plant cells is also sensitive to the drug (Beiderbeck 1972). Like E. coli, Agrobacterium
tumefaciens is insensitive to amanitin. However, the induction
of tumors by A. tumefaciens in plant cells is significantly decreased in the presence of 20 μg/ml of the drug. Both observations indicate that the process of tumor induction requires RNA
synthesis.

As reported by Gogala (1969) for the mesocotyls of avena, plant
cells are also susceptible to α-amanitin. They are, however,
less sensitive than mammalian cells (Seitz and Seitz 1971) as
17 μg/ml α-amanitin was required to inhibit the RNA synthesis

of parsley cells. At this concentration the ribosomal precursor (32S) RNA was not affected. De la Torre et al. (1974) observed that a concentration of 10 µg/ml had only a transient effect on the interphase nucleus and prolonged the cell division cycle of *allium* root tips. Mainly by prolongation of the G_1- and G_2 phase, the interphase lasted 33 h instead of 22.5 h in control cells. A concentration of 20-50 µg/ml finally caused necrosis of plant cells (Beiderbeck 1972).

Efforts have been made to facilitate the penetration of amatoxins into cultured cells. Kuwano and Ikehara (1973) detected that in cultured fibroblasts 2 µg/ml α-amanitin had no effect. However, the same toxin concentration together with 5 µg/ml amphothericin B caused a 100% inhibition of RNA polymerase II activity with decay of polysomes and a decline in protein synthesis. Foa-Tomasi et al. (1976) did not confirm this observation for HEp-2 and BHK cells, and found that amphothericin B alone was able to decrease RNA synthesis. They reported, however, that cell pretreatment with DEAE-dextrane significantly enhances the activities of both α- and β-amanitin. Another way to facilitate the penetration of amatoxins is their derivatization with lipophilic residues. Although the affinity of O-hexyl-α-amanitin to mammalian RNA polymerases II is lower than that of α-amanitin, its cytotoxicity for virus-transformed cultured lymphocytes proved to be four times that of the parent compound (Faulstich et al., in press).

Conjugation of amatoxins to proteins increases their toxicity and, moreover, confers specific activity towards protein consuming cells. β-amanitin-bovine serum albumin conjugates showed inhibitory capacity for in vitro RNA polymerase II, similar to that of the free toxin. However, its penetration into hepatocytes was much greater, causing an eight fold increase of cytotoxicity in comparison to α-amanitin. The toxicity could be increased further three-four times if the albumin was additionally derivatized with lipophilic residues like fluorescin (Fiume et al. 1971).

The protein consuming sinusoidal cells are two times more susceptible to the albumin conjugate than the hepatocytes. This is remarkable, since the liver sinusoidal cells are insensitive to low molecular weight amatoxins (Bonetti et al. 1976). Similarly, the β-conjugate causes lesions, that are typical for amatoxins, in the proximal tubules of rat kidney. These cells are also not affected by free amatoxins (Bonetti et al. 1974). The highly selective toxicity of β-amanitin-bovine serum albumin was also established with mouse peritoneal macrophages (Barbanti-Brodano and Fiume 1973). With these cells, amatoxin toxicity becomes about 50-fold greater when conjugated to albumin: 0.4 µg/ml kill 50% of the macrophages in a monolayer. By comparison, lymphocytes and fibroblasts were much less sensitive to this toxin: 25 µg/ml of the albumin derivative had no effect on lymphocytes, and mouse fibroblasts even withstood concentrations >100 µg/ml. Recent developments in chemical procedures have allowed the conjugation also of α-amanitin to proteins by a spacer molecule. Faulstich and Trischmann (in preparation) prepared conjugates of α-amanitin with γ-globulins, concanavalin A, histones,

ovalbumin, and other proteins. These derivatives have potential
for future biological application.

I. Resistant Cell Lines

Cell lines resistant to α-amanitin have been developed from
various mammalian cells, such as Chinese hamster ovary cells
(Chan et al. 1972; Ingles et al. 1976a), hamster BHK-T6-cells
(Amati et al. 1975), rat myoblasts (Somers et al. 1975a; Ingles
et al. 1976a), mouse myeloma MOPC 104 E cells (Wulf and Bautz
1976), human diploid fibroblasts (Buchwald and Ingles 1976),
and mouse lymphoblastoid cells L 51784 (Bryant et al. 1977).
Resistance was achieved by growing the cells in media containing
0.25 to 5 µg/ml α-amanitin ($2.5 \cdot 10^{-7}$ to $5 \cdot 10^{-6}$M). Owing to the
low inhibition constants of mammalian RNA polymerases II, only
a small portion of the toxin could have penetrated into the
cells.

Nearly all cell lines resistant to α-amanitin possessed RNA polymerases II activity with decreased sensitivity to the drug. The
inhibition constants (K_i) of the purified RNA polymerase II activities paralleled the relative resistance to growth inhibition.
The latter is, therefore, based on altered forms of DNA-dependent
RNA polymerases, which in some cases were 800- to 1000-fold less
sensitive than the wild-type enzymes.

The alterations of the RNA polymerases II seem to be based on
classical Mendelian mutations. The cell lines to be selected
were treated with the mutagen ethylmethanesulfonate. This improved the survival rate of cells grown with α-amanitin by a
factor of 20, and furthermore, the selected cells retained the
altered phenotype in the absence of the drug for many generations, e.g., rat myoblasts for more than 100 generations and
CHO-cells for more than 6 months. No other factors which might
be responsible for the resistance to growth inhibition could
be detected.

As the inhibition curve of their RNA polymerase II activity is
monophasic, CHO cells and human diploid fibroblasts probably
possess only the resistant type of the enzyme. Others, like rat
myoblasts, mouse myeloma, and mouse lymphoblastoid cells, show
a biphasic inhibition curve: part of their RNA polymerase activity is still of the wild type. Allele combinations have been
found with 1 wild type + 1 mutant corresponding to 50% of the
wild-type RNA polymerase activity, and others with 1 wild type
+ 3 mutant with 75% activity. In a few cases, resistant cell
lines had only RNA polymerase activities of the wild type.
This indicates that resistance can be based on conditions other
than possessing less sensitive RNA polymerases.

In several laboratories, hybrids of mutant cells with wild-type
cells, or with other mutant cells, have been produced. These
cells show inhibition curves of the RNA polymerase activities

that are biphasic and intermediate between their parents. The
allele of the genetic alteration, therefore, is expressed codom-
inantly. Then the mutant cells with 50 or 75% wild-type enzyme
were grown in medium containing α-amanitin the decline of wild-
type enzyme activity was readily compensated for by a subsequent
increase in activity of mutant RNA polymerase II. Therefore,
their RNA polymerase activity inhibition curve became monophasic,
and the total activity level remained nearly constant. As a result, the growth rate of these cells was maintained in presence
of the toxin (Somers et al. 1975a; Crerar and Pearson 1977). In
detail, there was a 50% reduction of the total activity during
the first 8 h, then the activity increased again after 16 h to
approximately its original value. In contrast, in mouse myeloma
cells the resistant enzymes did not totally compensate the decrease of the wild-type enzymes. The lower level of the total
RNA polymerase II activity was, hence, accompanied by a decrease
in the growth rate of the cells.

Since the blocked wild-type RNA polymerase II in rat myoblasts
under amanitin influence is rapidly substituted by the resistant
type of the enzyme, it has been deduced that regulation of RNA
synthesis is achieved by RNA polymerase II itself, acting as its
own repressor. Such regulation would be similar to that of the
corresponding enzyme in *E. coli*.

The low inhibition of RNA polymerase activity in mutant cells by
amanitin was accompanied by a low affinity of these enzymes for
the toxin. Association constants (K_A) were up to 100 times lower
for the mutant enzymes than for the wild type. The simplest explanation is that mutations in the structural gene coding for
the amanitin binding subunit B3 (Brodner and Wieland 1976b)
create the low affinity. This theory is consistent with the observation that most physical and functional properties of the
mutant enzymes, i.e., thermostability, K_m for UTP, salt optima,
or turnover numbers, remain unchanged in mutant cells. It cannot, however, be excluded that the mutation affects a subunit
other than B 3, altering the quaterny structure of the whole
complex and the affinity for the drug. Some evidence for this
was the twofold increase in activity of the mutant enzyme with
poly[d(AT)] as template rather than DNA, by comparison to the
wild-type enzyme (Bryant et al. 1977). It is also possible that
the affinity for α-amanitin depends on the subunit composition
of the RNA polymerase II. Arguments for such a possibility are
the sixfold increase in sensitivity to α-amanitin after the
removal of two subunits from (amanitin-sensitive) yeast RNA
polymerase I (Huet et al. 1975a) and the observation that
Drosophila RNA polymerase II loses its amanitin sensitivity
after chromatographic procedures (Phillips and Sumner-Smith 1977).
Some studies have shown that the elution patterns of mutant and
wild-type RNA polymerases II from DEAE-cellulose differ distinct-
ly (Chan et al. 1972; Amati et al. 1975).

The mutations seem to affect only one subunit. In hybrid cells
of two mutant cell lines with different resistance to growth
inhibition (Ingles et al. 1976b) no wild-type enzyme was detected. This would have been formed if different subunits had
been mutated in the parent cells rather than only one. A new

phase in the studies with amatoxin-resistant cells was started most recently by Greenleaf et al. (1979), who discovered a mutant of Drosophila melanogaster growing at concentrations of amanitin being lethal to the wild type. The RNA polymerase II of this mutant were approximately 250 times less sensitive to inhibition by amanitin than that of the wild type. The mutant represents the first organism - beside the toxic mushroom itself - that grows under high concentrations of α-amanitin.

J. Virus Research

α-Amanitin inhibits the replication of several RNA and DNA viruses. It has been generally concluded that viruses, whose replication is inhibited by α-amanitin, participate in the amanitin-sensitive RNA polymerase activities of the host cell. The virion-coded RNA polymerase of vaccinia virus was assayed with α-amanitin; it was found to be completely insensitive to the toxin.

One of the RNA viruses studied in great detail is influenza virus (Rott and Scholtissek 1970). At 10^{-5}M, α-amanitin produced 90% inhibition of virus replication in chick fibroblasts measured by several virus-specific activities. The toxin did not interfere significantly with the synthesis of the macromolecules of the host cell, and hence its activity can be regarded as virus-specific. These results were confirmed by Mahy et al. (1972), who concurrently observed the RNA polymerase II activity in nuclei of the chick embryo fibroblasts and the viral growth. They found that the decrease of the enzyme activity by α-amanitin was closely related to the decrease of viral replication. Recently, Lamb and Choppin (1977) furnished direct proof that cellular RNA polymerase II is involved in influenza virus replication. They used an α-amanitin-resistant cell line of Chinese hamster ovary cells (CHO-Ama 1). These cells synthesize influenza virus polypeptides in the presence of α-amanitin.

Recent experiments of Mark et al. (1979) made apparent, in what way the replication of influenza virus may depend on host RNA-polymerase II. By the complete inhibition of complementary RNA synthesis in MDCK cells, when α-amanitin was added at the beginning of the infection, they concluded that m-RNAs of the host, used as primers for the production of virus transcripts, may limit the viral replication.

Rous sarcoma virus in chick embryo fibroblasts also depends on cellular RNA polymerase II (Zanetti et al. 1971). The most significant decrease in virus production was observed when these cells were treated with 10^{-6}M α-amanitin 24 to 72 h after infection. This time element indicates that a later stage of viral development is inhibited. Indeed, the first step, the reverse transcription of viral RNA into proviral DNA, should be insensitive to α-amanitin. Other possibilities of transcription may exist: Perez-Bercoff et al. (1974) observed that an

α-amanitin-sensitive RNA polymerase of mouse L cells was bound
to virus-induced double-stranded RNA.

Using isolated chicken cell nuclei, Rymo et al. (1974) found
that a concentration as low as 10^{-8}M α-amanitin sufficed to
suppress the in vitro synthesis of Rous sarcoma virus-specific
RNA. Again, the inhibition is virus-specific. This is shown by
the 95% reduction in the synthesis of the virus-specific RNA,
identified by an hybridization assay, while the synthesis of
total RNA was decreased by only 50%. There was no such specifi-
ty found with actinomycin D, which reduced the viral and total
RNA synthesis equally. Dinowitz (1975) followed the in vivo incorporation of [^{3}H] uridine into virions in the presence of α-
amanitin, and found that after 5 h treatment with the toxin the
synthesis of the virions and the polyribosomal RNA were equally
inhibited (50%). At the same time there was only a slight inhibition (< 10%) of rRNA synthesis.

Studying the myeloblastosis virus, Rymo et al. (1974) measured
the synthesis of virus-specific RNA in the nuclei of chick mye-
loblasts. They found inhibition by α-amanitin, stimulation by
Mn^{2+} ions and high salt concentrations and transcription by calf
thymus RNA polymerase II; all of which strongly suggests that
here too most, if not all, of the virus-specific RNA is transcribed by cellular RNA polymerase II.

A similar, and complete dependance on host cell RNA polymerase
II must be expected also for viroids. These single stranded,
circular RNA molecules which infect plant cells, are too small
to code for a viroid-specific polymerase. Accordingly, it could
be demonstrated by Mühlbach and Sänger (1975) that viroid replication in tomato cells is sensitive to α-amanitin.

Among the other RNA viruses studied an early paper by Fiume et
al. (1966) reports that polio virus type 2 and para-influenza
virus type 3 are not inhibited by α-amanitin. Vesicular stoma-
titis virus (VSV) is also insensitive to α-amanitin and was
used by Zanetti et al. (1971) as a control experiment for α-
amanitin-sensitive viruses. Generally, all RNA viruses insensitive to α-amanitin replicate in the cytoplasm of the host cell
rather than in the nucleus.

Among the DNA viruses, the replication of adenovirus is inhibited
by α-amanitin. Viral activities were completely suppressed in
human embryonic kidney cells by 10^{-6}M α-amanitin (Ledinko 1971).
Chardonnet et al. (1972) reported that $5 \cdot 10^{-6}$M toxin reduced the
concentration of adenovirus released from HeLa cells into the
culture medium by a factor of 10^3. This was also confirmed by
Wallace and Kates (1972). At the same time the crystalline aggregates of the virus in the cell disappeared and only few
scattered virus particles were detected. As suggested by Price
and Penman (1972a), the transcription of adenoviral RNA needs
an α-amanitin-sensitive enzyme. It is very similar, although not
identical, to the host cell polymerase with respect to its dependence on ionic strength.

However, there may be small differences with respect to different templates and possible virus specific factors, which may influence the host cell polymerase II activities. These factors may alter the enzymes so that viral DNA is preferentially transcribed (Wallace and Kates 1972).

Weinmann et al. (1974) reported that in KB cells α-amanitin at high concentrations ($2 \cdot 10^{-5}$M) inhibits the synthesis of adenovirus-specific 5.5S RNA. This inhibition occurs along with the inhibition of cellular tRNA and 5S RNA, providing evidence that RNA polymerase III is also involved in the transcription of viral RNA. A similar RNA polymerase activity in adenovirus-infected HeLa cells had been detected by Price and Penman (1972a) and reported to be insensitive to α-amanitin; however, the enzyme was only tested with low concentrations of α-amanitin ($5 \cdot 10^{-7}$M). Jaehning et al. (1976) detected new species of low molecular weight RNAs in adenovirus-infected KB cells. The synthesis of these discrete RNAs was not inhibited by 1 µg/ml α-amanitin, however completely suppressed by 200 µg/ml. It has been suggested that these new low molecular weight RNAs are transcriptional products of RNA polymerase III.

Growth of herpes virus has been studied in various host cells in the presence of α-amanitin. Mannini-Palenzone et al. (1971) reported a significant impairment of growth for this virus in chick embryo fibroblasts with $3 \cdot 10^{-6}$M α-amanitin. In KB cells, growth of herpes virus was, however, less inhibited. This indicates that the effect of α-amanitin on virus growth depends on the type of host cell used. Alwine et al. (1974) reported that $3 \cdot 10^{-7}$ α-amanitin inhibited 90% viral RNA synthesis of herpes simplex type 1 in nuclei of HEp-2 and KB cells. Both laboratories conclude that herpes virus possibly utilized toxin-sensitive polymerases of the host cell at some stage of its replication.

Evidence for this assumption was also provided in experiments using mutant cell lines resistant to α-amanitin. Costanzo et al. (1977) followed herpes virus replication and the synthesis of virus α-, β- and γ-polypeptides in HEp-2-cells, and in AR1/9-5B CHO cells, a cell line resistant to α-amanitin. In the latter the activities of the virus were not affected by the drug. Similar results were obtained by Ben-Zeer and Becker (1977) who compared BSC-1 and BHK cells with the resistant cell line BHK-T6-G1. The mutant cells were bred with amanitin-sensitive 3T3 mouse cells to form a hybrid (Amati et al. 1975). When these hybrids were infected with polyoma virus, α-amanitin did not impair viral growth. This indicates that the RNA polymerase of the unsusceptible parental cell can participate in the transcription of the viral genome.

Another DNA virus sensitive to α-amanitin is the Simian virus 40, investigated by Hildebrandt-Jackson and Sugden (1972). They demonstrated that the synthesis of viral RNA in the nuclei of monkey kidney cells is achieved by an amanitin-sensitive enzyme, RNA polymerase II of the host cell, rather than by RNA polymerase I. Most recently, Monjardino (1978) reported that the trans-

cription of polyoma virus DNA in baby mouse kidney cells is
also dependent on host cell RNA polymerase II. He concluded
this after observation of inhibition by α-amanitin.

The growth of one species of DNA virus, vaccinia virus, is insensitive to α-amanitin (Fiume et al. 1966). In a more detailed
study of this virus Costanzo et al. (1970) report that it induces RNA polymerase activity which is completely insensitive
to α-amanitin (and to rifamycin). This RNA polymerase becomes
active only after treatment with detergents and is stimulated
by salt concentrations lower than those necessary for the activation of mammalian enzymes. It is probably a virion-coded
RNA polymerase, different from both mammalian and bacterial RNA
polymerases.

One DNA virus of the irido type, frog virus 3 (FV3), is by itself a strong inhibitor of RNA polymerase activity in the host
cell (Campadelli-Fiume et al. 1972). In this case the specific
inhibitory activity of α-amanitin was useful to characterize
the virus-inhibited enzyme as an RNA polymerase. Recently, one
of these authors (Campadelli-Fiume 1979) published a brief review on amatoxins and virus research.

In general, one can conclude from inhibition experiments with
both RNA and DNA viruses that α-amanitin inhibits those viruses
which replicate in the nucleus. Although currently little information is available on α-amanitin activity on viruses which
multiply predominantly in the cytoplasm of the host cells, these
species seem to be insensitive to the toxin.

K. Hormonal Research

Hormones and certain drugs like phenobarbital can induce de novo
synthesis of enzymes. α-Amanitin has been used to elucidate
whether such enzyme inductions are under the control of transcription. In cases where hormones induce de novo synthesis of
RNA, the toxin has been employed to differentiate the enzymatic
activities of RNA polymerases I, II, and III.

Administration of glucocorticosteroids in rats induces de novo
synthesis of the liver enzyme tyrosine-transaminase. In animals
pretreated with α-amanitin, the induction of the enzyme by cortisol is suppressed although overall protein synthesis is not
significantly affected. Sekeris et al. (1970) concluded that
the hormonal induction of tyrosine-transaminase depends on the
de novo synthesis of RNA. Jolicoeur and Labrie (1971) confirmed
these results and reported that α-amanitin also inhibited induction of this enzyme by dibutyryl-c-AMP. The toxin was active
when applied 1 h prior to administration of the c-AMP. Inhibition studies with α-amanitin afforded the estimation of the half
lives of tyrosine-transaminase and its mRNA to be 3 h, and 1 h,
respectively.

No inhibition effect by 10^{-6}M α-amanitin was found for the induction of tyrosine-transaminase in hepatoma cells (Sahib et al. 1971), while 10^{-6}M actinomycin or 10^{-4} cycloheximide were potent inhibitors in this system. Boctor and Grossman (1973) confirmed these findings and suggested that hepatoma cells may be impermeable to the drug. Other mechanisms are also possible: If the enzyme induction depends on an increase of rRNA, the process would be strongly inhibited by α-amanitin in rat liver cells but not in hepatoma cells, since in various cell lines unlike the cells of rat liver the drug affects the synthesis of mRNA only.

In chick embryo liver, the steroid etiocholanolone or certain barbiturates induce the synthesis of δ-aminolevulinate-synthetase. This enzyme controls the initial, rate-limiting, step of porphyrogenesis. Incefy and Kappas (1971) found that α-amanitin inhibits the synthesis of this enzyme and consequently also porphyrogenesis. They concluded that synthesis of new RNA is a necessary step 14-16 h prior to the enzyme induction. In a subsequent study Incefy et al. (1974) confirmed the effect in avian liver cell cultures. Since protein synthesis was grossly unaffected, is clear that the inhibited induction of one enzyme is not necessarily reflected by an impairment of total protein synthesis.

In bone marrow precursor cells, calf thymic factors cause the differentiation to T cells, a process which is inhibited by $5 \cdot 10^{-7}$M α-amanitin. Incefy and Good (1976) suggest that the toxin inhibits transcription, which must be essential for the differentiation to cells bearing T cell markers. In malignant trophoblast cells, $3 \cdot 10^{-6}$M α-amanitin prevented the release of human choriono-gonadotropin (HCG) induced by dibutyryl-c-AMP. Pattillo and Hussa (1975) conclude that c-AMP induces the synthesis of RNA coding for the hormone, rather than simply causing hormone release. This seems reasonable, because the increase in secreted HCG does not occur earlier than 3.5 h after stimulation, but is preceded by an increase in intracellular HCG.

α-Amanitin has also been tested in insects to study synthesis of hormone-induced enzymes. Shaaya and Sekeris (1971) investigated the enzyme Dopa-decarboxylase as induced by β-ecdyson in the larvae of *Calliphora*. When the drug was administered 3 h prior to the hormone, synthesis of the enzyme was inhibited along with that or pre-rRNA. Fong and Fuchs (1976b) had contradictory results when this enzyme was not inhibited in *Aedes*. However, β-ecdyson induced the enzyme in *Aedes* not earlier than 72 h after application, while in *Calliphora* the enzyme activity was raised significantly 5 h after the hormone treatment. Ecdyson also possibly regulates the ovarian development in *Aedes Aegypti*. If ovarian development is induced by administration of ecdyson, α-amanitin shows no effect, but if ovarian development is induced by a bloodmeal, the process is susceptible to the toxin. From these experiments Fuchs and Fong (1976) concluded that the ovarian development of these insects, as induced by a bloodmeal, may proceed via the de novo synthesis of the hormone.

In some cases enzymes can also be induced by drugs. Phenobarbital induces in rat liver the de novo synthesis of oxidases, among them P_{450} and ethylmorphine-N-demethylase. The induction of the latter enzyme is sensitive to α-amanitin for about 8 h after application of the agent. From this, Jacob et al. (1974) concluded that de novo synthesis of mRNA coding for these enzymes lasts for about 8 h. The synthesis of rRNA does not seem to be involved, because in other experiments by this group the activity of RNA polymerase I was found restored in rat livers 1-2 h after α-amanitin administration.

α-Amanitin was active in still another in vitro system, the organ-cultured intestine (Corradino 1973). Here, the drug inhibited the synthesis of Ca-binding proteins concomitantly with the uptake of ^{45}Ca. Both are events which are induced by 1,25 dihydroxycholecalciferol, a metabolite of vitamin D_3.

Most recently, Morrissey and Lovenberg (1978) reported that in rat pineal glands α-amanitin ($10^{-5}M$) inhibits the induction of acetyl-transferase by the β-agonist l-isoproterenol. Simultaneously the synthesis of mRNA is inhibited in line with the participation of mRNA synthesis in the process of enzyme induction.

Many steroid hormones are known to increase RNA synthesis. For example, when oestradiol is administered to immature female rats, the uteri nuclei exhibit 2 h later a distinct increase of RNA synthesis at low ionic strength. This increase in RNA synthesis is due to nucleolar activity, as was indicated by its insensitivity to α-amanitin (Raynoud-Jammet et al. 1971, 1972). A similar stimulation of RNA synthesis can also be achieved in vitro by the application of steroid-hormone receptor complexes. Davies and Griffiths (1974) reported that 5α-dihydro-testosterone receptor complex at low ionic strength stimulates both nucleolar and extra-nucleolar RNA polymerase activities which were differentiated by α-amanitin. Comparable results were obtained in immature chickens by Cox et al. (1973) who provided evidence that oestradiol stimulates both RNA polymerase I and II in nuclei of hormone-treated animals at different times. They claim that 12 h after hormone treatment there is a large increase in RNA polymerase I activity, while during the first 2 h there is a small, but distinct increase of α-amanitin-sensitive RNA polymerase II (1973).

Other steroid hormones, e.g., aldosterone, are also known to produce an increase of RNA in their target organ. Chu and Edelman (1972) reported that administration of aldosterone to rats increased in the following hours the ratio of RNA polymerase I:II in the kidneys of these animals. The aldosterone-induced-Na-transport of toad bladder was not affected by α-amanitin. This may, however, be a question of dosage, since amphibians are about ten times less sensitive to the toxin than rats.

A similar increase in rRNA synthesis was observed in rat liver nuclei after injection of bovine growth hormone (Smuckler and Tata 1971). It remains uncertain whether this increase in amanitin-resistant activity was accompanied by any change in RNA polymerase II. In another case it was reported that a hormone can induce increased activity of RNA polymerase III: in adrenal nuclei ACTH increases the synthesis of pre-4S and 5S RNA by a factor of 2, as assessed by the α-amanitin sensitivity of this process (Fuhrman and Gill 1976).

Smuckler and Tate (1971) tested the hormone triiodothyronine, and found that this enzyme also increases the RNA polymerase I activity of liver cells, although less than growth hormone does. Triiodothyronine activity can be studied by following the induction of two enzymes in the rat liver, α-glycerophosphate-dehydrogenase and cytoplasmic malic enzyme (Dillmann et al. 1977). These workers described that the hormone stimulus of enzyme synthesis could be suppressed by α-amanitin, but lasted for a long time, e.g., 72 h in the case of malic enzyme. They proposed the existence of a long-lived mediator of the hormone stimulus as a possible explanation of this observation. However, how could such a mediator be synthesized in the presence of α-amanitin, which completely blocks the transcription of mRNA?

L. Developmental Studies

Egg and embryonic development are based upon gene expression and hence involve RNA synthesis. α-Amanitin was used to ascertain at which stage of development the embryo starts its own RNA synthesis and, later, to distinguish the contributions of the various RNA polymerases (I, II, and III) to RNA synthesis in development. These studies were done with rabbit and mouse ova as well as with eggs of amphibia (*Xenopus laevis, Pleurodeles waltlii*), sea urchins (*Strongilocentrotus franciscanus*), and with insects such as the Colorado beetle (*Leptinotarsa decemlineata*). Similarly to results obtained after actinomycin D administration, a marked inhibition of embryo development was achieved by concentrations of $5 \cdot 10^{-7}$M to 10^{-6}M α-amanitin.

It became apparent in early experiments with rabbit zygotes (Manes 1973) and mouse ova (Golbus et al. 1973) that α-amanitin does not inhibit the first cleavages of the eggs, but rather subsequent ones. Thus, treatment with α-amanitine of 1-cell or 2-cell embryos arrests the egg development only at the 8-cell stage, or the morula, respectively. This may be due to the slow penetration of the toxin into cells, or more likely, to preformed mRNA sufficient for the first three cleavage steps.

It is reasonable to assume that α-amanitin acts on embryos through inhibition of the embryonal RNA polymerases, which can be detected as early as in the 2-cell stage (Warner and Versteegh 1974). After solubilization of the enzymes from mouse blastocysts, such RNA polymerase demonstrated maximum inhibition at $8 \cdot 10^{-7}$M amanitin. Apparently, these polymerases are less sensitive than the corresponding RNA polymerase of the adult mouse liver.

Mouse ova RNA polymerase activity, calculated per cell, remains nearly constant during development (Versteegh et al. 1975); while the contributions of the individual RNA polymerases I, II or III to the total activity may change. This was indicated by an analysis with $1.2 \cdot 10^{-6}$M and $3 \cdot 10^{-4}$M α-amanitin. However, the cellular concentrations of RNA polymerases do not reflect the cellular rates of synthesis of the corresponding RNAs. Although the total activity of RNA polymerase remains approximately constant, there is a tremendous increase in newly synthesized RNA during the development of embryos.

The rate of RNA synthesis decreases in the presence of α-amanitin. Warner and Hearn (1977) found all RNAs diminished in the presence of α-amanitin except the 4-5 S RNA. Interestingly, among the different species of high molecular weight RNA, the strongest inhibition was found for rRNA (28S, 18S), not for mRNA (6-17S). The impairment of rRNA synthesis is possibly most crucial in the developing cell and may eventually lead to cell death. This would be consistent with the observation that morphological alterations in mouse ova (Golbus et al. 1973) and amphibian eggs (Duprat 1973) were present predominantly in the nucleoli. Further support for this hypothesis is shown by Levey et al. (1977). They reported that the later stages of embryo development, in which synthesis of rRNA has become active, are about ten times more sensitive to α-amanitin than the earlier stages.

In the mature eggs of *Xenopus laevis* (Roeder 1974) all forms of RNA polymerases are present, but little or no RNA synthesis occurs. The pattern of the RNA polymerases was determined by utilizing the different sensitivities to α-amanitin. Unlike somatic cells, in oocytes type III and II$_A$ enzyme activities are dominant over those of type I and II$_B$. This proportion is constantly maintained until gastrulation, although the rates of RNA synthesis change. Similarly, in eggs of sea urchins (Morris and Rutter 1976) type III enzymes are also dominant, but they decrease in concentration until the blastula stage. Simultaneously, the types I and II enzymes increase in concentration. Also, part of their RNA polymerase II activity is present in the cytoplasm. The purified enzyme was inhibited by a two to four times lower concentration of α-amanitin than necessary with the crude enzyme present in the homogenate. This suggests that there may be additional proteins in the homogenate, which bind α-amanitin, or factors which regulate the sensitivity of RNA polymerase II against the drug.

In the embryos of the Colorado beetle RNA synthesis starts during the early cycles of the blastoderm (Schnetter 1975; Maisonhaute 1977). Treatment with α-amanitin indicated that the information for all cleavages before blastoderm is present in the egg, eventually, as a stable form of RNA inherited from the mother. Inhibition of insect protein synthesis by α-amanitin occurred after 20 h treatment, indicating a slow turnover of the newly synthesized mRNA.

M. Behavioral Research

Intracerebral ventricular (i.c.v.) administration of α-amanitin inhibits the consolidation of memory. Such experiments were carried out in the rat (Montanaro et al. 1971; Strocchi et al. 1971) and in the mouse (Thut et al. 1973; Thut and Lindell 1974).

In both species, the toxin affected the retention of active as well as passive avoidance tasks. When administered 6 h before training, the retention of the avoidance conditioning was significantly lower in amanitin-treated animals than in saline-treated controls. This was found when the retesting was done 4-6 h after training. There was no significant difference, when the animals were retested immediately after training. Evidently, the toxin only affects long term memory. Moreover, there is no effect of α-amanitin on the ability to recall behaviors learned several hours prior to treatment. Similar retarding effects on memory have been obtained by inhibitors of protein synthesis, like cycloheximide, leading to the hypothesis that protein or peptide synthesis is involved in the memory processes. It now appears that mRNA synthesis is the important step. Strocchi et al. (1971) showed that memory impairment by α-amanitin is strongest at that time when RNA synthesis is most strongly inhibited (6 h after intoxication) and not at the time when protein synthesis is at its lowest level (12 h after intoxication). Since at the time when retarding effects occur, protein synthesis is not significantly decreased, it appears that protein synthesis is not necessary for the formation of memory. However, the usual method for protein analysis, precipitation by trichloracetic acid, may not reveal a small portion of the protein as induced by the learning process. Moreover, the memory factor could escape the precipitation procedure, if it was a peptide rather than a protein. Scotophobin, for example, raised in rats under similar conditions (avoidance of dark) by Ungar et al. (1972), is indeed a pentadekapeptide.

There are some discrepancies between the results of studies with the mouse (Thut and Lindell 1974) and the rat (Strocchi et al. 1971). In the mouse, a virtually complete inhibition of RNA synthesis in the brain was necessary to impair memory. This was achieved with a dose 100 times the (i.c.v.) LD_{50} of α-amanitin. In contrast, in the rat only a 40% decrease of RNA synthesis was sufficient to impair memory, achieved with a 0.3-0.5 (i.c.v.) LD_{50} dose. These results may be due to different sensitivity to amatoxins, different permeabilities of the blood-brain barrier, and also by differences in brain content of RNA polymerases. Recently Montanaro et al. (1977) reported that such amnestic effects by α-amanitin in the rat can be antagonized by α-amphetamine.

α-Amanitin can retard still another effect associated with the brain, the masculinization of newborn female rats by testosterone. Salaman and Birkett (1974) demonstrated that s.c. application of testosterone propionate (80 μg) in combination with amanitin (0.15 mg/kg) increased the number of vaginal cycles to 89% (5% in control) and the number of recent corpora in the ovaries to 100% (10% in control). Similar to the memory process the masculinizing effect of testosterone can also be blocked by inhibitors of protein synthesis, e.g., puromycin.

Glasser and Spelsberg (1972) used α-amanitin to study the circadian rhythm of RNA polymerases I and II in vivo. They found that RNA polymerase II concentration increases with daylight and decreases during the night. Nucleolar RNA polymerase I showed complementary behavior. Although this phenomenon could be correlated to other biochemical processes with diurnal variations, the circadian rhythm of RNA polymerases is not necessarily linked to them.

In conclusion, it seems that in the future α-amanitin will be used more frequently as a tool in behavioral research. The possibility specifically to block RNA synthesis in distinct areas of the brain should help in analyzing and understanding mechanisms of this complex system.

References

Alwine, J.C., Steinhart, W.L., Hill, C.W.: Transcription of herpes simplex type 1 DNA in nuclei isolated from infected HEp-2 and KB cells. Virology 60, 302 (1974)

Amalric, F., Nicoloso, M., Zalta, J.-P.: A comparative study of "soluble" RNA polymerase activity of Zajdela hepatoma ascites cells and calf thymus. FEBS Lett. 22, 67 (1972)

Amati, P., Blasi, F., Di Porzio, U., Riccio, A., Traboni, C.: Hamster α-amanitin-resistant RNA polymerase II able to transcribe Polyoma virus genome in somatic cell hybrids. Proc. Natl. Acad. Sci. USA 72, 753 (1975)

Austoker, J.L., Beebee, T.J.C., Chesterton, C.J., Butterworth, P.H.W.: DNA-dependent RNA polymerase activity of Chinese hamster kidney cells sensitive to high concentrations of α-amanitin. Cell 3, 227 (1974)

Barbanti-Brodano, G., Fiume, L.: Selective killing of macrophages by amanitin-albumin conjugates. Nature New Biol. 243, 281 (1973)

Barsotti, P., Marinozzi, V.: A morphological study on the early changes in hepatocyte nuclei induced by low doses of α-amanitin. J. Submicr. Cytol. 11, 13 (1979)

Beermann, W.: Effect of α-amanitin on puffing and intranuclear RNA synthesis in Chironomus salivary glands. Chromosoma 34, 152 (1971)

Beiderbeck, R.: α-Amanitin hemmt die Tumorinduktion durch Agrobacterium tumefaciens. Z. Naturforsch. 27b, 1393 (1972)

Ben-Zeev, A., Becker, Y.: Requirement of host cell RNA polymerase II in the replication of herpes simplex virus in α-amanitin-sensitive and -resistant cell lines. Virology 76, 246 (1977)

Blair, D.G.R., Dommasch, M.: Nuclear DNA-dependent RNA polymerases or Ehrlich ascites tumor cells: Two discrete α-amanitin-sensitive forms. Biochem. Biophys. Res. Commun. 49, 877 (1972)

Boctor, A., Grossman, A.: Differential sensitivity of rat liver and rat hepatoma cells to α-amanitin. Biochem. Pharmacol. 22, 17 (1973)

Bonetti, E., Derenzini, M., Fiume, L.: Lesions in the cells of proximal convoluted tubules in rat kidney induced by amanitin-albumin conjugate. Virchows Arch. 16, 71 (1974)

Bonetti, E., Derenzini, M., Fiume, L.: Increased penetration of amanitine into hepatocytes when conjugated with albumin. Arch. Toxicol. 35, 69 (1976)

Bottomley, W., Smith, H.J., Bogorad, L.: RNA polymerases of maize: Partial purification and properties of the chloroplast enzyme. Proc. Natl. Acad. Sci. USA 68, 2412 (1971a)

Bottomley, W., Spencer, D., Wheeler, A.M., Whitefeld, P.R.: The effect of a range of RNA polymerase inhibitors on RNA synthesis in higher plant chloroplasts and nuclei. Arch. Biochem. Biophys. 143, 269 (1971b)

Brändle, E., Zetsche, K.: Zur Lokalisation der α-Amanitin sensitiven RNA-Polymerase in Zellkernen von Acetabularia. Planta 111, 209 (1973)

Brodner, O.G., Wieland, Th.: Die Isolierung eines Amatoxin-bindenden Proteins, das von der RNA-Polymerase B und C verschieden ist. Hoppe-Seyler's Z. Physiol. Chem. 357, 89 (1976a)

Brodner, O.G., Wieland, Th.: Identification of the amatoxin-binding subunit of RNA polymerase B by affinity labeling experiments. Subunit B3 - the true amatoxin receptor protein of multiple RNA polymerase B. Biochemistry 15, 3480 (1976b)

Brown, I.R.: RNA synthesis in isolated brain nuclei after administration of d-lysergic acid diethylamide (LSD) in vivo. Proc. Natl. Acad. Sci. USA 72, 837 (1975)

Bryant, R.E., Adelsberg, E.A., Magee, P.T.: Properties of an altered RNA polymerase II activity from an α-amanitin-resistant mouse cell line. Biochemistry 16, 4237 (1977)

Bucci, S., Nardi, I., Mancino, G., Fiume, L.: Incorporation of tritiated uridine in nuclei of triturus oocytes treated with α-amanitin. Exp. Cell Res. 69, 462 (1971)

Buchwald, M., Ingles, C.J.: Human diploid fibroblast mutants with altered RNA polymerase II. Somatic Cell Genet. 2, 225 (1976)

Campadelli-Fiume, G.: Amanitins in Virus Research. Arch. Virol. 58, 1 (1978)

Campadelli-Fiume, G., Costanzo, F., Mannini-Palenzona, A., La Placa, M.: Impairment of host cell ribonucleic acid polymerase II after infection with frog virus 3. J. Virol. 9, 698 (1972)

Capobianco, G., Farina, G., Pelella, E., Musella, V.: Sensitivity to α-amanitin of multiple DNA-dependent RNA polymerases from experimental tumors. Biochem. Biophys. Res. Commun. 77, 306 (1977)

Cessi, C., Fiume, L.: Increased toxicity of β-amanitin when bound to a protein. Toxicon 6, 309 (1969)

Chambon, P.: Eukaryotic nuclear RNA polymerases. Annu. Rev. Biochem. 44, 613 (1975)

Chambon, P., Gissinger, F., Mandel, J. Jr., Kedinger, C., Gniazdowski, M., Meilhac, M.: Purification and properties of calf thymus DNA-dependent RNA polymerases A and B. Cold Spring Harbor Symp. Quant. Biol. 35, 693 (1970)

Chambon, P., Gissinger, F., Kedinger, C., Mandel, J.L., Meilhac, M.: Animal nuclear DNA-dependent RNA polymerases. Cell Nucl. 3, 269 (1974)

Chan, V.L., Whitmore, G.F., Siminovitch, L.: Mammalian cells with altered forms of RNA polymerases II. Proc. Natl. Acad. Sci. USA 69, 3119 (1972)

Chardonnet, Y., Gazzolo, L., Pogo, B.G.T.: Effect of α-amanitin on Adenovirus 5 multiplication. Virology 48, 300 (1972)

Chesterton, C.J., Humphrey, St.M., Butterworth, P.H.W.: Comparison of the multiple deoxyribonucleic acid-dependent ribonucleic acid polymerase forms of whole rat liver and a minimal-deviation rat hepatoma cell line. Biochem. J. 126, 675 (1972)

Chu, L.L., Edelman, I.S.: Cordycepin and α-amanitin: Inhibitors of transcription as probes of aldosterone action. J. Membr. Biol. 10, 291 (1972)
Cochet-Meilhac, M., Chambon, P.: Animal DNA-dependent RNA polymerases.
11. Mechanism of the inhibition of RNA polymerases B by amaxotins. Biochim. Biophys. Acta 353, 160 (1974)
Cochet-Meilhac, M., Nuret, P., Courvalin, J.C., Chambon, P.: Animal DNA-
dependent RNA polymerases. 12. Determination of the cellular number of
RNA polymerase B molecules. Biochim. Biophys. Acta 353, 185 (1974)
Corradino, R.A.: 1,25-Dihydroxycholecalciferol: Inhibition of action in
organ-cultured intestine by actinomycin D and α-amanitin. Nature (London)
243, 41 (1973)
Costanzo, F., Fiume, L., La Placa, M., Mannini-Palenzona, A., Novello, F.,
Stirpe, F.: Ribonucleic acid polymerase inducted by Vaccinia virus: Lack
of inhibition by rifampicin and α-amanitin. J. Viro-. 5, 266 (1970)
Costanzo, F., Campadelle-Fiume, G., Foa-Tomasi, L., Cassai, E.: Evidence
that herpes simplex virus DNA is trancribed by cellular RNA polymerase
B. J. Virol. 21, 34 (1977)
Cox, R.F., Haines, M.E., Carey, N.H.: Modification of the template capacity
of chick-oviduct chromatin for form-B RNA polymerase by estradiol. Eur.
J. Biochem. 32, 513 (1973)
Crerar, M.M., Pearson, M.L.: RNA polymerase II regulation in α-amanitin-
resistant rat myoblast mutants. Changes in wild-type and mutant enzyme
levels during growth in α-amanitin. J. Mol. Biol. 112, 331 (1977)
Crerar, M.M., Andrews, St.J., David, E.S., Somers, D.G., Mandel, J.-L.,
Pearson, M.L.: Amanitin binding to RNA polymerase II in α-amanitin-resistant rat myoblast mutants. J. Mol. Biol. 112, 317 (1977)
Damle, S.P., Novack, J.: Biosynthesis of ribonucleic acid in L-1210 leukemic
cells: Lack of inhibition by α-amanitin in vivo. Fed. Proc. 31, 3608
(1972)
Davies, P., Griffiths, K.: Effects of α-amanitin on the stimulation of
prostatic ribonucleic acid polymerase by prostatic steroid-protein receptor complexes. Biochem. J. 140, 565 (1974)
De La Torre, C., Fernandez-Gomez, M.E., Gimenez-Martin, G., Gonzalez-Fernan-
dez, A.: Sensitivity of the cell division cycle to α-amanitin in Allium
root tips. J. Cell Sci. 14, 461 (1974)
Derenzini, M., Fiume, L., Marinozzi, V., Mattioli, A., Montanaro, L., Sperti,
S.: Pathogenesis of liver necrosis produced by amanitin-albumin conjugates.
Lab. Invest. 29, 150 (1973)
Derenzini, M., Marinozzi, V., Novello, F.: Effects of α-amanitin on chroma-
tin in regenerating rat hepatocytes. A biochemical and morphologic study.
Virchows Arch. B Cell Pathol. 20, 307 (1976)
Dezelee, S., Sentenac, A., Fromageot, P.: Study on yeast RNA polymerase
effect of α-amanitin and rifampicin. FEBS Lett. 7, 220 (1970)
Dillmann, W.H., Schwartz, H.L., Silva, E., Surks, M.I., Oppenheimer, J.H.:
Alpha-amanitin administration results in a temporary inhibition of hepatic
enzyme induction by triiodothyronine: Further evidence favoring a long-
lived mediator of thyroid hormone action. Endocrinology 100, 1621 (1977)
Dinowitz, M.: Inhibition of Rous sarcoma virus by α-amanitin: Possible role
of cell DNA-dependent RNA polymerase form II. Virology 66, 1 (1975)
Dinowitz, M., Lindell, T.J., O'Malley, A.: Altered sensitivity of Rous
sarcoma virus transformed cells to inhibition of RNA synthesis by α-
amanitin. Arch. Virol. 53, 109 (1977)
Doenecke, D., Pfeiffer, Ch., Sekeris, C.E.: Multiple forms of DNA-dependent
RNA polymerase from insect tissue. FEBS Lett. 21, 237 (1972)
Duprat, A.-M.: Etude des effets morphologiques de l'α-amanitine sur les
ribonucléoprotéines nucléaires de cellules embryonnaires d'Amphibiens
cultivées in vitro. C.R. Acad. Sci. Paris 277, 2393 (1973)

Duspiva, F., Scheller, K., Weiß, D., Winter, H.: Ribonucleinsäuresynthese in der telotroph-meroistischen Ovariole von Dysdercus intermedius Dist. (Heteroptera, Pyrrhoc.). Wilhelm Roux Arch. 172, 83 (1973)

Egyhazi, E., D'Monte, B., Edström, J.-E.: Effects of α-amanitin on in vitro labeling of RNA from defined nuclear components in salivary gland cells from *Chironomus tentans*. J. Cell. Biol. 53, 523 (1972)

Emanuilov, I., Nicolova, R.C., Dabeva, M.D., Hadjiolov, A.A.: Quantitative ultrastructural study of the action of α-amanitin on mouse liver cell nucleoli. Exp. Cell Res. 86, 401 (1974)

Faulstich, H., Fauser, U.: Untersuchungen zur Frage der Hämodialyse bei der Knollenblätterpilzvergiftung. Dtsch. Med. Wochenschr. 98, 2258 (1973)

Faulstich, H., Fauser, U.: Experimental intoxication with amanitin in the dog. (I). Death by hypoglycemia and importance of enterohepatic circulation. In preparation

Faulstich, H., Trischmann, H.: Toxicity of and inhibition of RNA polymerase by α-amanitin bound to macromolecules by an azo linkage. Hoppe Seylers Z. Physiol. Chem. 354, 1395 (1973)

Faulstich, H., Trischmann, H.: Toxicity, antigenic activity and capacity of RNA-polymerase II inhibition of amanitin conjugated to proteins or to polylysine. In preparation a

Faulstich, H., Georgopoulos, D., Bloching, M.: Quantitative chromatographic analysis of toxins in single mushrooms of *Amanita phalloides*. J. Chromatogr. 79, 257 (1973)

Faulstich, H., Georgopoulos, D., Bloching, M., Wieland, Th.: Analysis of the toxins of amanitin-containing mushrooms. Z. Naturforsch. 29c, 86 (1974)

Faulstich, H., Trischmann, H., Zobeley, S.: A radioimmunoassay for Amanitin. FEBS Lett. 56, 312 (1975)

Faulstich, H., Jahn, W., Wieland, Th.: Resorption and excretion rate of labeled amatoxins in the perfused rat liver. In preparation

Faulstich, H., Trischmann, H., Wieland, Th., Wulff, E., Zobeley, S.: Preparation of ether derivatives of α-amanitin for the introduction of radioactive lables, spacer moieties, and lipophilic residues. In preparation b

Faulstich, H., Zobeley, S., Fauser, U.: Experimental intoxication with amanitin in the dog (IV). Determination of lethal doses and studies with radioactivity labeled amatoxins. In preparation c

Faulstich. H., Wilbertz, C., Ungemach, B., Wieland, Th.: Growth inhibition of human lymphocytes by various synthetic amatoxins. In press

Fauser, U., Faulstich, H.: Beobachtungen zur Therapie der Knollenblätterpilzvergiftung. Dtsch. Med. Wochenschr. 98, 2259 (1973)

Fauser, U., Faulstich, H.: Knollenblätterpilzvergiftung und Dialysierbarkeit der Toxine. In: Aktuelle Probleme der Dialyseverfahren und der Niereninsuffizienz (eds. P.v. Dittrich, F. Skrabal, W.D. Stühlinger), p. 439. Friedberg/Hessen: 1975

Fauser, U., Faulstich, H.: Experimental intoxication with amanitin in the dog. II. Liver dystrophy after glucose treatment. In preparation

Fauser, U., Zimmermann, R., Faulstich, H.: Experimental intoxication with amanitin in the dog. III. Death later than 60 h after intoxication by hemorrhagia or uremia. In preparation

Fiume, L., Barbanti-Brodano, G.: Selective toxicity of amanitin-albumin conjugates for macrophages. Experienta 30, 76 (1974)

Fiume, L., Laschi, R.: Lesioni ultrastrutturali prodotte nelle cellule parenchimali epatiche dalla falloidina e dalla α-amanitina. Sperimentale 115, 288 (1965)

Fiume, L., Stirpe, F.: Decreased RNA content in mouse liver nuclei after intoxication with α-amanitin. Biochim. Biophys. Acta 123, 643 (1966)

Fiume, L., Wieland, Th.: Amanitins. Chemistry and action. FEBS Lett. 8, 1 (1970)

Fiume, L., La Placa, M., Portolani, M.: Ricerche sul meccanismo dell'azione citopatogena della α-amanitina. Sperimentale 166, 15 (1966)

Fiume, L., Marinozzi, V., Nardi, F.: The effects of amanitin poisoning on mouse kidney. Br. J. Exp. Pathol. 50, 270 (1969)

Fiume, L., Campadelli-Fiume, G., Wieland, Th.: Facilitated penetration of amanitin-albumin conjugates into hepatocytes after coupling with fluorescein. Nature New Biol. 230, 219 (1971)

Fiume, L., Busi, C., Campadelli-Fiume, G., Franceschi, C.: Production of antibodies to amanitins as the basis for their radioimmunoassay. Experientia 31, 1233 (1975)

Foa-Tomasi, L., Costanzo, F., Campadelli-Fiume, G.: Enhanced inhibition of RNA synthesis by amanitins in vitro cultured cells. Experentia 32, 45 (1976)

Fong, W.-F., Fuchs, M.S.: The long term effect of α-amanitin on RNA synthesis in adult female *Aedes aegypti*. Insect Biochem. 6, 123 (1976a)

Fong, W.-F., Fuchs, M.S.: Studies on the mode of action of ecdysterone in adult female *Aedes aegypti*. Mol. Cell. Endocrinol. 4, 341 (1976b)

Frederiksen, S., Hellung-Larsen, P., Gram Jensen, E.: The differential inhibitory effect of α-amanitin on the synthesis of low molecular weight RNA components in BHK cells. FEBS Lett. 87, 227 (1978)

Fuchs, M.S., Fong, W.-F.: Inhibition of blood digestion by α-amanitin and actinomycin D and its effects on ovarian development in *Aedes aegypti*. J. Insect. Physiol. 22, 465 (1976)

Fuhrman, Sh.A., Gill, G.N.: Adrenocorticotropic hormone regulation of adrenal RNA polymerases. Stimulation of nuclear RNA polymerase III. Biochemistry 15, 5520 (1976)

Gallerani, R., Saccone, C., Cantatore, P., Gadaleta, M.N.: DNA dependent RNA polymerase from rat liver mitochondria. FEBS Lett. 22, 37 (1972)

Glasser, St.R., Spelsberg, T.C.: Mammalian RNA polymerases I and II: Independent diurnal variations in activity. Biochem. Biophys. Res. Commun. 47, 951 (1972)

Gogala, N.: Amanitin and Phalloidin - Wachstumshemmstoffe für höhere Pflanzen. Biol. Vestn. 17, 27 (1969)

Golbus, M.S., Calarco, P.G., Epstein, Ch.J.: The effects of inhibitors of RNA synthesis (α-amanitin and actinomycin D) on preimplantation mouse embeyogenesis. J. Exp. Zool. 186, 207 (1973)

Govindan, M.V.: Diploma work, University of Heidelberg 1969

Grant, W.D.: The effect of α-amanitin and $(NH_4)_2SO_4$ on RNA synthesis in nuclei and nucleoli isolated from *Physarum polycephalum* at different times during the cell cycle. Eur. J. Biochem. 29, 94 (1972)

Greenleaf, A.L., Krämer, A., Bautz, E.K.F.: Polymerases from *Drosophila melanogaster* larvae. In: RNA Polymerases (eds. R. Losick, M. Chamberlin), p. 793. Cold Spring Harbor Laboratory 1976

Greenleaf, A.L., Borsetti, L.M., Jiamachello, P.F., Coulter, D.E.: α-Amanitin-resistant D melanogaster with an altered RNA polymerase II. Cell 18 (1979)

Guialis, A., Beatty, B.G., Ingles, C.J., Crerar, M.M.: Regulation of RNA polymerase II activity in α-amanitin-resistant CHO hybrid cells. Cell 10, 53 (1977)

Habermann, E., Cheng-Raude, D.: Central neurotoxicity of apamin, crotamin, phospholipase A and α-amanitin. Toxicon 13, 465 (1975)

Hadjiolov, A.A., Dabeva, M.D., Mackedonski, V.V.: The action of α-amanitin in vivo on the synthesis and maturation of mouse liver ribonucleic acids. Biochem. J. 138, 321 (1974)

Hager, G., Holland, P., Valenzuela, F., Weinberg, F., Rutter, W.: RNA poly-
merases and transcriptive specificity in *Saccharomyces cerevisiae*. In:
RNA Polymerase (eds. R. Losick, M. Chamberlin), p. 745. New York: Cold
Spring Harbor Laboratories 1976

Hastie, N.D., Mahy, B.W.J.: Effects of α-amanitin in vivo on RNA polymerase
activity of cultured chick embryo fibroblast cell nuclei: Resistance of
ribosomal RNA synthesis to the drug. FEBS Lett. 32, 95 (1973)

Hastie, D., Armstrong, S.J., Mahy, B.W.J.: Effects of α-amanitin on protein
and nucleic acid synthesis in chick-embryo fibroblast cells. Biochem. J.
130, 28 (1972)

Higashinakagawa, T., Mita, T.: DNA-dependent RNA polymerase of eukaryotic
cells: A study with protozoon *Tetrahymena pyriformis*. Gunma Symp. Endocrinol. 10, 41 (1973)

Higashinakagawa, T., Muramatsu, M.: In vitro effect of cycloheximide on
the nucleolar and extranucleolar nuclear RNA polymerases of rat liver.
Biochem. Biophys. Res. Commun. 47, 1 (1972)

Hildebrandt, A., Sauer, H.W.: DNA-dependent RNA polymerase from *Physarum
polycephalum*. FEBS Lett. 35, 41 (1973)

Hildebrandt-Jackson, S., Sugden, B.: Inhibition by α-amanitin of Simian
virus 40-specific ribonucleic acid synthesis in nuclei of infected monkey
cells. J. Virol. 10, 1086 (1972)

Hodo III, H.G., Blatti, St.P.: Purification using poyethylenimine precipitation and low molecular weight subunit analyses of calf thymus and wehat
germ DNA-dependent RNA polymerase II. Biochemistry 16, 2334 (1977)

Holt, Th.K.H., Kuijpers, A.M.C.: Induction of chromosome puffs in *Drosophila
hydei* salivara glands after inhibition of RNA synthesis by α-amanitin.
Chromosoma 37, 423 (1972a)

Holt, Th.K.H., Kuijpers, A.M.C.: Effects of α-amanitin on nucleolar structure and metabolism in *Drosophila hydei*. Erperientia 28, 899 (1972b)

Horgen, P.A., Griffin, D.H.: Specific inhibitors of the three RNA polymerases
from the aqua tic fungus *Blastocladiella emersonii*. Proc. Natl. Acad. Sci.
USA 68, 338 (1971)

Hossenlopp, P., Wells, D., Chambon, P.: Animal DNA-dependent RNA polymerases.
Partial purification and properties of three classes of RNA polymerases
from uninfected and Adenovirus-infected HeLa cells. Eur. J. Biochem. 58,
237 (1975)

Huet, J., Buhler, J.-M., Sentenac, A., Fromageot, P.: Dissociation of two
polypeptide chains from yeast RNA polymerase A. Biochemistry 72, 3064
(1975a)

Huet, J., Buhler, J.-M., Sentenac, A., Fromageot, P.: Dissociation of two
polypeptide chains from yeast RNA polymerase. Proc. Natl. Acad. Sci. USA
72, 3034 (1975b)

Incefy, G.S., Good, R.A.: The need for transcription and translation for
differentiation of bone marrow cells by thymic factors in man. In: Immune
Reactivity of Lymphocytes (eds. M. Feldman, A. Globerson). New York:
Plenum Publishing Corporation 1976

Incefy, G.S., Kappas, A.: Inhibitory effect of α-amanitin on the induction
of δ-aminolevulinate synthetase in chick embryo liver. FEBS Lett. 15,
153 (1971)

Incefy, G.S., Kappas, A.: Enhancement of RNA synthesis in avian liver cell
cultures by a 5β-steroid metabolite during induction of δ-aminolevulinate
synthase. Proc. Natl. Acad. Sci. USA 71, 2290 (1974)

Incefy, G.S., Rifkind, A.B., Kappas, A.: Inhibition of δ-aminolevulinate
synthetase induction by α-amanitin in avian liver cell cultures. Biochim.
Biophys. Acta 361, 331 (1974)

Ingles, C.J., Beatty, B.G., Buialis, A., Pearson, M.L., Crerar, M.M., Lobban, P.E., Siminovitch, L., Somers, D.G., Buchwald, M.: α-amanitin-resistant mutants of mammalian cells and the regulation of RNA polymerase II activity. In: RNA Polymerase (eds. R. Losick, M. Chamberlin). New York: Cold Spring Harbor Laboratory 1976a

Ingles, C.J., Guialis, A., Lam, J., Siminovitch, L.: α-amanitin resistance of RNA polymerase II in mutant Chinese hamster ovary cell lines. J. Biol. Chem. 251, 2729 (1976b)

Jacob, S.T., Muecke, W., Sajdel, E.M., Munro, H.N.: Evidence for extranucleolar control of RNA synthesis in the nucleolus. Biochem. Biophys. Res. Commun. 40, 334 (1970a)

Jacob, S.T., Sajdel, E.M., Muecke, W., Munro, H.N.: Soluble RNA polymerases of rat liver nuclei: Properties, template specificity, and amanitin responses in vitro and in vivo. Cold Spring Harbor Symp. Quant. Biol. 35, 681 (1970b)

Jacob, S.T., Scharf, M.B., Vesell, E.S.: Role of RNA in induction of hepatic microsomal mixed function oxidases. Proc. Natl. Acad. Sci. USA 71, 704 (1974)

Jacquet, M., Groner, Y., Monroy, G., Hurwitz, J.: The in vitro synthesis of avian myeloblastosis vrial RNA sequences. Proc. Natl. Acad. Sci. USA 71, 3045 (1974)

Jaehning, J.A., Weinmann, R., Brendler, Th.G., Raskas, H.J., Roeder, R.G.: Function and regulation of RNA polymerase II and III in Adenovirus-infected KB cells. In: RNA Polymerase (eds. R. Losick, M. Chamberlin), p. 819. New York: Cold Spring Harbor Laboratories 1976

Jendrisak, J., Guilfoyle, T.I.: Eukaryotic RNA polymerases: Comparative subunit structures, immunological properties, and α-amanitin sensitivities of the class II enzymes from higher plants. Biochemistry 17, 1322 (1978)

Jensen, E.G., Hellung-Larsen, P., Frederiksen, S.: Synthesis of low molecular weight RNA components A, C, and D by polymerase II in α-amanitin resistant hamster cells. Nucleic Acid Res. 6, 321 (1979)

Johnson, B.E.C., Preston, J.F.: Quantitation of amanitins in *Galerina autumnalis*. Mycologia 68, 1248 (1976)

Johnson, B., Preston III, J.F.: Unique amanitin resistance of RNA-synthesis in isolated nuclei from Amanita species accumulating amanitins. Arch. Microbiol. 122, 161 (1979)

Jolicoeur, P., Labrie, F.: Induction of rat liver tyrosine aminotransferase by dibutyryl cyclic amp and its inhibition by actinomycin D and α-amanitin. FEBS Lett. 17, 141 (1971)

Kaufmann, R., Voigt, H.-P.: Soluble RNA polymerases from human placenta. Functional and structural properties. Hoppe Seylers Z. 345, 1438 (1973)

Kedinger, C., Chambon, P.: Animal-dependent RNA polymerases. 3. Purification of calf thymus BI and BII enzymes. Eur. J. Biochem. 28, 283 (1972)

Kedinger, C., Simard, R.: The action of α-amanitin on RNA synthesis in Chinese hamster ovary cells. J. Cell Biol. 63, 831 (1974)

Kedinger, C., Niazdowski, M., Mandel, J.L. Jr., Gissinger, F., Chambon, P.: α-Amanitin: A specific inhibitor of one of two DNA-dependent RNA polymerase activities from calf thymus. Biochem. Biophys. Res. Commun. 38, 165 (1970)

Kedinger, C., Nuret, P., Chambon, P.: Structural evidence for two α-amanitin-sensitive RNA polymerases in calf thymus. FEBS Le-t. 15, 169 (1971)

Kostansek, E.C., Lipscomb, W.N., Yocum, R.R., Thiessen, W.E.: The crystal structure of the mushroom toxin β-amanitin. J. Am. Chem. Soc. 99, 1273 (1977)

Küntzel, H., Schäfer, K.P.: Mitochondrial RNA polymerase from Neurospora crassa. Nature New Biol. 231, 265 (1971)

Kuwano, M., Ikehara, Y.: Inhibition by α-amanitin of messenger RNA formation in cultured fibroblasts: Potentiation by amphotericin B. Exp. Cell Res. 82, 454 (1973)

Lamb, R.A., Choppin, P.W.: Synthesis of influenza vuris polypeptides in cells resistant to alpha-amanitin: Evidence for the involvement of cellular RNA polymerase II in virus replication. J. Virol. 23, 816 (1977)

Lampert, A., Feigelson, P.: A short lived polypeptide component of one of two discrete functional pools of hepatic nuclear α-amanitin-resistant RNA polymerases. Biochem. Biophys. Res. Commun. 58, 1030 (1974)

Ledinko, N.: Inhibition by α-amanitin of Adenovirus 12 replication in human embryo kidney cells and of Adenovirus transformation of hamster cells. Nature New Biol. 233, 247 (1971)

Levey, I.L., Troike, D.E., Brinster, R.L.: Effects of α-amanitin on the development of mouse ova in culture. J. Reprod. Fertil. 50, 147 (1977)

Lindell, Th.J.: Evidence for an extranucleolar mechanism of actinomycin D action. Nature (London) 263, 347 (1976)

Lindell, Th.J., Weinberg, F., Morris, P.W., Roeder, R.G., Rutter, W.I.: Specific inhibition of nuclear RNA polymerase II by α-amanitin. Science 170, 447 (1970)

Lobban, P.E., Siminovitch, L.: α-amanitin resistance: a dominant mutation in CHO cells. Cell 4, 167 (1975)

Lobban, P.E., Siminovitch, L.: The RNA polymerase II of an α-amanitin-resistant chinese hamster ovary cell line. Cell 8, 65 (1976)

Mahy, B.W.J., Hastie, N.D., Armstrong, S.J.: Inhibition of influenza virus replication by α-amanitin: Mode of action. Proc. Natl. Acad. Sci. USA 69, 1421 (1972)

Maisonhaute, C.: Comparaison des effects de deux inhibiteurs de la synthèse d'ARN (actinomycine D et α-amanitine) sur le developpement embryonnaire d'un insecte: déterminisme génique du debut de l'embryogénèse de *Leptinotarsa decemlineata* (Coleoptera). Wilhelm Roux Arch. 183, 61 (1977)

Mancino, G., Nardi, I., Corvaja, N., Fiume, L., Marinozzi, V.: Effects of α-amanitin on triturus lampbrush chromosomes. Exp. Cell Res. 64, 237 (1970)

Mandel, J.L., Chambon, P.: Purification of RNA polymerase B activity from rat liver. FEBS Lett. 15, 175 (1971)

Manes, C.: The participation of the embryonic genome during early cleavage in the rabbit. Dev. Biol. 32, 453 (1973)

Mannini-Palenzona, A., Costanzo, F., La Placa, M.: Impairment of herpesvirus growth in chick embryo fibroblast cultures by α-amanitin. Arch. gesamte Virusforsch. 34, 381 (1971)

Marinozzi, V., Fiume, L.: Effects of α-amanitin on mouse and rat liver cell nuclei. Exp. Cell Res. 67, 311 (1971)

Mark, G.E., Taylor, J.M., Broni, B., Krug, R.M.: Nuclear accumulation of influenza viral RNA transcripts and the effects of cycloheximide, actinomycin D and α-amanitin.

Meilhac, M., Kedinger, C., Chambon, P., Faulstich, H., Govindan, M.V., Wieland, Th.: Amanitin binding to calf thymus RNA polymerase B. FEBS Lett. 9, 258 (1970)

Menon, I.A.: Differential effects of α-amanitin on RNA polymerase activity in nuclei and mitochondria. Can. J. Biochem. 49, 1395 (1971)

Monjardino, J.P.: The effect of α-amanitin on the synthesis of Polyoma specific RNA. Biochem. Biophys. Res. Commun. 80, 1049 (1978)

Montanaro, N., Novello, F., Stirpe, F.: Effect of α-amanitin on ribonucleic acid polymerase II of rat brain nuclei and on retention of avoidance conditioning. Biochem. J. 125, 1087 (1971)

Montanaro, L., Novello, F., Stirpe, F.: Inhibition of ribonucleic acid and of protein synthesis in the organs of rats and mice poisoned with α-amanitin. Biochim. Biophys. Acta 319, 188 (1973)

Montanaro, N., Strocchi, P., Dall'Olio, R., Gandolfi, O.: Recovery of amnestic effect of α-amphetamine in rat. Neuropsychobiology 3, 1 (1977)

Montecuccoli, G., Novello, F., Stirpe, F.: Effect of α-amanitin poisoning on the synthesis of deoxyribonucleic acid and of protein in regenerating rat liver. Biochim. Biophys. Acta 319, 199 (1973)

Moore, G.P.M., Ringertz, N.R.: Localization of DNA-depen-ent RNA polymerase activities in fixed human fibroblasts by autoradiography. Exp. Cell Res. 76, 223 (1973)

Moreno Diaz de la Espina, S., Risueno, M.C.: Effect of α-amanitine on the nucleolus of meristematic cells of Allium cepa in interphase and mitosis: an ultrastructural analysis. Cytobiology 12, 175 (1976)

Morris, P.W., Rutter, W.J.: Nucleic acid polymerizing enzymes in developing Strongylocentrotus franciscanus embryos. Biochemistry 15, 3106 (1976)

Morris, P.W., Venton, D.L., Kelley, K.M.: Biochemistry of the amatoxins: Preparation and characterisation of stably iodinated α-amanitin. Biochemistry 17, 690 (1978)

Morrissey, I.I., Lovenberg, W.: Synthesis of RNA in pineal-gland during N-acetyltransferase induction. Effects of actinomycin D, α-amanitin and Cordycepin. Biochem. Pharmacol. 27, 557 (1978)

Mühlbach, H.P., Sänger, H.L.: Viroid replication is inhibited by α-amanitin. Nature 278, 185 (1979)

Niessing, J., Schnieders, B., Kunz, W., Seifart, K.H., Sekeris, C.E.: Inhibition of RNA synthesis by α-amanitin in vivo. Z. Naturforsch. 25, 1119 (1970)

Pattillo, R.A., Hussa, R.O.: Early stimulation of human chorionic gonadotropin secretion by dibutyryl cyclic AMP and theophylline in human malignant trophoblast cells in vitro: Inhibition by actinomycin D; α-amanitin and cordycepin. Gynecol. Invest. 6, 365 (1975)

Paweletz, N., Hoffmann, H.: Patterns of junctionally active chromatin in α-amanitin-treated chicken fibroblasts. Naturwissenschaften 59, 368 (1972)

Perez-Bercoff, R., Carrara, G., Dolei, A., Conciatore, G., Tita, G.: In vitro binding of a cellular, α-amanitin sensitive, RNA polymerase to infectious, mengovirus-induced double-stranded RNA. Biochem. Biophys. Res. Commun. 56, 876 (1974)

Petrov, P., Sekeris, C.E.: Early action of α-amanitin on extranucleolar ribonucleoproteins, as revealed by electron microscopic observation. Exp. Cell Res. 69, 393 (1971)

Phillips, J.P., Sumner-Smith, M.: Form III RNA polymerase from Drosophila nuclei: multiple forms and loss of α-amanitin sensitivity. Insect Biochem. 7, 323 (1977)

Polya, G.M., Jagendorf, A.T.: Wheat leaf PNA polymerases I. Partial purification and characterization of nuclear, chloroplast and soluble DNA-dependent enzymes. Arch. Biochem. Biophys. 146, 635 (1971)

Pong, Sh.-Sh., Loomis, W.F. Jr.: Multiple nuclear ribonucleic acid polymerases during development of Dictyostelium discoideum. J. Biol. Chem. 218, 3833 (1973)

Prasanna, H.R., Gupta, S.R., Viswanathan, L., Venkitasubramanian, T.A.: A comparative study of the effects of aflatoxin B_1, α-amanitin and actinomycin-D on RNA synthesis by rat liver. Environ. Physiol. Biochem. 5, 341 (1975)

Preston, J.F., Stark, H.J., Kimbrough, J.W.: Quantitation of amanitins in Amanita verna with calf thymus RNA polymerase B. Lloydia 38, 153 (1975)

Price, R., Penman, S.: A distinct RNA polymerase activity, synthesizing 5.5s, 5s and 4s RNA in nuclei from Adenovirus 2-infected HeLa cells. J. Mol. Biol. 70, 435 (1972a)

Price, R., Penman, S.: Transcription of the Adenovirus genome by an α-amanitin-sensitive ribonucleic acid. J. Virol. 9, 621 (1972b)

Raynaud-Jammet, C., Bieri, F., Baulieu, E.-E.: Effects of oestradiol, α-amanitin and ionic strength on the in vitro synthesis of RNA by uterus nuclei. Biochim. Biophys. Acta 247, 355 (1971)

Raynaud-Jammet, C., Catelli, M.G., Baulieu, E.-E.: Inhibition by α-amanitin of the oestradiol-induced increase in α-amanitin insensitive RNA polymerase in immature rat uterus. FEBS Lett. 22, 93 (1972)

Ro-Choi, T.S., Raj, N.B.K., Pike, L.M., Busch, H.: Effects of α-amanitin, cycloheximide, and thiocetamide on low molecular weight nuclear RNA. Biochemistry 15, 3823 (1976)

Roeder, R.G.: Multiple forms of deoxyribonucleic acid-dependent ribonucleic acid polymerase in *Xenopus laevis*. J. Biol. Chem. 249, 249 (1974)

Roeder, R.G.: Eukaryotic nuclear RNA polymerases. In: RNA Polymerase (eds. R. Losick, M. Chamberlin), p. 285. New York: Cold Spring Harbor Laboratory 1976

Roeder, R.G., Rutter, W.J.: Multiple forms of DNA-dependent RNA polymerase in eukaryotic organisms. Nature (London) 224, 234 (1969)

Roeder, R.G., Rutter, W.J.: Multiple ribonucleic acid polymerases and ribonucleic acid synthesis during sea urchin development. Biochemistry 9, 2543 (1970)

Roeder, R.G., Reeder, R.H., Brown, D.D.: Mutliple forms of RNA polymerase in *Xenopus laevis*: Their relationship to RNA synthesis in vivo and their fidelity of transcription in vitro. In: Transcription of Genetic Material. Cold Spring Harbor Symp. Quant. Biol. 35, 727 (1970)

Romen, W., Knobloch, U., Altmann, H.-W.: Vergleichende Untersuchungen der Kernveränderungen von Rattenhepatocyten nach Actinomycin D- und α-Amanitin-Vergiftung. Virchows Arch. B Cell Pathol. 23, 93 (1977)

Rott, R., Scholtissek, C.: Specific inhibition of influenza replication by α-amanitin. Nature (London) 228, 56 (1970)

Rymo, L., Parsons, J.T., Coffin, J.M., Weissmann, C.: In vitro synthesis of Rous sarcoma virus-specific RNA is catalyzed by a DNA-dependent RNA polymerase. Proc. Natl. Acad. Sci. USA 71, 2782 (1974)

Saccone, C., Gallerani, R., Gadaleta, M.N., Greco, M.: The effect of α-amanitin on RNA synthesis in rat liver mitochondria. FEBS Lett. 18, 339 (1971)

Sahib, M.K., Jost, Y.-C., Jost, J.P.: Role of cyclic adenosine 3', 5'-monophosphate in the induction of hepatic enzymes. J. Biol. Chem. 246, 4539 (1971)

Salaman, D.F., Birkett, S.: Androgen-induced sexual differentiation of the brain is blocked by inhibitors of DNA and RNA synthesis. Nature (London) 247, 109 (1974)

Santelli, R.V., Machado-Santelli, G.M., Lara, F.J.S.: In vitro transcription by isolated nuclei of *Rhynchosciara americana* salivary glands. Characteristics of incorporation and inhibition by α-amanitin. Chromosoma 56, 69 (1976)

Schmid, W., Sekeris, C.E.: Possible involvement of nuclear DNA-like RNA in the control of ribosomal RNA synthesis. Biochim. Biophys. Acta 312, 549 (1973)

Schnetter, W.: Zur Proteinsynthese in der frühen Embryonalentwicklung von *Leptinotarsa decemlineata* (Coleoptera): Rolle der mütterlichen und der embryoeigenen Messenger-RNS. Verh. Dtsch. Zool. Ges., S. 197 (1975)

Schultz, L.D., Hall, B.D.: Transcription in yeast: α-amanitin sensitivity and other properties which distinguish between RNA polymerases I and III. Proc. Natl. Acad. Sci. USA 73, 1029 (1976)

Seifart, K.H., Benecke, B.J.: DNA-dependent RNA polymerase C. Occurrence and localization in various animal cells. Eur. J. Biochem. 53, 293 (1975)

Seifart, K.H., Sekeris, C.E.: Amanitin, a specific inhibitor of transcription by mammalian RNA-polymerase. Z. Naturforsch. 24b, 1538 (1969)

Seifart, K.H., Benecke, B.J., Juhasz, P.P.: Multiple RNA polymerase species from rat liver tissue: Possible existence of a cytoplasmic enzyme. Arch. Biochem. Biophys. 151, 519 (1972)

Seitz, U., Seitz, R.: Selektive Hemmung der Synthese der AMP-reichen RNS durch α-Amanitin in Zellen höherer Pflanzen. Planta 97, 224 (1971)

Sekeris, C.E., Schmid, W.: Action of α-amanitin in vivo and in vitro. FEBS Lett. 27, 41 (1972)

Sekeris, C.E., Niessing, J., Seifart, K.H.: Inhibition by α-amanitin of induction of tyrosine transaminase in rat liver by cortisol. FEBS Lett. 9, 103 (1970)

Serfling, E., Wobus, U., Panitz, R.: Effect of α-amanitin on chromosomal and nucleolar RNA-synthesis in *Chironomus thummi* polytene chromosomes. FEBS Lett. 20, 148 (1972)

Seyberth, H.W., Schmidt-Gayk, H., Hackenthal, E.: Toxicity, clearance and distribution of endotoxin in mice as influenced by actinomycin D, cycloheximide, α-amanitin and lead acetate. Toxicon 10, 491 (1972)

Shaaya, E., Clever, U.: In vivo effects of α-amanitin on RNA synthesis in *Calliphora erythrocephala*. Biochim. Biophys. Acta 272, 373 (1972)

Shaaya, E., Sekeris, C.E.: Inhibitory effects of α-amanitin on RNA synthesis and induction of DOPA-decarboxylase by β-ecdysone. FEBS Lett. 16, 333 (1971)

Sinclair, G.D., Brasch, K.: Reversible action of α-amanitin on nuclear structure and molecular composition. Exp. Cell Res. 111, 1 (1978)

Sklar, U.E.F., Roeder, R.R.: Purification, characterisation and structure of class III RNA polymerases. Fed. Proc. Fed. Am. Soc, Exp. Biol. 34, 650 (1975)

Smuckler, E.A., Hadjiolov, A.A.: Inhibition of hepatic deoxyribonucleic acid-dependent ribonucleic acid polymerases by the exotoxin of *Bacillus thuringiensis* in comparison with the effects of α-amanitin and cordycepin. Biochem. J. 129, 153 (1972)

Smuckler, E.A., Tata, J.R.: Changes in hepatic nuclear DNA-dependent RNA polymerase caused by growth hormone and triiodothyronine. Nature (London) 234, 37 (1971)

Somers, D.G., Pearson, M.L., Ingles, C.J.: Isolation and characterization of an α-amanitin-resistant rat myoblast mutant cell line possessing α-amanitin-resistant DNA polymerase II. J. Biol. Chem. 250, 4825 (1975a)

Somers, D.G., Pearson, M.L., Ingles, C.J.: Regulation of RNA polymerase II activity in a mutant rat myoblast cell line resistant to α-amanitin. Nature (London) 253, 5490 (1975b)

Sperti, S., Montanaro, L., Fiume, L., Mattioli, A.: Dissociation constants of the complexes between RNA-polymerase II and amanitins. Experientia 29, 33 (1973)

Stirpe, F., Fiume, L.: Studies on the pathogenesis of liver necrosis by α-amanitin. Effect of α-amanitin on ribonucleic acid synthesis and on ribonucleic acid polymerase in mouse liver nuclei. Biochem. J. 105, 779 (1967)

Strain, G.C., Mullinix, K.P., Bogorad, L.: RNA polymerase of maize: Nuclear RNA polymerases. Proc. Natl. Acad. Sci. USA 68, 2647 (1971)

Strocchi, P., Montanaro, N., Dall'Olio, R., Novello, F., Stirpe, F.: Effect of α-amanitin on brain RNA and protein synthesis and on retention of avoidance conditioning. Pharmacol. Biochem. Behav. 6, 433 (1971)

Suzuki, Y., Giza, P.E.: Accentuated expression of silk fibroin genes in vivo and in vitro. J. Mol. Biol. 107, 183 (1976)

Tata, J.R., Hamilton, M.J., Shields, D.: Effects of α-amanitin in vivo on RNA polymerase and nuclear RNA synthesis. Nature New Biol. 238, 161 (1972)

Thut, P.D., Lindell, T.J.: α-amanitin inhibition of mouse brain form II ribonucleic acid polymerase and passive avoidance retention. Mol. Pharmacol. 10, 146 (1974)

Thut, P.D., Hruska, R.E., Kelter, A., Mizne, J., Lindell, T.J.: The effect of α-amanitin on passive and active avoidance acquisition in mice. Psychopharmacologia 30, 355 (1973)

Tocchini-Valentini, G.P., Grippa, M.: Ribosomal RNA synthesis and RNA polymerase. Nature (London) 228, 993 (1970)

Tsai, M.-J., Michaelis, G., Griddle, R.S.: DNA-dependent RNA polymerase from yeast mitochondria. Natl. Acad. Sci. 68, 473 (1971)

Ungar, G., Desiderio, D.M., Parr, W.: Isolation, identification and synthesis of a specific behaviour-inducing brain peptide. Nature (London) 238, 198 (1972)

Vaisius, A.C., Horgen, P.A.: Purification and characterisation of RNA polymerase II resistant to α-amanitin from the mushroom Agaricus Bisporus. Biochemistry 18, 796 (1979)

Versteegh, L.R., Hearn, T.F., Warner, C.M.: Variations in the amounts of RNA polymerase forms I, II and III during preimplantation development in the mouse, Dev. Biol. 46, 430 (1975)

Villa, L., Agostoni, A., Jean, G.: Investigations on the toxic action of Amanita phalloides Fr. on the hepatic cell. Experentia 24, 576 (1968)

Voigt, H.-P., Kaufmann, R., Matthaei, H.: Solubilized DNA-dependent RNA polymerase from human placenta: A Mn^{2+}-dependent enzyme. FEBS Lett. 10, 257 (1970)

Wallace, R.D., Kates, J.: State of Adenovirus 2 deoxyribonucleic acid in the nucleus and its mode of transcription: Studies with isolated viral deoxyribonucleic acid-protein complexes and isolated nuclei. J. Virol. 9, 627 (1972)

Warner, C.M., Hearn, T.F.: The effect of α-amanitin on nucleic acid synthesis in preimplantation mouse embryos. Differentiation 7, 89 (1977)

Warner, C.M., Versteegh, L.R.: In vivo and in vitro effect of α-amanitin on preimplantation mouse embryo RNA polymerase Nature (London) 248, 678 (1974)

Weil, P.A., Blatti, St.P.: HeLa cell deoxyribonucleic acid-dependent RNA polymerases: Function and properties of the class III enzymes. Biochemistry 15, 1500 (1976)

Weinmann, R., Raskas, H.J., Roeder, R.G.: Role of DNA-dependent RNA polymerase II and III in transcription of the Adenovirus genome late in productive infection. Proc. Natl. Acad. Sci. 71, 3426 (1974)

Wieland, Th.: Poisonous principles of mushrooms of a genus Amanita. Science 159, 946 (1968)

Wieland, Th.: Struktur und Wirkung der Amatoxine. Naturwissenschaft 59, 225 (1972)

Wieland, Th., Brodner, O.: Herstellung von [6^{ind}-^{3}H]-Amanin, einem radioaktiven Amatoxin mit Carboxyfunktion. Liebigs Ann. Chem. 1976, 1412 (1976)

Wieland, Th., Dose, K.: Veränderungen der Proteinverteilung im Blutserum bei der Amanitinvergiftung. Biochem. Z. 325, 439 (1954)

Wieland, Th., Fahrmeir, A.: Oxydation und Reduktion an der γ,δ-Dihydroxy-isoleucin-Seitenkette des O-Methyl-α-amanitins. Methyl-aldoamanitin, ein ungiftiges Abbauprodukt. Liebigs Ann. Chem. 736, 95 (1970)

Wieland, Th., Faulstich, H.: Amatoxins, Phallotoxins, Phallolysin and Antamanide, the biologically active components of poisonous *Amanita* mushrooms. Crit. Rev. Biochem. 1978

Wieland, Th., Faulstich, H.: The phalloidin story. In: Frontiers in Bioorganic Chemistry and Molecular Biology (eds. Yu. A. Orchinnikov, M.N. Kolosov); Amsterdam-New York: Elsevier North-Holland 1979

Wieland, Th., Wieland, O.: Chemistry and toxicology of the toxins of *Amanita phalloides*. Pharmacol. Rev. 11, 87 (1959)

Wieland, Th., Wieland, O.: The toxic peptides of *Amanita* species. In: Microbial Toxins (eds. S. Kadis. A. Ciegler, S.J. Ajl), Vol. XIII, p. 249. New York: Academic Press 1972

Wilhelm, I., Dina, D., Crippa, M.: A special form of deoxyribonucleic acid-dependent ribonucleic acid polymerase from oocytes of *Xenopus laevis*. Isolation and characterisation. Biochemistry 13, 1200 (1974)

Wintersberger, E.: DNA-dependent RNA polymerase from mitochondria of a cytoplasmic "petite" mutant of yeast. Biochem. Biophys. Res. Commun. 40, 1179 (1970)

Wobus, U., Panitz, R., Serfling, E.: α-Amanitin: its effect on RNA synthesis in polytene chromosomes. Experientia 27, 1202 (1971)

Wulf, E., Bautz, L.: RNA polymerase B from an α-amanitin-resistant mouse myeloma cell line. FEBS Lett. 69, 6 (1976)

Yagura, T., Yanagisawa, M., Iwabuchi, M.: Evidence for two α-amanitin-resistant RNA polymerases in vegetative amoebae of *Dictyostelium discoideum*. Biochem. Biophys. Res. Commun. 68, 183 (1976)

Yocum, R.R., Simons, D.M.: Amatoxins and phallotoxins in *Amanita* species of the Northeastern United States. Lloydia 40, 178 (1977)

Young, H.A., Whiteley, H.R.: Deoxyribonucleic acid-dependent ribonucleic acid polymerases in the dimorphic fungus *Mucor rouxii*. J. Biol. Chem. 250, 479 (1975)

Zanetti, M., Foa, L., Costanzo, F., La Placa, M.: Specific inhibition of rous sarcoma virus by α-amanitin. Arch. gesamte Virusforsch. 34, 255 (1971)

Zybler, E.A., Penman, Sh.: Products of RNA polymerases in HeLa cell nuclei. Proc. Natl. Acad. Sci. USA 68, 2861 (1971)

Specialized Genetic Recombination Systems in Bacteria: Their Involvement in Gene Expression and Evolution

D. J. Kopecko

A. Introduction

A variety of phenomenal DNA units, aided by specialized recombination processes, are responsible for a major proportion of the
"spontaneous" chromosomal alterations observed in bacteria. These
structurally and genetically distinct DNA segments (i.e., bacterial viruses, insertion sequence elements, and transposons)
can be inserted within many chromosomal loci. In addition to
causing insertion mutations and encoding new genetic potential,
these discrete units act as supernumerary genetic regulatory
switches capable of enhancing or eliminating the expression of
nearby genes. Furthermore, recombination promoted by these DNA
elements can result in various chromosomal rearrangements affecting large or small chromosomal DNA regions and involving the
joining of unrelated DNA regions that lack apparent nucleotide
sequence homology. Though this review is aimed at describing
the intensive study of recombination mediated by viruses and
transposable elements in bacteria, there is considerable evidence to suggest that transposable elements are also significantly involved in genetic reorganization and regulation in higher
organisms (McClintock 1957; Bukhari et al. 1977; Kleckner 1977;
Starlinger 1977; See Sect. F). Intermolecular exchange of a DNA
segment(s) (i.e., genetic recombination or crossing over) between homologous parental chromosomes, resulting in the formation of a hybrid molecule, has been recognized in eukaryotic
systems since the early days of classical genetics (Hayes 1968).
This marvelous process is important in providing us with the
breadth of phenotypic diversity that one sees within a single
plant or animal species. More fundamentally, recombination promotes new genetic combinations upon which the forces of natural
selection can act, eventually leading to the evolution of an
organism more suited to the environment. Unfortunately, the mul-
tichromosomal organization of eukaryotic hereditary information
as well as the absence of experimental methods to manipulate
this material have precluded, for the most part, molecular analyses of either mutations or various recombinational events in
higher organisms. The recent "genetic engineering" techniques
lay the foundation for fine structure study of eukaryotic chromosomes, but little has yet been accomplished along these lines.
However, the eugonic bacteria and their viruses, each containing
one relatively simple chromosome, seem to have been specially constructed for the molecular geneticist. Bacteria are relatively
simple, undifferentiated organisms that reproduce by asexual
fission, a process characterized by the doubling of the cellular

contents followed by the equipartitioning of the replicated hereditary information at cell division. Thus, each daughter cell is essentially a genetic replica of the parent. Despite the absence of sexual reproduction in bacteria, intercellular exchange of genetic information can occur readily, not only between different bacterial species, but also intergenerically (for review see Hayes 1968; Lewin 1977; Kopecko et al. 1979). Beyond the apparent evolutionary significance of this intercellular genetic exchange, bacterial genetic transfer processes (e.g., conjugation) have been successfully manipulated to obtain our present understanding of the genetic and molecular organization and expression of hereditary material.

Bacterial evolution, until recently, was thought by many to occur by a very slow process encompassing the induction of a small alteration in a chromosomal DNA sequence (i.e., a mutation), environmental selection for the desirable mutations, and the accumulation of beneficial mutations through intercellular genetic exchange and general recombination (see Fig. 1). This concept was fostered by the results of early genetic studies in which bacteria were found to mutate relatively infrequently (i.e., one mutant per 10^6-10^8 cells for any given trait) and each "spontaneous" mutation appeared to represent an alteration in only one

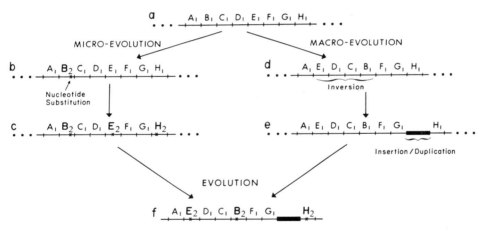

Fig. 1. Schematic representation of evolution. The *horizontal lines* represent a portion of the bacterial chromosome and arbitrary genes are labeled A, B, ... and H. The subscript, 1 or 2, after each gene designates an allelic form of that gene. The chromosomal segment shown in (a) can undergo small single nucleotide base changes·or micro-evolutionary events, such as that shown in (b) and (c). The mutated genes are indicated by a subscript 2 and pronounced lettering, and the mutations are located by an X. In addition, large chromosomal or macro-evolutionary rearrangements such as the inversion of a DNA segment (d), the deletion of a DNA segment (not shown) or the insertion of a DNA sequence (e) can occur. Overall bacterial evolution appears to result from the accumulation, via genetic recombination, of both micro- and macro-evolutionary chromosomal alterations, as shown in (f)

or a few adjacent mucleotide base pairs of DNA. It seems appropriate to refer to these mutational events, which involve the
addition, deletion or substitution of only one or a few nucleotides, as *micro-evolutionary* (Dobzhansky 1955, see p. 165; Cohen
et al. 1978). Within the past 10-15 years, however, it has become
apparent that large chromosomal rearrangements (e.g., duplications, deletions, inversions, and transpositions of multinucleo-
tide DNA segments) occur in bacteria, oftentimes at relatively
high frequencies. These events which result in the rearrangement, gain, and/or loss or large DNA segments can accordingly
be described as *macro-evolutionary* (Dobzhansky 1955, see p. 166;
Cohen et al. 1978) and certainly must account for a major proportion of bacterial evolution. It appears that bacterial evolution results from the temporal accumulation of both micro-
and macro-evolutionary DNA alterations (see Fig. 1).

Genetic recombination in bacteria, described in more detail below, can be divided into two broad categories: (1) general and
(2) specialized. In short, *general* recombination mechanisms mediate genetic interchange at random points between largely homologous deoxyribonucleotide segments. Following the intercellular transfer of mutated DNA segments, both micro- and macro-
evolutionary changes can be stably incorporated into the recipient chromosome via general recombination (Fig. 1f). However,
whereas micro-evolutionary mutational events are thought to be
caused in vivo by errors in DNA metabolism, sometimes induced
by intermediary metabolites or radiation exposure, macro-evolutionary DNA alterations appear to be generated via a variety
of special recombinational processes. Moreover, these macro-
evolutionary genetic exchange processes are functionally independent of known general recombination systems. In addition,
the distinctive behavior (i.e., the lack of a requirement for
extended homology between interacting DNA regions) of these
macro-evolutionary processes allows them to be categorized as
specialized recombination systems.

Research on macro-evolutionary chromosomal alterations has only
recently shifted from the purely descriptive to the mechanistic
level. However, the recent results of electron microscope and
DNA sequencing studies of various macro-evolutionary events
have suggested that several distinct specialized recombination
systems are functionally related, a finding that appears yet to
be appreciated by many. Thus, my primary intent in this review
is to describe for the non-specialist the various types of bacterial specialized recombination systems and to relate these
processes to the overall scope of bacterial heredity (i.e.,
mutation, gene regulation, genetic exchange, and evolution).
Although this review is slanted toward a general introduction
to this rapidly unfolding area of specialized genetic recombination, I hope that the coverage is in adequate depth and breadth
to be of value also to the genetic specialist. For the novice,
several recent cursory and detailed surveys of bacterial heredity and genetic nomenclature are recommended for reference
(Demerec et al. 1966; Hayes 1968; Hershey 1971a; Falkow 1975;
Novick et al. 1976; Watson 1976; Lewin 1977; Luria et al. 1978;
Kopecko et al. 1979). Because of the broad scope of this review

only representative research reports and review articles have been referenced. In fact, there are several excellent detailed literature reviews of various specific aspects of specialized genetic recombination and these references have been included in the appropriate sections below.

This review is organized as follows. General and specialized recombination are defined in the following section. Then, I have briefly reviewed the currently perceived genetic and molecular bases of general recombination in order to ensure that the significance of specialized genetic interchange is fully appreciated. Afterwards, several specialized recombination systems are discussed in detail and, finally, the inherent evolutionary significance and practical experimental application of these processes is discussed.

B. Genetic Recombination: An Overview

I. General vs Specialized Recombination

"Genetic recombination" refers to a variety of DNA exchange of processes, in different organisms, that result in heritable altered linkage relationships of genes or parts of genes. For instance, reciprocal exchange of DNA segments between largely homologous regions of similar chromosomes, a recombinational capability of most organisms, can be represented by ...abCDe... X ... ABcdE... = ...ABCDE... + ...abcde..., where similar upper and lower case letters denote, respectively, dominant and recessive forms (alleles) of the same gene on separate chromosomes. In addition to this simple reciprocal exchange between homologous DNA regions, presently recognized recombination processes encompass such diverse events as the chromosomal integration of entire extrachromosomal genetic elements such as bacterial plasmids or viruses (e.g., the F plasmid and phage λ) as well as chromosomal gene duplication (e.g., ...ABABCD...), inversions (e.g., ...ABDCEF...), delections (e.g., ...ABDBCD...), and transpositions (e.g., ...ADEFBC...). Albeit varied, these events are all mediated by one of two classes of recombinational activities. *General* recombination mechanisms mediate random genetic exchanges between largely homologous segments on the same or on different genomes and require certain host general recombination functions. As a result of general recombination, virtually any DNA segment can be exchanged, but only between homologous DNA regions. In contrast, *specialized* recombinational activities act independently of general recombination functions and do not require large regions of homology between interacting DNA segments. Certain *specialized* recombination processes catalyze the integration and/or excision at a limited number of chromosomal sites of physically defined genetic units (e.g., λ phage integration into the *E. coli* genome), or, in other cases, insertion or deletion of discrete DNA segments at seemingly random chromosomal loci (e.g., Mu phage or transposon insertion into the *E. coli* genome). In addition to the currently described specialized

recombination systems that mediate the genetic exchange of recognizable, defined DNA units (e.g., λ or Mu bacteriophage; insertion sequence elements), present evidence would lead one to
believe that these or similar specialized recombination systems
promote the occasionally observed exchanges, at variable loci,
between bacterial DNA segments that lack apparent nucleotide sequence homology (e.g., "spontaneous" chromosomal deletion or
transposition events; and specialized transducing phage formation). General recombination and the various specialized recombination systems appear to be mediated by separate overall processes, but may share common components of DNA metabolism, such
as winding/unwinding enzymes, ligase, polymerases, various nucleases, and DNA binding proteins.

II. Genetic Aspects of General Recombination

Particularly noteworthy prerequisites to our current conceptua-
lizations of genetic recombination in bacteria were the disco-
veries of DNA structure and the informational organization of
DNA, as well as the various mechanisms whereby bacteria can
exchange their hereditary material intercellularly (Hayes 1968;
Watson 1976). Upon the introduction into recipient bacterial
cells of large segments of a similar (donor) bacterial chromosome via conjugation, or of smaller bacterial DNA fragments via
transformation or transduction, normally haploid recipient cells
become partially diploid (i.e., merodiploid) for the DNA segment
transferred. Merodiploid cells are usually genetically unstable
and characteristically lose the newly inherited trait. One can
observe this loss if the donor DNA segment, ...ABDdef..., is
phenotypically distinguishable from the analogous recipient
chromosomal region, ...abcDEF... . However, with a probability
of ∿0.5, these merodiploid cells can undergo a genetic recombination event(s) in which a random segment of the newly inherited
material is exchanged for an analogous portion of the recipient
chromosome. If, for instance, the dominant ABC alleles were inserted in place of the recessive abc alleles of the recipient,
the resulting recombinant bacterial genome would be dominant
for the region ...ABCDEF..., and all progeny bacteria would in-
herit this recombinant genotype (Hayes 1968). This random recombinational exchange of DNA segments between largely homologous
interacting DNA regions is thought to occur universally among
bacteria. However, most intensive genetic analyses of this general recombination phenomenon have been conducted in the well
characterized *Escherichia coli* K-12 system.

The insightful genetic studies of A.J. Clark and others (see
Clark 1973; Lewin 1977; Mahajan and Datta 1979) have established
the specific involvement of several recombination *(rec)* genes in
this process. Through genetic complementation analyses of various
mutants in recombination deficient (Rec⁻) *E. coli*, Clark (1973)
has deduced the normal existence of two general recombination
pathways, both of which require the 40,000 mol. wt. *recA* protein.
One pathway that utilizes, in addition to the *recA* protein product, the *recB, C* exonuclease, normally accounts for almost all
general recombination. In cells deficient in the *recB, C* exo-

nuclease, a second pathway involving the uncharacterized *recF* gene mediates recombination at about 1% of wild-type levels. There is now considerable evidence to suggest that the *recB, C* recombinational pathway is involved primarily in double-strand genetic exchanges, while the *recF* pathway mediates mostly single-strand DNA exchanges (Mahajan and Datta 1979). In addition, separate studies have revealed similar phage-specified general recombination systems in λ, T4, T7 and P22 (Lewin 1977).

III. Molecular Mechanisms of Homologous Recombination Processes

Little is factually known about the molecular events involved in recombination. However, there is considerable evidence to indicate that general recombination encompasses physical breakage of parental molecules and reunion of exchanged DNA segments (see Lewin 1977). Stahl and co-workers have recently provided evidence for the existence of *recB, C*-dependent, randomly located (about every 5000 base pairs) recombination sequences, called Chi, on the *E. coli* chromosome (Stahl et al. 1975; Malone et al. 1978; Chattoraj et al. 1979). Although not yet proven, physical pairing of interacting DNA regions (i.e., recombinational synapse) may be catalyzed by the *recA* protein (Shibata et al. 1979), while the Chi sequences may be involved as enzyme recognition sites in the final resolution of the hybrid structure. Molecular models for general recombination are necessarily speculative, but several have been included here to generate a general concept of events likely to be involved in recombination and for later comparison and contrast to specialized recombination mechanisms. Figure 2 diagrammatically depicts events likely to be involved in the integration of a single linear DNA strand, acquired by conjugation or transformation, into the bacterial chromosome. Experimental results suggest that entering donor single-stranded DNA quickly pairs with a homologous region on the chromosome. The donor single-strand is exchanged with the recipient molecule at a gap either created by general recombination enzymes or remaining from DNA replication or repair activities. Following nuclease trimming of the non-exchanged ends of the donor strand, covalent closure of the newly constructed recombinant might involve repair synthesis or simply ligation. The exchange of a single DNA strand between two double-stranded (duplex) DNA molecules, as shown in Figure 1E, could occur by a minor variation of the scheme shown for single-strand integration. If the complementary strands in the hybrid region of the recombinant molecule differ, DNA repair mechanisms might remove any mispaired bases. Alternatively, replication of the hybrid molecule would generate daughter chromosomes that differ in the region of the original recombinational event. Following single-strand DNA transfer to a recipient cell via conjugation, segments as large as 500,000 nucleotide bases in length have been detected by genetic means to be incorporated by general recombination. Although hybrid molecules are formed in the absence of DNA synthesis, the final covalent linkage of exchanged strands in the hybrid molecule appears to require cell growth, but the specific requirements for DNA, RNA, or protein synthesis are unknown (reviewed by Lewin 1977).

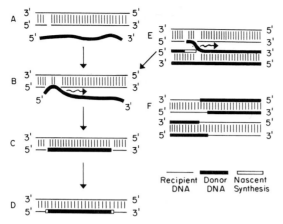

Fig. 2. Molecular models for general recombination. Interacting DNA strands
are indicated by *horizontal lines*; hydrogen bonds between complementary nucleotides are shown as *short vertical lines*. Steps A-E describe events likely to occur during the integration of a single DNA strand into a recipient
double-strand molecule. (A) Entering single-strand donor DNA quickly pairs
with a homologous region of the chromosome. Genetic exchange is initiated
on the recipient genome at a single-strand gap created by a recombination
enzyme(s) or some other DNA metabolic activity. (B) Extension of the exchanged region (see *wavy arrow*) following displacement and/or degradation
of the corresponding recipient strand. (C) Termination of the genetic exchange may occur at a gap introduced by a specific recombination enzyme or
by some other event. Unincorporated donor sequences are exonucleolytically
removed. (D) The gaps on each side of the incorporated DNA segment are repaired by DNA polymerase and ligase. Any differences (base mispairings, nucleotide additions or deletions) between the strands of these recombinant
DNA molecules are either corrected by DNA repair processes or expressed following replication. (E) The exchange of a single DNA strand between two
double-stranded DNA molecules could occur in a manner similar to that described above, except that exchange between paired regions would require, at
least, single phosphodiester bond cleavages in corresponding strands of both
donor and recipient molecules. Incorporation of the exchanged DNA segment
would occur as shown in steps B-D. Complementary donor strand synthesis could
occur subsequent to or simultaneously with displacement of the donor single
strand. Alternatively, (not shown) a single strand from each molecule could
exchange with the opposite molecule giving rise to a reciprocal exchange of
single DNA strands. (F) This diagram depicts the products expected from a
single reciprocal, double-strand exchange between two linear DNA molecules.
It should be noted that chromosomal integration of the F plasmid or phage
lambda occurs by a specific, single, reciprocal, double-strand exchange between two circular molecules and results in a large composite circle. Two
reciprocal exchanges would have to occur in order to exchange any DNA segment between two circular genomes. Reciprocity in genetic recombination is
explained in the text

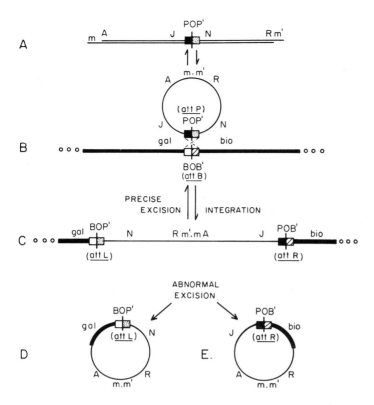

Fig. 3. Integration and excision of phage λ. This unscaled diagram depicts various molecular states of the temperate bacteriophage λ. The phage chromosome, normally ∿45 kilobase pairs in length, is represented by a *thin line(s)* while the *thicker lines* correspond to segments of the bacterial chromosome. The *rectangles* descriptively labeled POP' and BOB' represent the analogous attachment sites on the phage *(att*P*)* and bacterial *(att*B*)* chromosomes, respectively. The common core sequence, O, is depicted by a *vertical line* in the middle of each *att* site whereas the unique flanking sequences P, P', B and B' are represented by the *differently marked boxes*. The DNA sequences marked with A, J, N, and R are phage structural genes while *m·m'* is the site on λ which is endonucleolytically cleaved giving rise to single-stranded, 12 nucleotide long complementary (cohesive) 5' ends, *m* and *m'*. The only bacterial genes shown are those for utilization of galactose *(gal)* and the synthesis of biotin *(bio)*, which are situated nearby to the primary *att*B site in *E. coli*. (A) The linear λ genome. This is the form in which λ DNA is packaged in the virion. During packaging, phage monomers are excised from tandem oligomeric replicative forms by specific endonucleolytic cleavage at symmetrically identical sites that are separated by 12 nucleotides in opposite strands at the *cos (m·m')* locus. Both strands of λ DNA are depicted in this section to emphasize the cohesive termini. In steps B through E, both λ and bacterial DNA are depicted by single lines of different width. (B) Phage circularization and lysogenization. Once injected intracellularly, linear λ circularizes by ligation of the complementary termini, *m* and *m'*. Phage integration occurs after pairing of the analogous attachment regions on both λ (POP') and the host (BOB') genomes, by a specific recombinational crossover mediated by the λ-coded integration *(int)* protein. (C) λ prophage. The normal linear arrangement

Exchange of a double-stranded segment between two duplex DNA
molecules, although more complex, might follow a course of
events similar to that described above. However, the requirements of the exchange would depend upon the physical state (i.e.,
circular or linear) of the interacting molecules. Also, the exchange could be *reciprocal*, an event in which all DNA ends created by recombinational cleavage are rejoined to new sequences,
or *non-reciprocal*, an event in which new DNA ends are generated
by the recombinational event. The end product of a single reciprocal double-strand exchange is shown in Fig. 2F. Note that two
reciprocal, double-strand crossover events are required to exchange a single contiguous DNA segment via generalized recombination between two circular molecules, which is the normal phy-
sicochemical structure of bacterial, plasmid, and many virus genomes. In contrast, one reciprocal, double-strand exchange between two circular molecules would produce one larger composite
circular molecule, the product observed for the chromosomal integration of viruses or plasmid (see Fig. 3B, C). Additionally,
the extent of homology between the F plasmid and the bacterial
chromosome that results in *recA*-dependent integration of F is
now known to be approximately 1000 nucleotide base pairs (Davidson et al. 1974). Thus, only a relatively small amount of sequence homology between interacting molecules is needed for homologous recombination.

General recombination mechanisms, then, mediate the physical exchange of single or double-stranded DNA at random points between
paired and largely homologous DNA segments. In *E. coli*, general
recombination is specifically dependent upon the *recA* gene product, but other DNA metabolic activities are also involved
(Radding 1973; Clark 1974; Lewin 1977). The recent cloning of
the *recA* gene and purification of its protein product (McEntee
and Epstein 1977) in conjunction with the findings of Stahl and
co-workers (Malone et al. 1978) of the Chi sequences and the recent direct visualization by electron microscopy of synapsed DNA
regions (Potter and Dressler 1979) would suggest that some of
the events involved in generalized genetic exchange may soon be
deciphered. Hopefully, this brief overview of general recombination will better enable the reader to comprehend the catalog of
"aberrant" or specialized recombination events described in the
following sections.

---

of the prophage genome within the host chromosome is shown. This integration
process is reversible, and upon induction, precise excision of the λ genome is
effected by both *int* and the excision *(xis)* proteins, resulting in the production of normal λ phage (see step B). Occasionally, however, prophage excision
is inexact and creates a defective specialized transducing phage (see D and E).
(D) Circular λ d*gal*. During excision of the λ prophage, the recombinational
crossover took place between a site within λ and a site in the *gal* region of
the chromosome giving rise to a defective phage particle which has lost a set
of phage genes, but gained a corresponding length of bacterial DNA, in this
case the *gal* genes. (E) A similar abnormal excision involving the opposite
end of the λ prophage can create λ d*bio* transducing phages

C. Specialized Recombination Systems in Bacteria

I. Introduction

Atypical recombination events in bacteria were first characterized in depth during the study of the temperature bacterial viruses. For example, in contrast to the randomness of general (or
legitimate) recombination events, bacteriophage λ was observed
to integrate itself physically as a linear DNA addition into a
specific site on the *E. coli* K-12 chromosome. Although upon induction, prophage λ excision from the integrated state was most
often precise, occasionally an imprecise event would take place
in which part of the λ sequences would excise along with some
adjacent bacterial DNA giving rise to a specialized transducing
phage (Weisberg et al. 1977; see Fig. 3). Furthermore, entirely
separate studies conducted during the 1960's revealed that reshuffling of large segments of the *E. coli* chromosome through
duplication, deletion, inversion, or transposition could occur
(see reviews by Starlinger and Saedler 1976; Starlinger 1977).
Although seemingly different, all of the above events required
little or no homology between interacting DNA regions and could
occur in bacteria deficient in general recombination ability
(e.g., *recA*-deficient *E. coli* K-12). This *recA*-independent, physical joining of two apparently non-homologous DNA segments, once
thought to be a rare, aberrant event, has previously been termed
"illegitimate" recombination (Franklin 1971). These processes result in the formation of a novel joint as two grossly unrelated
DNA regions are fused (Hershey 1971a).

More recently, a series of discrete DNA segments, called transposable elements, have been identified which can transpose independently of host *recA* function, intra- or inter-molecularly.
These elements have been found to promote many of the macro-evolutionary events described above, as well as to affect gene expression by causing insertion mutations or by carrying DNA sequences that act as genetic transcriptional promoter and/or termination signals (Starlinger and Saedler 1976).

"Illegitimate" recombination was an amorphous discipline with
limited examples at the time of Franklin's comprehensive review
(1971). During the last decade, due specifically to the development of electron microscope heteroduplexing and denaturation
mapping procedures (Westmoreland et al. 1969; Inman and Schnos
1970; Davis et al. 1971), the discovery of many site-specific
DNA endonucleases (reviewed by Roberts 1976), and advances in DNA
sequencing techniqes (Maxam and Gilbert 1977), we have uncovered
and defined a series of what appear to be different specialized
recombination systems that are responsible for some of the events
heretofore termed "illegitimate". This section will entail general
descriptions of representative examples of various specialized
recombination systems. However, some macro-evolutionary rearrangements will be described for which no known specialized genetic
exchange mechanism has yet been implicated. These events which
occur in the absence of general recombination and extended DNA
homology will be referred to as "aberrant" or "illegitimate".

II. Bacteriophage λ

1. Site-specific Recombination System

The temperate bacteriophage λ is normally packaged in the virion as a linear, double-stranded DNA molecule, as diagrammed in Fig. 3A, which has been enzymatically cleaved by a specific endonuclease that creates 12 nucleotide-long complementary (cohesive) 5' ends, *m* and *m'*. Following bacteriophage infection of a bacterial cell, the injected phage DNA molecule stably circularizes by ligation of its reannealed cohesive ends. Either lytic replication can ensue or, by definition, a temperate bacterial virus can exist intracellularly in a quiescent (i.e., lysogenic) state from which it can later be induced to undergo lytic/vegetative growth. During lysogenization the functions needed for λ lytic growth are repressed. In addition, a site-specific reciprocal recombination event occurs between a specific attachment/recognition site, called *att*P, on the circular phage DNA and an analogous receptor site, *att*B, on the bacterial genome, resulting in the ordered linear insertion of λ into the *E. coli* chromosome (see Fig. 3B,C). This event requires a phage-encoded integration (*int*) protein, which has a subunit molecular weight of approximately 40,000 daltons, that binds to specific sites within *att*P and is known to have DNA nicking-sealing activity (Nash 1977; Kikuchi and Nash 1978; Nash, personal communication). The λ prophage, which is now replicated as an integral part of the bacterial genome, is bounded by hybrid attachment/recognition sites which have been designated *att*L (left) and *att*R (right; see Fig. 3C). No extended regions of homology could be detected among *att*P, *att*B, *att*L and *att*R by electron microscope heteroduplex procedures (Hradecna and Szybalski 1969; Davis and Parkinson 1971), but recent direct DNA sequence analysis has shown that each of these four *att* sites has the following 15 deoxyribonucleotide base common core: 5' -GCTTTTTTATACTAA- 3' (Landy and Ross 1977). Despite the presence of the common core region, however, the sequences on either side of the core in *att*B or *att*P are different from one another. Because λ integration is a reversible process, this common core region must be the physical locus at which integrative recombinational DNA breakage, exchange, and reunion occur. As a result of the different core-adjacent sequences, each *att* site is genetically distinct, i.e., each displays a unique set of affinities for the other *att* sites during integrative/excisive recombination events (Parkinson 1971; Nash 1977). The overall organization of the *att* sites and exact size of the core-adjacent sequences which affect site-specific recombination are not yet known. Although λ normally integrates into a primary *att*B locus located at 17 min on the *E. coli* genetic map (Bachmann et al. 1976), in bacterial hosts containing a deletion of this primary bacterial *att* site λ will integrate less efficiently into a variety of secondary bacterial *att* sites. These secondary sites appear, in all respects, to be natural variants of *att*B (Shimada et al. 1972) and when λ is integrated at a secondary site that lies within a detectable gene, the λ prophage acts as a large insertion mutation within that gene.

At induction λ prophage repressor is inactivated, possibly through proteolytic cleavage by the host *recA* protein (Roberts et al. 1979), allowing λ gene expression which is normally followed by excisive recombination between the hybrid *att* sites, an event that appears essentially to be the reverse of integration (see Fig. 3B,C). Excision requires the action of two phage-encoded products, the *int* protein and an excision *(xis)* gene product (encoded by < 250 base pairs; Nash 1977), and results in the generation of circular bacterial and λ chromosomes. Neither phage nor host general recombination abilities or phage genes other than *int* and *xis* appear to be involved in this site-specific specialized recombination system. Furthermore, λ integration and excision can occur in the absence of DNA, RNA or protein synthesis. Thus, in contrast to our current knowledge of general recombination mechanisms, these integration/excision events must not involve exonucleolytic trimming of exchanged DNA, the random creation of single-strand gaps, or gap-filling by nascent DNA synthesis (Kikuchi and Nash 1978). Recently, mutant λ phage that simultaneously carry *att*P and *att*B, or *att*L and *att*R, have been constructed. These mutant viruses participate in both in vivo and in vitro inter- and intra-molecular site-specific recombination reactions as illustrated in Fig. 4 (Engler and Inman 1977; Nash 1977; Nash et al. 1978). Data obtained with

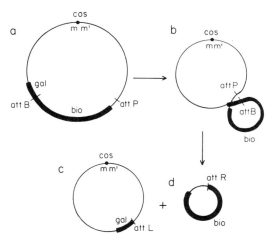

Fig. 4. λ *att*B–*att*P, substrate for integrative recombination in vitro. (a) The unusual transducing phage carrying both *att*P and *att*B was reported by Nash 1974. Lambda sequences *(att*P; *cos)* are represented by a *thin line* and the bacterial sequences *(gal, bio, att*B) carried by the phage are indicated by a *thicker line*. (b) Intramolecular recombination between *att*P and *att*B generates two smaller circular molecules, c and d. (c) Viable λ transducing phage carrying the bacterial *gal* genes plus the hybrid *att*L site. (d) A small non-self-replicating circle carrying the *bio* gene segment plus the hybrid *att*R site. Other λ transducing phage carrying one or more of the *att* sites *(att*P, *att*B, *att*R and *att*L) have since been isolated and are extremely useful in biochemical analyses of site-specific integration and excision (for details see Nash et al. 1978)

these and other mutant phage indicate that λ, through the action of *E. coli* ligase and gyrase, must exist as a covalently-sealed, negatively supertwisted circle in order for integration to occur (Mizuuchi et al. 1978).

Beyond the requirement for host ligase and gyrase, several *E. coli* genes have been identified by mutations that affect the λ site-specific integrative recombination process. The host integration mutations *(him)* have been mapped at 38 min *(himA)*, 84 min *(himB)*, and at 20 min *(hip* or host integration protein) on the *E. coli* genetic map (Miller and Friedman 1977; Nash et al. 1978). Interestingly, these mutations also affect other specialized recombination processes and these genes may comprise a series of proteins common to DNA metabolism (i.e., repair, replication, and recombination). Although the mechanisms of λ integration/excision events have not been physically defined, Landy and Ross (1977) have pointed out several features of the *att* sites (e.g., direct and inverted repeat DNA sequences, as well as adenine plus thymine rich regions) that might influence their specific recombinational behavior (see later section on mechanism of transposition).

Thus, bacteriophage λ encodes a specialized recombination mechanism that enables λ, as a discrete genetic unit, to integrate site-specifically into and excise from one or a limited number of sites on the *E. coli* genome. In addition, λ codes for a random genetic exchange system, specified by the λ *red* genes, that is analogous to the bacterial *rec* system and is not required for λ phage integration or excision. Though not mentioned above, complex genetic regulation of λ gene expression controls the fate of the infecting phage (i.e., lytic or lysogenic state). Several recent excellent reviews of these regulatory controls as well as of λ integration/excision are available (Gottesman 1974; Campbell 1976; Schwesinger 1977; Nash 1977; Weisberg and Adhya 1977; Weisberg et al. 1977; Nash et al. 1978).

In addition to performing precise integration/excision functions, the λ site-specific recombination system is able to promote recombination between two autonomous phage chromosomes (Table 1). Such events are *int*-dependent, occur only at the *att* sites (presumably via integrative recombination) and can take place in the absence of λ Red or host Rec general recombination (Weil and Signer 1968; Echols et al. 1968). For example, in a cell doubly infected with two λ derivatives that are genetically marked by mutations on opposite sides of *att*P in each phage type, one can detect reciprocal recombinant phage which carry approximately half of each parent phage sequence, with the recombination event occuring at *att*P on each molecule, presumably between linear phage molecules. Furthermore, this specialized recombination system also appears to be involved in the occasional formation of certain site-specific phage deletion mutants. Formation of these phage deletions, in which one deletion end point always occurs at *att*P, has been shown to require the λ *int* protein. Davis and Parkinson (1971) proposed two mechanisms to explain this infrequent deletion formation: (1) exonucleolytic digestion from a nicked *att*P site followed by joining the ends of the

Table 1. Specialized recombination events involving λ phage

Type of event	Functions and sites Required	Not required	Physical state of phage Initial	Final	Final State of recipient chromosome	Frequency	References
Site-specific phage integration	*int*, *attP*, *att*B	*xis*, *red*, *recA*	Autonomous phage	Prophage inserted in one orientation at *att*B	Linear, ordered insertion of the phage at *att*B		Hershey (1971a) Gottesman (1974) Campbell (1976) Nash (1977) Nash et al. (1978)
Precise prophage excision	*int*, *xis*, *att*R, *att*L —	*red* *recA*	Prophage	Autonomous phage	Original host sequences restored upon prophage excision	All prophage can be excised after induction	Hershey (1971a) Gottesman (1974) Campbell (1976) Nash (1977)
Site-specific exchange between λ phages	*int*, *attP* and/or other *att* sites —	*red* *recA*	Two autonomous linear phages?	Two recombinant phages, each carrying reciprocal halves of the parental molecules, with the exchange beginning at *attP*	—		Echols et al. (1968) Weil and Singer (1968) Lewin (1977)
Site-specific phage deletions	*int*, *attP* —	*red*, *xis?* *recA*	Autonomous phage	Autonomous phage containing a deletion with one end point at *attP*: the other end point can be fixed (e.g., λ*b2*) or random	—	5×10^{-6} events/ cell	Davis and Parkinson (1971) Parkinson and Huskey (1971) Weisberg and Adhya (1977)

Event	Functions/sites not necessary	Functions/sites required	Pre-recombinational state	Post-recombinational state	Effect on bacterial chromosome	Frequency	References
Imprecise prophage excision or transducing phage formation	– –	*int*, *xis*, *red*, attL, attR –	Prophage	Specialized transducing phage carrying bacterial sequences from one or both sides of attB	Chromosomal deletion; part or all of prophage is removed along with adjacent bacterial sequences	10^{-5}–10^{-3} events/cell, after induction	Hershey (1971a) Weisberg and Adhya (1977)
Aberrant λ deletion formation	? –	*red*, *int*, *xis*, att sites *recA*	Prophage	Prophage containing a non-specific deletion encompassing λ sequences involved with cell lysis & also some adjacent bacterial sequences	Chromosomal deletion; part of the prophage and some adjacent host sequences are deleted	10^{-7}/cell after a single phage growth cycle	Weisberg and Adhya (1977)
Internal λ deletion or λdv formation	*int?*, *xis?* –	*red* *recA*	Replicating autonomous phage; e.g., λ *vir*	Small defective phage containing large internal deletions of normal λ sequences; e.g., λdv	–		Matsubara and Kaiser (1968) Berg (1974) Chow et al. (1974)
DNA sequence duplication within λ phage	– –	*red* *recA*	Autonomous phage containing a deletion	Autonomous phage containing a tandem direct duplication	–	10^{-4} events/cell	Emmons et al. (1975) Weisberg and Adhya (1977)

Occurrence of the above events in the absence of phage *red* or host *rec* general recombination systems is the unifying characteristic of these genetic exchanges. Those functions/sites that are known to be required or that are not necessary for the above events include: phage integration (*int*), excision (*xis*), and *red* functions and the attachment sites (attP, attR and attL); bacterial *recA* function and the attachment site (attB). The pre- and post-recombinational states of the phage and the effect, if any, on the bacterial chromosome is given for each type of event. The recombinational systems involved in the latter four events are unknown

digested strand as shown in Fig. 5a-c, and (2) an *int*-promoted
unequal crossing-over event in which a region beginning at *att*P
on one molecule was exchanged at a non-homologous secondary
site on another phage molecule (Fig. 5d-f). The involvement, if
any, of *xis* in λ site-specific deletion formation is unknown,
but one can envision *int* and *xis*-dependent deletions which might
occur by intramolecular deletion of a sequence between *att*P and
certain specific secondary sites in λ. Such an intramolecular
deletion is exemplified by the exchange between *att*P and *att*B
on the special λ *att*P-*att*B phage, as illustrated in Fig. 4. Current evidence seems to favor this latter mechanism: Shimada et
al. (1972) have observed *int*-promoted inefficient recombination
between *att*P and secondary *att* sites in bacteria lacking the
primary *att*B locus; several apparently identical *int*-dependent
deletions of λ, call *b*2, have been isolated independently (Weisberg and Adhya 1977); and, finally, *int* and *xis*-dependent recombination events have been reported between each *att* site and
any of the other sites (i.e., *att*B, *att*P, *att*L or *att*R), though
the recombination frequencies varied widely (Parkinson 1971;
Guarneros and Echols 1973; Lewin 1977; Nash 1977).

2. Transducing Phage Formation and Other Illegitimate Recombination

Imprecise or aberrant excision of a λ prophage occurs occasionally by a process(es) that takes place independently of *int* or *xis*
functions, the prophage *att* sites, and the λ *red* or host *recA*
genes (Weisberg and Adhya 1977). This process(es) yields specialized transducing viruses, at a frequency of ∿1 per 10^6 normal

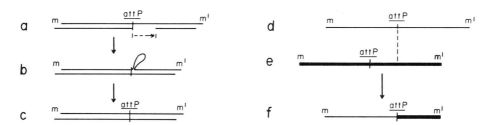

Fig. 5. Proposed mechanisms for *int*-dependent site-specific deletion formation in λ. Linear phage chromosomes are represented by *single* (d, e, f) or
double (a, b, c) *horizontal lines*. The cohesive ends, m and m', as well as
*att*P are shown on each molecule. As proposed by Davis and Parkinson (1971),
int-dependent deletion formation can occur in two ways. Shown in steps a-c,
exonucleolytic digestion - depicted by *dashed line* in (a) - from a specific
single strand cleavage in *att*P is followed by joining of the ends of the
digested strand (b). Revomal of the unpaired sequences, shown as a loop in
(b), would generate a λ deletion mutant, as shown in c. Alternatively, unequal crossing-over between two λ phages (d, e) in which a region beginning
at *att*P on molecule was exchanged (see *dashed line*, d and e) at a non-homologous secondary site on another phage, resulting in a shorter recombinant
molecule (f)

phage released following induction, whose genomes are composed
of part of the original phage sequences and some of the adjacent
bacterial genes (see Fig. 3D,E). Specialized transducing phage
can be serially propagated, but sometimes only in the presence
of coinfecting helper phage to supply missing essential phage
functions. Each originally isolated transducing particle appears
to be distinct from other transducing particles with respect to
the extent of both phage genes remaining and bacterial genes substituted. This finding indicates that the recombinational crossover occurs at random points between the interacting phage and
bacterial sequences. However, weak evidence exists to suggest
that there are preferred sites in the bacterial DNA sequences
adjacent to the prophage at which abnormal excision/transducing
phage formation occurs (Weisberg and Adhya 1977). The low frequency of specialized transducing phage formation is probably
a composite of (1) an inefficient recombination event(s), (2)
a requirement that the transducing phage contain the cohesive
ends, *m* and *m'*, in order to be packaged, (3) the fact that
fusion must occur between the phage and bacterial sequences
at the abnormal site of recombinational crossover to generate
a circle (or else scission at *cos (m·m')* during phage packaging
would generate two fragments that could not be serially propagated), and (4) the necessity for the excised product to be between 0.73 and 1.09 λ length in order to be packaged (Weisberg
and Adhya 1977). No phage or bacterial genes have yet been identified that affect this abnormal excision process. It has been suggested that specialized transducing phages are formed at or after
lysogenic induction, but the molecular bases for these apparently illegitimate events are currently unknown (Campbell 1963;
Weisberg and Adhya 1977). However, since no extended region of
homology exists between the interacting phage and bacterial DNA
sequences, during aberrant excision, one could envision a phenotypically cryptic, chromosomally-determined specialized recombinational systems as being responsible for, at least, some transducing phage production (see later section on *recA*-independent
bacterial recombination system). Therefore, transducing phage
might be produced continuously at a low frequency by such a
system, but detection would occur mainly following induction,
when helper phage are expressed.

It is worth noting that a class of defective transducing λ phages,
called λ*doc*L and λ*doc*R (*doc* = defective carrying one cohesive end),
have been characterized (Little and Gottesman 1971). These phages,
which cannot be serially propagated, carry bacterial sequences
from the left or right side of *att*B plus half of λ and are generated after lysogenic induction, in part, by the site-specific
cleavage of λ prophage at *cos (m·m')*. Because the free left cohesive end of λ appears to be packaged first, the bacterial end
of λ*doc*L is probably generated by the action of DNAses on the
bacterial DNA protruding from the filled phage head. The process
which results in the cleavage in the bacterial sequences of λ*doc*R
is not adequately understood, but is thought to involve a non-
specific DNAse (for more detail see Weisberg and Adhya 1977).
Thus, λ*doc* phages do not appear to be generated by recombinational
mechanisms.

Lambda phage appears to be involved in several other types of
illegitimate recombinational events, besides transducing phage
formation (see Table 1). By examining bacterial survivors following induction of a heat-inducible λ lysogen, one can isolate
bacterial deletion mutants which have lost part of the λ prophage (i.e., at least those λ genes involved in cell death) and
some neighboring regions of the host chromosome. It is not
known when (i.e., before or after phage induction) or how these
deletions occur, but their formation is rare ($\sim 10^{-7}$/cell after
one phage growth cycle), is not site-specific, and does not require λ *red*, *int*, *xis*, or *att* genes/sites (Weisberg and Adhya
1977). Again, a host-mediated, *recA*-independent recombination
system may be responsible (see later section).

Vegetatively replicating λ*vir* or similar phage that express constitutive replication functions have been observed on occasion
to undergo internal deletion of a large contiguous region creating λ*dv* (defective virulent) molecules (Matsubara and Kaiser
1968; Matsubara and Otsuji 1978). These non-integrative defective phage molecules, which can be formed in the absence of
recA and *red* functions (Berg 1974), retain basically that part
of λ which is normally essential for replication and responsible
for immunity. Consequently, λ*dv* molecules lack most phage properties and exist intracellularly as multicopy circular plasmids
comprising 50-250 copies/cell. Each separate λ*dv* isolate contains
only from $\sim$3 to 6 kilobase pairs of original λ information. However, λ*dv*'s often exist as dimers or higher multimers, consisting
entirely of direct or inverted large tandem duplications (e.g.,
3'...ABCABC...5' or 3'...ABCC'B'A'...5', respectively), where
the primed letters represent the nucleotide sequence complements
of the corresponding unprimed letters), sometimes interspersed
by a unique DNA region (e.g., 3'...ABCDC'B'A'...5'). Based on
electron microscope heteroduplex analyses, Chow et al. (1974)
have hypothesized that λ*dv*'s arise from a partially replicated
λ chromosome through recombinational events as depicted in Fig. 6.

In contrast to λ*dv* deletion mutants, other λ phage deletion mutants lose only short stretches of DNA. Because of the minimum
DNA length requirements of λ packaging, it is not surprising
that some λ deletion mutants often undergo partial genetic duplication which allows them to be packaged. Reiteration of some
of the sequences in a λ phage deletion mutant does occur at a
relatively high frequency ($\sim 2 \times 10^{-4}$ phage derivatives/cell after
a single phage growth cycle; Emmons et al. 1975) compared to λ
deletion formation. These duplication derivatives can easily be
selected by their increased density in cesium chloride density
gradients, by the increased concentration of a gene product, or
by various genetic means (see Weisberg and Adhya 1977). Duplication occurs independetly of *red* or *recA* gene products, possibly
by an intramolecular recombination event between daughter arms
of a partially replicated molecule or via intermolecular exchange
(Weisberg and Adhya 1977). Many regions of the λ genome have
been duplicated and the tandem direct repeat (3'...ABCBCDE...5')
has been observed most often. Duplication mutants with direct
repeats are genetically unstable and easily detected because the
reiterated sequence can be lost by Rec- or Red-promoted recombi-

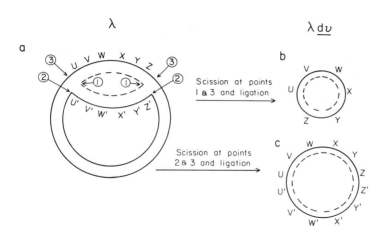

Fig. 6. Models for λ*dv* formation. Phage double-stranded DNA is represented by the two *parallel circular lines*. *Dashed lines* represent newly synthesized DNA. Arbitrary DNA sequences U through Z and the complementary DNA sequences, given in *primed letters*, are indicated. These illustrations are based on the models for λ*dv* formation proposed by Chow et al. (1974). By this proposal λ*dv* molecules are formed via scission immediately within or just outside of the termini of the replication fork and subsequent *recA*-independent recombination, resulting in the joining of parental to parental and progeny to progeny phage strands. (a) Bidirectionally replicating λ molecule. Breakage at the points marked by *arrows* 1 and 3 would generate a small linear fragment carrying the sequences U through Z. (b) λ*dv* formation could involve *recA*-independent recombination between the ends of this linear fragment and creation of a small circular molecule containing unique DNA sequences. (c) Scission of the replicating λ, shown in (a), at points 2 and 3 and subsequent joining of the original phage strands between sequences U and U', and Z and Z' would create a λ*dv* comprising a tandem inverted duplication. Note that single-strand interruptions may already exist at certain points in the replication fork due to the replication process, e.g., at points marked 1. Though not shown, a λ*dv* comprising a tandem-inverted dimer with interposed unique sequences at one or both ends of the repeated sequence could be generated by staggering the recombinational breakpoints at one or both ends of the replication fork (for details see Chow et al. 1974)

nation between the homologous regions (Bellett et al. 1971). Collectively, these data indicate that λ is involved in a series of seemingly aberrant recombinational events which occur independently of site-specific (i.e., *int/xis*), Red-mediated, or Rec-promoted recombination and that create specialized transducing viruses and internal phage sequence duplications, as well as deletions in autonomous or integrated λ and in adjacent bacterial sequences. As pointed out by Weisberg and Adhya (1977) in their recent review of illegitimate recombination events, more than one mechanism may be responsible for these aberrant exchanges involving λ. The various specialized and aberrant recombinational events in which λ phage are involved have been summarized in Table 1. The molecular maps of wild type λ and various derivatives as described above are included in the review of Szybalski and Szybalski (1979).

Albeit the bacteriophage λ integration/excision system is the best understood, other temperate bacterial viruses (e.g., Ø80, 434, 21, P2 and P22) are known to integrate via different specialized recombination systems at one or a few sites on the host chromosome and also to participate in many aberrant chromosomal rearrangements similar to those described above (Lewin 1977; Luria et al. 1978; Susskind and Botstein 1978). Therefore, when considering the causes of "illegitimate" recombinational events, one must be cognizant of the seemingly large assortment of phage-mediated specialized recombination systems and also of the types of genetic rearrangements that are promoted by these systems. Defective or non-inducible prophages or prophage remnants specifying specialized recombination systems may play a large role in bacterial macro-evolution. On the other hand, however, many genomic rearrangements observed in λ or other phages could be mediated by recombinational processes encoded by the bacterial host.

In contrast to the site-specific phage integration systems exemplified by λ, the unique temperate phage Mu can insert randomly into many chromosomal sites often causing detectable insertion (e.g., auxotrophic) mutations. Furthermore, this phage appears to be able to integrate into practically any phage, plasmid, or bacterial chromosome and can promote a variety of chromosomal reshufflings. Present knowledge of the Mu specialized recombination system, which represents the opposite end of the spectrum in recipient site specificity, is related in Sect. III.

III. The Mutator Bacteriophage, Mu

1. Genetic Definition

The novel bacterial virus, Mu, behaves as a temperate phage lysogenizing 5-10% of the cells which it infects or generating 50-100 infective virions per cell following the lytic cycle. Unlike most temperate viruses which integrate at one or a few specific host chromosomal sites, Mu can integrate at many, if not all, chromosomal loci. Consequently, about 2% of the cells lysogenized by Mu concomitantly acquire a new nutritional requirement or other recognizable mutation (Taylor 1963). This represents a frequency of mutation within a single gene of 50-100-fold higher than the spontaneous frequency observed in the absence of Mu. Both genetic and physical evidence indicates that these mutagenic events occur by the recA-independent, linear insertion of Mu into the affected gene (reviewed by Howe and Bade 1975). A series of studies on Mu integrated into several E. coli operons, mainly the lactose and tryptophan regions, has resulted in the following generalizations: (1) Mu-induced mutations are strongly polar, probably due to termination of transcription distal to the inserted phage (Jordan et al. 1968; Toussaint 1969; Daniell and Abelson 1973); (2) the mutations induced by Mu are extremely stable, with an apparent reversion frequency of $< 10^{-9} - 10^{-10}$ per viable cell (Taylor 1963; Jordan et al. 1968); (3) during lysogenization 10-15% of the integrating Mu phage concomitantly cause deletions in bacterial sequences (see Howe and Bade 1975; discussed later in section on Mu-promoted deletions); (4) there do not appear to be preferred sites for Mu insertion within a gene, indicating that, if insertion is not absolutely random, the chromosomal attachment/recognition receptor sequence must only be two to three nucleotides long (discussed by Bukhari 1976; Couturier 1976); (5) non-transcribed genes or repressed operons are mutated by Mu approximately five times more often than actively transcribed DNA regions (see Howe and Bade 1975); (6) Mu phage and prophage genomes are collinear, i.e., all Mu prophage genomes have the same gene order as that found for the linear vegetative phage molecule (Howe and Bade 1975); and, finally, (7) Mu can integrate in both possible orientations within a given gene (reviewed in Howe and Bade 1975).

In further contrast to the λ-like integration/excision systems, Mu prophage is not induced by ultraviolet light or other agents known to induce λ prophage. Instead, exponentially growing cultures of Mu lysogens generally contain 10^5-10^6 plaque-forming units of spontaneously induced phage per milliliter, with some evidence suggesting that prophage derepression is caused by normal transcription of a Mu-infected DNA segment (Howe and Bade 1975). Even though Mu prophage induction occurs spontaneously at a moderate level, in an apparent contradiction, reversion to prototrophy is practically undetectable. Why? Howe and Bade (1975) correctly reasoned that this low reversion frequency might be because prophage are excised infrequently and/or excision usually results in cell death. The isolation and experimental use of temperature-sensitive Mu mutants, called Mucts, which can be induced at 42°C to excise from the prophage state and are probably affected in a prophage repressor gene, have been instrumental in solving this problem. For example, utilizing $E.$ $coli$ lysogenized in the $lacZ$ gene with a temperature-inducible Mu mutant, Bukhari (1976) has isolated, upon Mu induction, non-conditional phage mutants, called X, that are unable to replicate their DNA or express other lytic functions, but which allow for Mu excision and subsequent lac gene expression. Precise excision of induced MuX, cts prophage resulting in Lac$^+$ revertants occurs at a frequency of 10^{-6}, while imprecise excision occurs ten times more often and yields defective revertants that remain LacZ$^-$ but which express the more operator-distal $lacY$ gene. Precise reversion of MuX, cts-induced bacterial mutations, although infrequent, demonstrates in these instances that Mu is inserted without the alteration of any bacterial sequences adjacent to its insertion site and suggests that integration/excision involves specific recognition of sequences at the Mu termini (Bukhari 1976). The recent isolation of deletion mutants of Mu that have lost the terminal sequences on one Mu end and which fail to lysogenize, lend support to the concept of specific terminal recognition sequences (Bukhari 1976; Chow and Bukhari 1977). Phage genes A and/or B, which are thought to be involved in integration and replication, respectively, also appear to be required for other Mu-promoted recombinational events such as deletion formation and chromosomal genetic inversion or translocation (Faelen et al. 1977, 1978;

O'Day et al. 1978). Mu phage growth, following induction of a thermoinducible lysogen or infection of *E. coli* cells with Mu, appears to require the host *dnaC* replication initiation function, the host DNA elongation factor encoded by *dnaB*, and host DNA polymerase III (Toissant and Faelen 1974). Aside from these replication requirements, no bacterial genes have been identified that are absolutely required for the various Mu-promoted recombinational events. However, some of the bacterial host mutations that inhibit λ integration also inhibit Mu induction and growth, and probably these host genes code for common DNA metabolic proteins that are necessary for a variety of processes (Kleckner 1977; Miller and Friedman 1977).

2. *Molecular Organization*

Physical analysis of Mu phage DNA has revealed several unexpected complexities (see Fig. 7a and reviews by Howe and Bade 1975; Bukhari 1976; Couturier 1976). Mu is packaged as a linear, double-stranded DNA molecule of slightly variable size averaging 25 megadaltons (i.e., ∿37,000 nucleotide base pairs or ∿37 kilobase pairs). Since Mu DNA contains neither detectable cohesive ends nor terminal redundancy, it lacks any obvious means to circularize (Couturier 1976). DNA molecules released from Mu phage that were isolated from a single plaque have been found to vary in size from about 36-38 kilobase pairs (Martuscelli et al. 1971; Daniell et al. 1973a,b). When the phage DNA molecules originating from a single plaque are completely denatured and allowed to reanneal, structures like those depicted in Fig. 7b and c are observed. The resulting molecules are predominantly double-stranded and generally contain variable length (0.5-3.2 kilobase pairs) heterogeneous terminal sequences represented by the long single-stranded (split) ends at one terminus (Fig. 7b,c). Short variable length sequences of 100 base pair average size have been identified at the opposite Mu terminus more recently (see Chow and Bukhari 1977). Additionally, some reannealed molecules contain an internal 3.0 kilobase pair non-renaturable region (generally termed a substitution bubble) called the G segment, which is located at a constant position within these molecules and always proximal to the longer split ends. The cogent features of Mu DNA that have emerged from various molecular and genetic analyses (see recent reviews: Bukhari 1976; Couturier 1976; Chow and Bukhari 1977) have been summarized below and in Fig. 7a. The heterogeneous terminal regions of Mu are comprised of seemingly random bacterial sequences that differ among phage molecules. The physical map of Mu has been divided into the α, G, and β segments, as shown in Fig. 7a. Prophage immunity functions map close to one end of Mu (now termed the immunity end) and lie adjacent to the majority of known Mu genes, which are located in the 31 kilobase pair α segment. The 3.0 kilobase pair G segment frequently undergoes genetic inversion so that this region is distinguished as an internal substitution bubble in some reannealed phage molecules (Fig. 7c). The remaining 1600 base pairs of actual Mu DNA, the β segment, is situated immediately adjacent to the long heterogeneous terminal sequences, referred to as the variable end (Chow and Bukhari 1977). Therefore, disregarding the

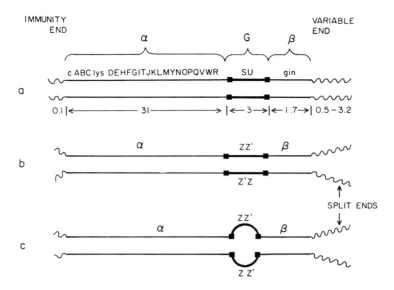

Fig. 7. Physical structure of Mu molecules. Each *horizontal line* represents
a single-strand of phage DNA. (a) Linear double-stranded mature phage DNA
molecule of approximately 25 megadaltons (not drawn to scale). The heterogeneous terminal sequences, represented by a *squiggled line*, consist of
random bacterial sequences that vary in composition and length from molecule to molecule. True phage DNA sequences are divided into α, G, and β segments. The size in kilobase pairs for each region is given (Chow and Bukhari
1977; Kahmann et al. 1977). The invertible G segment is bounded by small
(∿20 base pair) inverted repeat DNA sequences represented by solid blocks,
which are thought to be involved in the inversion of this segment. The
middle one-third of the G segment contains a less stable pair of inverted
repeat DNA sequences, identified in the electron microscope, that are represented in steps b and c as (Z·Z'). The map order of known genes on Mu
phage is shown in (a) Bukhari et al. 1977; M. Howe, personal communication).
The immunity gene *c*, integration gene *A*, and replication gene *B*, are located at one end of Mu, termed the immunity end. Genes *S* and *U* are located
in the G segment, the inversion of which is controlled by the *gin* function
that is encoded in the β segment. (b,c) Double-stranded heteroduplex structure resulting from the reannealing of different denatured Mu molecules.
One end of reannealed Mu molecules, termed the variable end, was observed
to contain long heterogeneous terminal sequences which have been referred
to as split ends. More refined techniques have also revealed the existence
of short heterogeneous terminal sequences at the immunity end. In the molecule shown in (c), the G segment is inserted in opposite orientations in
each strand, creating a substitution bubble

heterogeneous termini, the actual phage DNA sequences in all Mu
molecules are identical barring only the inverted G segment in
some molecules.

3. *Integration/Excision*

Electron microscope and DNA sequence analyses of plasmids or phages harboring a Mu prophage have shown that none of the heterogeneous terminal sequences associated with autonomous Mu molecules is inserted during integration (Hsu and Davidson 1974; Allet 1978). It seems reasonable at this point to assume that Mu somehow sheds its terminal host sequences prior to or during integration. Randomly isolated Mu*X*, *cts* prophage-mediated auxotrophs have been observed to revert to prototrophy, probably by precise prophage excision, indicating that most Mu prophage exist as point insertions (Bukhari 1976). Furthermore, since Mu prophage and phage maps are collinear, this implies that insertion requires the specific recombinational interaction between the true Mu termini. One might logically hypothesize, then, that the mechanism of Mu integration is similar to that of λ or P22, except that the bacterial *att*Mu receptor sites are numerous and the attachment/recognition sequences on Mu are located immediately adjacent to the heterogeneous host sequences present in mature phage, as illustrated in Fig. 8. Unlike cohesive-ended λ or terminally redundant P22, however, Mu molecules have no obvious physical means to circularize, a requirement for the chromosomal integration of many phages and plasmids (Campbell 1976). As expected, infecting Mu molecules have never been observed to form covalently sealed circles (Bukhari 1976). However, the predominant Mu form observed after infection sediments in neutral sucrose gradients twice as fast as linear Mu monomers. This faster sedimenting form might represent a Mu DNA-protein complex, similar to that shown in Fig. 8B, where a protein accomplishes the non-covalent fusion of the true Mu termini. Recent evidence indicates that intracellular infecting Mu phage are assimilated by the host very slowly, and that Mu integration appears to require both Mu and bacterial DNA replication (Ljungquist et al. 1979). Perhaps the infecting Mu molecule acts as a template upon which only the true Mu phage sequences are replicated; this newly replicated single- or double-stranded structure might form short-lived covalently sealed circles or, possibly, undergo fusion of the Mu termini through the assistance of a protein (see Fig. 8B). Regardless of the exact intermediate, Mu integration apparently involves a specific recombinational exchange between the Mu termini and the host receptor site, resulting in linear Mu insertion, as illustrated in Fig. 8D. The gene *A*-encoded integration function and the terminal recognition sequences are the only presently known phage functions or sites needed for Mu integration and possibly excision (Bukhari 1975; Faelen et al. 1978; O'Day et al. 1978). Although infrequent, precise prophage excision can be detected in induced Mu*X*, *cts* lysogens. Restoration of the original host sequences at the receptor site might involve essentially a reversal of the integration process, i.e., recognition of the Mu termini by phage-encoded functions (Fig. 8C,D). However, recent DNA sequence analyses have revealed that a 5 base pair bacterial sequence at the insertion site is apparently duplicated in direct order during insertion, so that a 5 base pair repeat sequence flanks the inserted phage (Allet 1979). Thus, reversion of a Mu prophage-induced mutant would necessitate excision of one 5 base pair repeat plus the entire

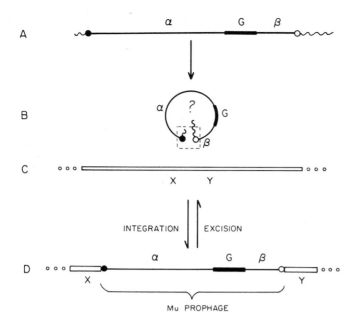

Fig. 8. Hypothetical illustration of precise integration and excision of Mu.
(A) Linear double-stranded Mu DNA molecule as packaged in the virion. The
heterogeneous bacterial DNA associated with each mature phage molecule is
represented by *squiggled lines*. The true phage sequences α, β and G, are
shown as *solid lines* of different width. The Mu attachment/recognition sequences are indicated by the *open and filled circles*. (B) Presumptive integration intermediate structure in which the true ends (i.e., *att* sites) of
Mu are brought within close proximity (enclosed within *dashed lines*), perhaps by covalent linkage or via protein. The infecting Mu phage might undergo replication only of its true Mu sequences (i.e., excluding the heterogeneous ends) and this newly replicated structure (single- or double-stranded) could be the integration intermediate. (C) Portion of a bacterial, plasmid, or phage recipient chromosome, represented by an *open rectangle*. (D)
Physical integration of Mu results in the linear insertion of only the true
α, β, and G segments of Mu into the recipient chromosome. Although Mu can
be inserted in either of two physical orientations, the prophage and vegetative phage maps are always collinear. Precise excision of Mu may simply
involve the reversal of the integration process (discussed in detail by
Bukhari 1976; Ljungquist et al. 1979; Couturier 1976) or may be mediated
by some Rec-independent bacterial recombination process, as described in
the text

prophage, an event that may be mediated by some bacterial process which recognizes short, directly repeated sequences (see
Sect. C.V.). Relative to λ phage specialized recombination, our
level of understanding of Mu integration/excision is very ele-
mentary. Furthermore, Mu phage induction constitutes a biological paradox as described below.

If only true Mu sequences are chromosomally inserted during prophage formation, the question arises, how do the phage particles isolated from an induced single Mu lysogen acquire many different terminal host sequences? The observations that Mu*cts* prophage can be induced independently of host Rec-ability and that Mu does not encode a general genetic exchange system (see Howe and Bade 1975) would lead one to believe that the heterogeneous phage termini are generated during replication and/or packaging, but not via recombination. Remember that in bacteria lysogenized by various other temperate viruses (e.g., λ, P22), induced prophage in virtually every cell excise and replicate as autonomous units. However, upon Mu*X, cts* prophage induction, only one out of every 10^5 cells appears to lose the prophage by exact or inexact excision, as detected by mutant reversion or loss of polarity, although every prophage is seemingly induced. Assuming that this situation closely mimics wild-type Mu induction, then most induced Mu prophage do not excise from the original integration site, yet they are replicated and give rise to progeny phage containing heterogeneous terminal sequences. This is the Mu paradox! Apparently, an induced Mu prophage generally replicates in situ at the original site of integration and transposes either single or double-stranded replicas of itself to other chromosomal loci where they insert, probably via the Mu integration system (Ljungquist and Bukhari 1977; see Fig. 10b,d). The newly inserted prophage can continue the lytic replication/transposition process giving rise to multiple lysogens containing ten or more Mu prophages (Toussaint et al. 1977; Ljungquist et al. 1979). In striking contrast to other temperate phages' integration functions which are not utilized during lytic growth, Mu integration functions (e.g., gene *A*) are necessarily expressed during productive lytic growth (Bukhari 1976).

Completion of the unique Mu lytic cycle encompasses encapsulation of mature phage, cell death, and virion release (Howe and Bade 1975). Compelling evidence indicates that Mu is packaged by a headful mechanism with encapsulation beginning at the phage immunity end. It is not currently known if the mature phages are excised and packaged directly from the host chromosome, from the large supertwisted circular molecules of varied sizes comprised of both Mu and host DNA that have been observed during lytic growth (see legend Fig. 10k), or in some alternate manner (Bukhari 1976, 1977; Waggoner et al. 1977). However, the packaging enzyme responsible for generating the immunity end of mature Mu molecules apparently recognizes a specific Mu site, but cleaves at a variable distance from the recognition site, averaging about 100 nucleotides into the adjacent host sequences (similar to the action of Type I restriction endonucleases). Beginning with the immunity end of Mu, the entire phage is packaged along with 500-3200 base pairs of adjacent bacterial DNA which is attached to the opposite Mu terminus. Recent experiments have shown that Mu phages containing sizable internal deletions or additions are packaged with longer or shorter stretches, respectively, of bacterial DNA attached to the β end of the phage molecules (Chow and Bukhari 1977; Chow et al. 1977). Therefore, the heterogeneous Mu termini seemingly result from the encapsulation of Mu molecules that are inserted within many different host sequences.

4. *Internal DNA Inversion Regulates Phage Viability*

The G segment of Mu has the remarkable and somewhat perplexing ability to undergo *recA*-independent inversion. This relatively high frequency inversion event is thought to occur by recombination between short inverted repeat DNA sequences (∿ 20 base pairs in length) found at the ends of the G region (Hsu and Davidson 1974; Chow and Bukhari 1977). The importance of these events is just being realized. Recently, a Mu gene (termed *gin*, for G inversion) that controls G inversion and which probably encodes a recombinational enzyme has been mapped in the β segment (Chow et al. 1977). Further studies have shown that in monolysogens a Mu *gin⁻* prophage with either G orientation appears to be produced equally well upon induction, but that only mature phage containing G in one specific orientation, the lytic orientation, can successfully infect similar bacterial hosts (Kamp et al. 1978). Several lines of evidence suggest that phage-coat borne adsorption functions, encoded by genes *S* and *U* which are located on the G segment, are not expressed when G is in the opposite orientation and, thus, resulting phage can not adsorb to similar host cells. However, Howe (1978) has intuitively noted that there may be two sets of adsorption functions on the G segment, each on opposite DNA strands and each of which express different adsorption proteins. Thus, G segment inversion may act as an on-off switch controlling phage viability, as supported by present data, or G inversion may change the host range of Mu.

It is very noteworthy that phages D108, P1, and P7 contain an invertible sequence that is virtually identical, by heteroduplex analysis, to the Mu G segment (Hull et al. 1978; Kamp et al. 1978, 1979). P1 and P7 are similar temperate viruses, which are circularly permuted and terminally repetitious, that exist as plasmids in the prophage state and, in most respects, are very different from Mu. Moreover, in P1 and P7 the G segment is bordered by larger (620 base pair) inverted repeat sequences (Chow et al. 1978b). However, G segments in Mu *gin⁻* phages can be inverted in the presence of P1 phage, demonstrating functional relatedness between these inversion systems (Chow and Bukhari 1977). The lack of homology between Mu and P1 or P7 suggests that these invertible G segments are capable of independent translocation from one molecule to another and may have arisen in these diverse systems by such an event (see Howe 1978). More important thant origin, what is the significance of genetic inversion to the phage or host bacterium? Recently, Zeig et al. (1978) have reported an inversion event that controls the alternating but exclusive expression of one of two flagellar antigenic types in *Salmonella*. In conjunction with these data, G inversions in phage demonstrate that inversion of DNA segments is a general mechanism, at least in bacteria and their viruses, for the control of gene expression (mechanism discussed in Sect. C).

5. *Mu-promoted "Illegitimate" Genetic Exchanges*

As discussed above and outlined in Table 2, Mu phages participate in what appears to be a variety of specialized recombination

Table 2. Specialized recombination events involving Mu phage

Type of event	Functions and sites		Physical state of phage	
	Required	Not required	Initial	Final
Phage integration	MuA, att sites	MuB	Infecting autonomous phage with random bacterial sequences at its terminal	Prophage in either orientation at any site of a recipient genome. The random bacterial sequences attached to mature Mu are lost upon integration
	(host replication functions)[a]	host recA, recB, recC, recF		
Phage transposition	Mu A, B, att sites	-	Replicating prophage	Multiple prophages with Mu in either orientation at two or more chromosomal loci
	?	host recA		
Precise prophage excision	Mu A	Mu B	Prophage	(Autonomous phage without heterogeneous termini)
	?	host(recA), recB, recC, recF		
Imprecise prophage excision	Mu A (Mu packaging enzymes)	Mu B	Prophage	Autonomous phage with heterogeneous termini
	?	Host(recA)		
G segment inversion	Mu gin, specific G segment termini	-	Prophage with G segment in either orientation; orientation of G regulates phage viability	Unaltered
	-	host recA, recB, recC		
Mu-promoted chromosomal deletion	Mu A, att sites	Mu B	Prophage or infecting phage	Prophage
	-	host recA		
Mu-promoted chromosomal inversion	Mu A or gin att or G segment termini	-	Prophage or infecting phage	Two prophages, separated by the inverted DNA segment, that are in opposing orientations or that have G segments in opposing orientations
	-	host recA		

Final state of recipient chromosome	Frequency	References
Insertion mutation, i.e., the Mu prophage	1-10% of the cell survivors of a single cycle of Mu infection are lysogens	Howe and Bade (1975) Couturier (1976) O'Day et al. (1978) Ljungquist et al. (1979)
Multiple lysogen	Occurs apparently for all prophages following induction	Toussaint and Faelen (1974) Bukhari (1976) Faelen et al. (1978)
Original host sequences restored upon prophage excision	Detected only upon induction of Mu X, cts mutants at 10^{-8} -10^{-6} events/cell	Bukhari (1975, 1976) Couturier (1976)
Chromosomal deletion; entire prophage and adjacent bacterial sequences are deleted together	Detected only upon induction of Mu X, cts mutants at $10^{-7}-10^{-5}$ events/cell	Bukhari (1975, 1976) Couturier (1976) Toussaint et al. (1977)
Unaltered	50% of the Mu released from an induced Mu cts lysogen have G segment inverted	Daniell et al. (1973b) Howe and Bade (1975) Kamp et al. (1978, 1979)
Lysogenic and containing deletions of host sequences that were originally adjacent to prophage insertion site	10-15% of Mu lysogens contain adjacent deletions	Howe and Bade (1975) Toussaint et al. (1977) Faelen et al. (1978)
The inverted host DNA is flanked on each side by a Mu prophage	10^{-4} events/cell in induced Mu X lysogen	Faelen et al. (1978) Toussaint (personal communication)

Table 2 (continued)

Type of event	Functions and sites		Physical state of phage	
	Required	Not required	Initial	Final
Mu-promoted integration of circular DNA	Mu *A*, *att* sites	Mu *B*	Infecting phage, or prophage, in the host or plasmid chromo- some	Two prophages with same orien- tation, flanking the inserted DNA
	Circular DNA mole- cule	host *recA*		
Mu-promoted transposition of chromosomal DNA	Mu *A*, *B* *att* sites	–	Prophage or infecting phage	Two prophages with same orien- tations, flanking the transposed sequences
	–	host *recA*		
Internal Mu deletion	?	?	Prophage	Prophage con- taining internal deletions of up to 78% of Mu se- quences. Resulting phage can be de- fective or viable
	?	?		

Functions or sites required or non-essential for these events are listed with the Mu functions/sites above the host functions/sites for each type of event. Although not required, Mu *B* increases the frequency of phage integration (Toussaint, personal communication)

Information shown in parentheses is conjectural

att = specific Mu termini

[a]host *dnaA* is not required, but host *dnaB*, *dnaC* and *dnaE* are required for Mu replication, which may preceed some of the above events

Final state of recipient chromosome	Frequency	References
Infecting phages promote plasmid integration at random chromosomal loci. In induced Mu lysogen, the plasmid is integrated at the original prophage site	10^{-4} events/cell in induced Mu X, *cts* lysogens	Howe and Bade (1975) Faelen et al. (1975) Toussaint et al. (1977) Faelen et al. (1978)
Transposed sequence flanked by 2 prophages and inserted into a new chromosomal locus	For any given gene, $2 \cdot 10^{-4}$ events/cell in induced Mu *cts* lysogen	Toussaint et al. (1977) Faelen et al. (1978)
Unaltered	?	Chow et al. (1977) Faelen et al. (1978)

events. Mu phage can insert as a discrete unit in either of two
physical orientations within practically any locus, be it on a
plasmid, phage or bacterial chromosome. However, due to the in
situ replication/transposition lytic process characteristic of
induced Mu *cts* prophages, reversion of Mu-induced mutations occurs only infrequently. Be that as it may, both precise and imprecise excision of Mu *X* prophages have been detected and these
events require at least Mu gene *A* function (Bukhari 1975; 1977;
Toussaint, personal communication). Also, phage viability or
host range is controlled by the orientation of the internal invertible G segment relative to the surrounding Mu sequences. Mu
integration/transposition and G loop inversion events are mediated by specialized recombinational processes that are encoded by
Mu. Mu prophage excision involves an as yet uncharacterized recombination event which appears to be enhanced by the presence
of *recA* protein (Bukhari 1975).

As a consequence of the ability of Mu phage to replicate in situ
and continuously transpose to different chromosomal sites during
lytic growth, Mu causes a variety of aberrant host chromosomal
rearrangements that ordinarily do not occur or which are detected
at a much lower frequency in the absence of Mu (Toussaint et al.
1977, see Table 2). These events, all of which can occur in the
absence of host Rec ability, include host chromosomal deletions,
transpositions, and inversions, as well as the Mu-mediated chromosomal integration of autonomous circular DNA. Though the specific
molecular mechanisms involved are not known, Mu gene *A* function
together with the terminal Mu recognition/attachment sites are required to promote all of these events except Mu-mediated transposition of host sequences which apparently also requires the
"replication" function specified by Mu gene *B* (Faelen et al.
1978; O'Day et al. 1978). Furthermore, besides these and possibly
other Mu-encoded proteins, the Mu genome is a direct physical
participant in both host Rec-independent and Rec-dependent chromosomal alterations. In addition to mediating host chromosomal
rearrangements, Mu phage can undergo internal deletions entirely
within the phage genome (Chow et al. 1977; Faelen et al. 1978).
Therefore, besides promoting phage integration/transposition
events, the Mu-mediated specialized recombination system(s) is
involved in mediating macro-evolutionary chromosomal alterations,
as described below and illustrated in Figures 9 and 10.

a) Deletions. Mu-promoted host chromosomal deletions can be generated by different mechanisms (see Table 2). Imprecise prophage
excision results in the deletion of host sequences at the host
chromosomal insertion site of Mu. This inexact excision of Mu
prophages occurs at a relatively low spontaneous frequency
(10^{-7}-10^{-5} events/cell) in a lysogenic bacterial population and
involves the removal of the Mu DNA accompanied by some adjacent
bacterial DNA sequences, frequently from both sides of the prophage insertion site (Bukhari 1976; Toussaint et al. 1977). Since
there are 10^5-10^6 viable phage per milliliter of an exponentially
growing, non-induced culture of a Mu lysogen (Howe and Bade 1975),
there must be a constant low level spontaneous induction of phage
functions. Though entirely speculative, inexact Mu prophage excision may sometimes involve the premature packaging of chromo-

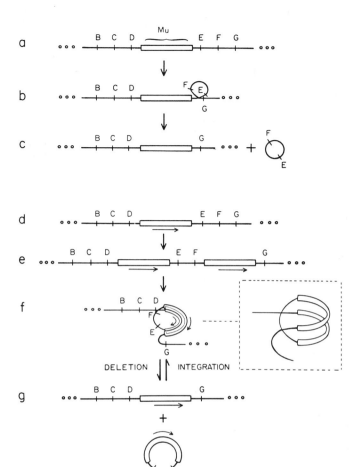

<u>Fig. 9.</u> Simplified hypothetical scheme for Mu-promoted deletion of host
chromosomal DNA. (a) The *thin black horizontal line* represents a portion of
a double-stranded bacterial chromosome which contains a Mu prophage, depicted by the *open rectangle*. Several hypothetical host genes are denoted
alphabetically and the Mu prophage is inserted between host genes D and E.
Many Mu-promoted host DNA deletions apparently occur from either Mu prophage
terminus to random points within the adjacent chromosomal DNA and the prophage always remains intact. These Mu-mediated deletions might result from
a recombinational exchange between either Mu terminus and nearby host sequences. (b) Deletion of host genes E and F could result from a reciprocal
genetic exchange between a Mu terminus and a chromosomal site situated between genes F and G. (c) The resulting chromosome would contain a deletion
of some host sequences immediately adjacent to one prophage terminus. The
deleted material might exist as a non-replicating small circle which would
be diluted out of the cell population during growth. Although this scheme
allows one to conceptualize how Mu might mediate deletion formation, the
more sophisticated model illustrated in Fig. 10 (b,h,k and l) is a more
probable mechanism for this reaction.

somally linked prophage with concomitant closure of the host chromosome.

In contrast to these low frequency deletion events, Mu can promote deletions at much higher frequencies both during lysogenization or following partial induction of a thermoinducible prophage. Normally, during lysogenization about 15% of the inserted Mu prophage cause host chromosomal deletions that occur immediately adjacent to the prophage (see Howe and Bade 1975). Similarly, some of the survivors of partial induction of a Mu *cts* lysogen have been found to contain deletions of host sequences to either side of the prophage. In these instances, the prophage always remains intact and the deleted sequences span from either prophage end to a seemingly random point on the host chromosome. These recombinational events require the physical presence of a Mu prophage and can occur in the absence of host general recombination. It is probable that these latter events involve a common mechanism in which a Mu prophage(s) somehow undergoes an exchange between a Mu terminus and some nearby bacterial DNA sequence (Howe and Bade 1975; Toussaint et al. 1977). The mechanism of this deletion formation has not been elucidated. However, the simple reciprocal exchange between one Mu terminus and some adjacent bacterial sequences as illustrated in Fig. 9b,c offers, at least, a visual conception of this event. A less simple scheme proposed by Faelen and Toussaint (1976) is more likely to direct the deletions described above (see Fig. 10b,h,k,l). Their propo-

---

(Legend to Fig. 9. continued)
It is important to note that although the above events do not require participation of the host Rec system, deletion formation can nevertheless occur between nearby Mu prophage via host chromosome-mediated general recombination. (d,e) Partial induction of a Mu monolysogen could result in the formation of a dilysogen. (f) Recombinational crossover between the nearby prophages could occur at any point along the paired Mu genomes. (g) In this example, general recombination between the identically-oriented, paired, nearby prophage resulted in the same chromosomal product as shown in (c). However, the deleted material now consists of all of the host DNA sequences initially located between the prophages (i.e., host genes E and F) as well as one entire Mu genome.

As shown in diagrams (g) to (f), the Rec-dependent integration of circular DNA into the chromosome can occur if both circular DNA's contain a Mu prophage. Though not shown, deleted host sequences like those depicted in step (g) could be reinserted at a new chromosomal prophage locus, an event that would constitute site-specific, Rec-dependent translocation of host DNA. Thus, Rec-dependent integration occurs by the apparent reversal of the deletion process and Rec-dependent transposition encompasses a deletion event followed by reintegration of the deleted sequences at a new chromosomal locus.

Imprecise prophage excision (not shown above) can also generate deletions of adjacent host sequences, frequently from both sides of the prophage insertion site. In contrast to the above events, the prophage is always removed during inexact excision. Although entirely speculative, the inexact excision may involve the premature packaging of chromosomally linked prophage

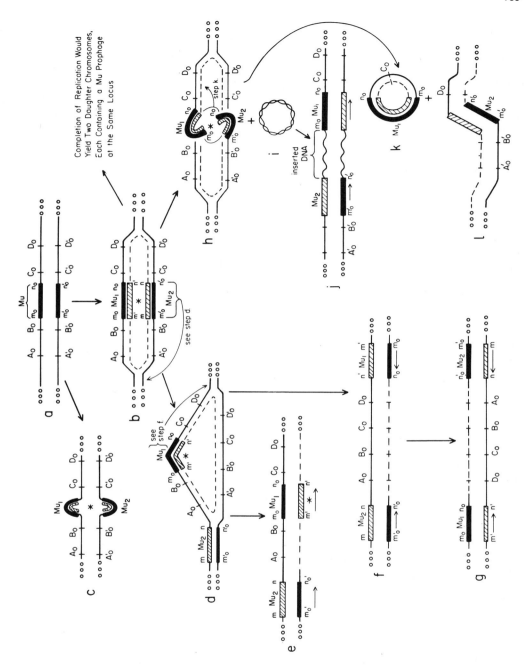

Fig. 10. Mu-mediated aberrant chromosomal rearrangements. Bacterial or plasmid single-stranded DNA sequences are represented by *thin horizontal lines* and Mu prophages are depicted by *rectangular boxes*. *Dashed lines* or *rectangles* with enclosed *slanted lines* indicate newly synthesized bacterial or phage DNA strands, respectively. Bacterial genes A, B, C and D, shown on one

170

(Legend to Fig. 10 continued)
DNA strand, and the complementary sequences (A', etc.) are labeled with *sub-script zero* to indicate the initial sequences of the unreplicated parental molecule. Similarly marked are the specific Mu termini at the immunity and variable ends, which have been labeled *m* and *n*, respectively, in keeping with the terminology of Faelen and Toussaint (1976). As described below, most chromosomal rearrangements promoted by Mu can be explained on the basis of in situ prophage duplication coupled with subsequent prophage recombination events. The hypothetical scheme illustrated here is an adaptation and extension of the models previously proposed by Faelen and Toussaint (1976) and Toussaint et al. (1977). (a) A portion of bacterial chromosomal double-stranded DNA with a Mu prophage inserted between bacterial genes B and C. (b) During chromosome replication the segment carrying the prophage can be duplicated, giving rise to two prophages in opposite arms of the replication fork. Newly replicated Mu DNA appears to be the "active" form upon which the Mu integration function can act and these recombinationally active forms of Mu are denoted with an *asterisk*. Perhaps during growth, Mu DNA is specifically modified (e.g., by nucleotide methylation) at some point in both strands and the modified Mu cannot be attacked at the specific termini by the Mu integration enzyme. However, following semi-conservative replication of a Mu prophage, the integration enzyme might be able to recognize and act upon the termini of Mu that are modified in only one strand.

The prophages in (b) are depicted as having been replicated during normal chromosome duplication. Host-determined replication is sufficient to create recombinationally active prophages that can promote most genomic rearrangements. Completion of a round of replication would generate two daughter molecules, each containing prophages in the same location. In addition to Mu being replicated along with the host genome, specific Mu-determined prophage replication can not only occur, but is required for Mu-mediated transposition of host DNA. (c) Mu-specified replication of the Mu prophage would generate recombinationally active Mu prophage, which could also cause the same chromosomal rearrangements as shown in the following steps. However, the structures shown in (b) and (h) more easily convey a picture of the "recombinationally active", newly replicated, daughter Mu prophages. (d) This example illustrates the specific translocation of the lower prophage, Mu_2, shown in (b) to a point outside of the replication fork. Mu transposition might occur, as shown, by the recombination of the Mu_2 termini with a DNA sequence to the left of gene A. Although Mu could be inserted in either orientation, I have chosen to show a direct transposition. (e) Following either degradation of the replication fork from which Mu_2 was deleted or completion of a round of replication, the chromosome would be dilysogenic with both prophages in the same physical orientation as depicted by the *horizontal arrows*. It is important to recognize that recombinational activation of the Mu termini apparently accompanies prophage replication. However, Mu transposition rarely causes a reversion of the original Mu insertional mutation, evidently because only one daughter prophage is commonly transposed. Alternatively, transposition may occur by single-strand exchange and always leaves Mu at the original site. (f) In addition to the transposition event shown in (d), precise prophage excision might result from the subsequent transposition of the upper prophage, Mu_1 (also shown in d) to some other chromosomal site. In this case, Mu is transposed in inverted orientation, rightward to a point adjacent to the replication fork. Upon dissolution of one strand of the replication fork or completion of host genome replication, the chromosome would be dilysogenic with the two prophages in opposite orientations (f). As a consequence of precise prophage excision, the host sequences at the initial prophage insertion site, as

(Legend to Fig. 10 continued)
shown in (a), are restored. (g) Host sequences flanked by nearby prophages in physical apposition, as shown in (f), are substrates for inversion. This diagram illustrates the product of an inversion which occurred between the outer termini of the prophages. Inversion of the enclosed host and phage sequences has also been observed between the oppositely oriented G segments of two identically oriented, nearby prophages (Faelen et al. 1977; not shown).

"Active" Mu termini, although attached to random bacterial DNA during integration or transposition, can insert into virtually any chromosomal locus (see Fig. 8). Because of the proximity to each other of the newly duplicated prophages shown in (b), the integration enzyme complex may recognize the m end of one prophage together with the n terminus of the opposite prophage, and recombine these opposing ends as shown in the following steps. (h) This schematic shows the interacting opposite ends of the two prophages in a "fused" state, ready to recombine with any DNA sequence. In order to emphasize the interacting ends of the opposing prophages, one end of each prophage is illustrated as being disconnected from its corresponding arm of the replication fork. However, the interaction of the two opposing prophages may not initially require such disconnections. Toussaint, Faelen and co-workers have ingeniously deduced that "fused", newly replicated prophages, as shown in (h), can react with circular DNA to promote its insertion or can cause the deletion of adjacent host bacterial sequences (Faelen and Toussaint 1975; Toussaint et al. 1977). (i) A covalently closed, circular double-stranded DNA molecule, such as $\lambda dgal$ or a bacterial plasmid, is represented by the *squiggled line*. (j) Linear insertion of the circular DNA in any permutation would occur by recombination between the "fused" prophage termini, shown in (h), and the circular DNA. The resulting chromosomal insertion would occur at the original prophage insertion site on the host chromosome and the inserted DNA would be flanked by identically-oriented Mu prophages. Note that similarly oriented, nearby prophages are susceptible to host Rec-dependent recombination and could result in deletion of one phage and the interposed host sequences, as shown in Fig. 9g. Instead of recombining with an autonomous circle, if the "fused" prophage termini (h) had recombined with some nearby chromosomal sequence, a site-specific deletion would be generated. (k,l) Recombination between bacterial sequences C_O and D_O would result in the deletion of a circle (k) containing one prophage, Mu_1, and the DNA sequences immediately surrounding C_O. Similar circular molecules have been observed following Mu prophage induction and may be the structures from which mature phage containing heterogeneous termini are obtained for virion packaging. Following the deletion of Mu_1 and bacterial gene C_O (h), chromosomal circularity would be reestablished by the joining of Mu_2 to a point close to the sequences labeled D_O (l). Exonucleolytic degradation of the $-(A_O-B_O)-$ and $-(C_O'-D_O')-$ arms of the replication fork (h) would result in a bacterial chromosome containing a deletion of host sequences occurring immediately adjacent to the remaining prophage, as shown in (l). Though not shown, recombination between the "fused" Mu termini (h) and a point outside of the replication fork would generate circles with tails, similar to those observed following prophage induction (Toussaint et al. (1977).

Transposition of host sequences might require two of the above described recombination events. First, newly replicated prophages could cause the deletion of a circle comprised of one prophage and some adjacent host sequences, as shown in (k) and (l). Secondly, specific Mu-determined replication of the prophage in this circle would generate a Mu-Mu structure similar to that depicted in (c). "Fusion" of the m and n termini of opposite

(Legend to Fig. 10 continued)
prophages and recombination with the host chromosome at any locus would
result in the transposition of bacterial sequences. The transposed sequences
would be flanked by identically-oriented Mu prophages just as shown in (j)

---

sal takes into account the currently known properties of Mu, especially its ability to integrate randomly, and the fact that Mu-
mediated deletions do not require the Mu B "replication" function.
Simply stated, host chromosome-determined replication of a prophage would generate two prophages on daughter arms of the replication fork, as shown in Fig. 10b. "Fusion" of these prophages
into a structure resembling a directly repeated tandem Mu dimer
would generate a "recombinationally active site", the adjacent
Mu termini (see Fig. 10h). This "active site" could recombine
with any nearby host chromosomal sequence, always resulting in
the deletion of one Mu genome plus some adjacent host DNA and
leaving one intact Mu prophage at the site of the deleted sequences (see Fig. 10k,l).

All of the above deletion events can occur via specialized recombination. However, it is very important to emphasize that
due to the relatively large size of Mu DNA, host Rec-dependent
exchange between Mu prophage can also generate host chromosomal
deletions (Faelen et al. 1977). Spontaneous induction of a Mu
prophage could generate identically oriented, nearby prophage
which flank some host genomic sequences (see Fig. 9l). Rec-dependent random genetic exchange between paired, nearby prophage
genomes (Fig. 9f) would result, like the phage termini-specific
exchange described in Fig. 10k,l, in the deletion of one Mu genome and any interposed bacterial sequences (see Fig. 9g). Thus,
the integration via specialized recombination of two or more Mu
into nearby regions of the bacterial chromosome results in the
prerequisites (i.e., large homologous DNA regions) for Rec-dependent recombination. In other words, specific DNA segments like
Mu phage or transposons (see later section) which are recognized
and exchanged by specialized recombination enzymes can also be
substrates for homologous recombination systems.

Aside from mediating deletions in recipient chromosomes, Mu prophages have been observed to undergo internal deletions of Mu
sequences. Recently, Faelen et al. (1978) have isolated, by thermal induction of Mu cts lysogens, mini-Mu mutants which contain
large internal deletions of up to 78% of the Mu genome. These
mini-Mu phage still carry the Mu terminal recognition/attachment
sequences and can be propagated in the presence of helper phage.
Furthermore, when induced these mini-Mu phage can transpose and
promote chromosomal rearrangements. Another thermoinducible Mu
lysogen, from which non-defective internal deletion mutants have
been obtained, contains an unusual 2.6 kb insertion, which has
been identified as insertion sequence elements IS2 and IS5 (Chow
and Broker 1978). Although the specific mechanism of Mu phage
internal deletion formation is not known, all such deletion mutants, to date, have been found to contain insertion sequence
elements that may be responsible for these deletion events (see
Sect. IV.2.b.). The involvement of Mu genes in the formation of
these internal deletions is presently unknown (Chow et al. 1977).

b) Integration of Circular DNA. Host chromosomal integration of circular DNA molecules, such as λ*dgal* or bacterial plasmids, can be mediated by Mu (Faelen et al. 1975; Howe and Bade 1975; Toussaint et al. 1977). This process, outlined in Table 2, can occur independently of host Rec ability, but requires Mu gene *A* function, as well as the physical participation of the Mu genome itself. Mu-mediated integration occurs at frequencies as high as one per 10^4 Mu *cts* phage in an induced population of monolysogens. The product is always the linear insertion within the recipient chromosome of the circular DNA molecule flanked on each side by one entire Mu genome, with both flanking prophages in the same orientation (Faelen et al. 1978). Considering that induced Mu *cts*, B^- (non-self-replicating) monolysogens have been observed to promote the integration of circular DNA (Faelen et al. 1975), how is the resultant second flanking prophage synthesized? Faelen et al. (1975) have proposed that two daughter Mu prophages, formed by normal chromosome replication of a preexisting prophage, can interact with one another to promote the integration of circular DNA, as illustrated in Fig. 10b,h-j. When integration is promoted by an induced Mu prophage in a monolysogen, the resulting "Mu-inserted DNA-Mu" chromosomal complex is always located at the original prophage insertion site. In contrast, infecting Mu phage can mediate the linear insertion of circular DNA into virtually any chromosomal site. However, in both cases the integrated DNA is inserted with any circular permutation. Although not described in Fig. 10, the above chromosomal integration event could also be mediated by a plasmid-borne Mu prophage. In this case, however, the duplicated daughter prophages, formed during plasmid replication, would promote the insertion of the plasmid with only one permutation into any host chromosomal locus.

Though Mu-mediated integration of circular DNA may occur by the process described in Fig. 10, an alternative mechanism has been proposed. Faelen and Toussaint (1976) previously theorized that infecting Mu phage might somehow dimerize by the fusion of two vegetative phage genomes. The resulting directly-repeated, tandem, Mu, circular dimer would contain two sets of hyperactive, fused Mu termini which could mediate the co-integration of two circles at random points. Finally, one should be cognizant of the fact that if both the bacterial chromosome and any circular DNA molecule simultaneously carry Mu prophages, Rec-dependent recombination between the paired prophages can promote chromosomal integration of the circular DNA (see Fig. 9g and legend).

c) Mu-mediated Transpositions. Mu phage has the remarkable ability to mediate the transposition of any DNA segment from one location to another on the same or a different molecule (see Table 2). Either an induced lysogen or an infecting phage can effect transposition of a chromosomal segment. The frequency of transposition for any specific chromosomal gene can be as high as 2×10^4 events/induced Mu *cts* lysogen (Faelen and Toussaint 1976). Assuming that the *E. coli* genome comprises about 3000 genes, then in an induced, Mu-lysogenic population a phenomenal one out of every two cells should contain a transposed DNA sequence. The end product is identical to that obtained in Mu-mediated plasmid integration, i.e., the transposed DNA segment is flanked by two

Mu prophages in the same orientation. Additional evidence indicates that: only closely linked markers are cotransposed; all
genes situated between two cotransposed genes are simultaneously
cotransposed; the cotransposition frequency of an unselected
marker is related to its distance from the selected marker; and
transposition of segments as large as 3.5 min of the *E. coli*
chromosome or approximately 90 megadaltons (135 kilobase pairs)
has been detected (Faelen and Toussaint 1976). Although transposition events, for all practical purposes, would be undetectable within a single cell, transposition of chromosomal genes
has been readily assessed in cells carrying the F plasmid. Following induction of a cell population lysogenic for Mu *cts*, various segments of the host chromosome are transposed onto the F
plasmid. Specific transposed genes can then be detected after
conjugal plasmid transfer to suitably marked recipient cells.
Unlike most other Mu-mediated specialized recombinational events
which require only Mu *A* function in addition to the specific Mu
termini, transposition of host sequences also requires the Mu
gene *B* (presumptive replication) function.

Albeit the exact mechanism of Mu-mediated transposition has not
been elucidated, Toussaint et al. (1977) have developed a model
to explain this process. Accordingly, Mu-mediated transposition
of host sequences is a two step recombinational process. Initially, newly replicated prophages interact, as shown in Fig. 10h,
and promote the deletion of a circle composed of one prophage
and some adjacent host sequences (see Fig. 10k,l). Subsequently,
specific Mu-determined replication of the prophage in this circle
(Fig. 10k) would generate a forked Mu-Mu structure similar to
that illustrated in Fig. 10c. The second recombinational event
would entail "fusion" of one *m* and one *n* terminus, each from opposite daughter prophages on the plasmid, and reciprocal genetic
exchange of this recombinationally active site (i.e., the adjacent Mu termini) with the host chromosome at any locus. This
process would result in the integration of the entire circle into the chromosome. The resulting transposed bacterial sequences
would be flanked by identically oriented Mu prophages just as
shown in Fig. 10j. Although newly transposed or integrated
DNA segments are hereditarily stable in a Rec-deficient host,
it is interesting to note that because of the large size of
the flanking homologous Mu genomes, these insertions are
deleted at a frequency of about 1% from Rec[+] hosts (Faelen
et al. 1975; see Fig. 9e-g). In the absence of Mu specialized
recombination systems, the host general recombination system
may potentiate Mu-mediated transposition as described in the
legend to Fig. 9. In contrast to specific transposition of the
Mu genome that occurs following Mu induction and in which one
Mu prophage generally remains at the original insertion site
(Fig. 10d,e), Mu-mediated transposition of host DNA appears to
involve the absolute deletion of host sequences from one area
of the chromosome followed by their insertion into a different
chromosomal locus. However, current evidence does not eliminate
other explanations. Transposition of bacterial sequences that
are flanked by directly repeated Mu prophages might occasionally
involve recognition of the opposite ends of these prophages so
that the recombinational exchange would mimic that observed in

specific Mu phage transposition in which one copy of the transposed segment always remains at the initial site.

DNA sequence transposition, in general, allows for the movement of relatively large, complex DNA segments to other chromosomal sites, sometimes placing them under different genetic regulatory controls. Also, transposition of large DNA segments to plasmid or phage vectors can promote rapid intercellular dissemination of hereditary information. Fusion of various genes to novel genetic promotors (Casadaban et al. 1977) and the in vivo cloning of a gene via Mu-mediated transposition onto a plasmid (Denarie et al. 1977; Faelen et al. 1977) have been successfully employed in recent genetic studies.

d) Mu-mediated DNA Inversions. Mu can mediate the inversion of adjacent chromosomal DNA sequences by two different mechanisms. Inversion in an induced Mu X lysogen occurs at a frequency of 10^{-4}, requires Mu gene A product, and the inverted chromosomal sequence is usually flanked by two Mu prophages in opposite orientations (Toussaint, pers. commun.; Faelen et al. 1978; depicted in Fig. 10f,g). The prophages observed on both sides of the inverted host DNA sequence may not be a prerequisite for inversion, but may in some way be a consequence of the inversion process, as is the case for Mu-mediated insertion or transposition. Also, in addition to promoting G loop inversion, the Mu *gin* function can apparently promote inversion of all host and phage sequences located between two opposing G segments in nearby prophages (Faelen et al. 1977).

6. *Effect of Mu on Bacterial Evolution*

Temperate bacterial viruses are usually limited in host range to one or two genera because of limited bacterial species carrying the proper cellular surface receptors and/or the inability of the phage, once injected, to lysogenize or replicate. Though Mu will infect *E. coli* K-12, *Citrobacter freundii, Shigella dysenteriae,* and some strains of *Klebsiella pneumoniae,* Mu phage do not form plaques on *E. coli* C, B, S, or W, *Salmonella typhimurium* or other related enterobacteria. However, following integration of Mu into the conjugally promiscuous RP4 plasmid, this plasmid has been successfully employed to introduce Mu via conjugation into many different hosts. Recently, Mu has been observed to replicate, to lysogenize, and to promote chromosomal rearrangements in many diverse gram-negative bacteria which are not ordinarily susceptible to Mu infection (Denarie et al. 1977).

Temperate viruses that can physically integrate into the host chromosome are apparently each recognized by a specialized recombination system. Following virus integration as discrete DNA units at one (e.g., λ), a limited number of (e.g., P22), or many (e.g., Mu) chromosomal recognition/attachment sites, prophages can subsequently cause localized mutagenesis (e.g., host DNA deletion). Unlike other temperate phage, however, Mu and phage D108, which share most of their DNA sequences, can be inserted in either physical orientation at any chromosomal site, an event causing either simple gene inactivation or a strongly polar mutation in an operon (Hull et al. 1978; Kamp et al. 1979). In addition, Mu specialized recombination systems mediate a variety of host genomic rearrangements. Therefore, Mu and λ may represent opposite ends of a spectrum of site-specific or specialized recombination systems, each of which recognizes a specific class of DNA elements (e.g., λ, insertion sequence elements, or Mu) and which can effect their specific insertion into only one, a few, or many recipient DNA sites. For the interested reader, many of the properties of Mu have been reviewed in detail elsewhere (Howe and Bade 1975; Bukhari 1976, 1977; Couturier 1976; Bukhari et al. 1977). One wonders how many more Mu-like viruses exist in nature. Certainly these viruses or their remnants within a chromosome must be responsible for a significant proportion of the genetic flexibility of many bacteria.

Recently, an increasingly large group of transposable genetic entities has been detected in many diverse bacterial genera. These elements do not appear to replicate in a physically autonomous state like a plasmid and cannot exist extracellularly like a virus, but they can promote a variety of host genetic alterations. As described in the following sections, these transposable elements appear to be functionally related to λ and Mu viruses in that all are recognized as discrete DNA units by specialized recombination processes.

IV. Transposable Genetic Elements

1. Definition

During the past decade, a variety of unique transposable DNA segments have been identified in the chromosomes of bacteria, their plasmids, and viruses. Apparently common constituents of chromosomes, these elements have been detected by their transposition to and inactivation of a known gene. Beyond their simple transposition to a chromosomal locus, these remarkable elements cause a variety of macro-evolutionary chromosomal rearrangements, similar to those promoted by the bacterial virus Mu. Furthermore, some of these transposable DNA units encode transcriptional initiation and termination signals and can act as supernumerary regulatory switches affecting gene expression (see reviews by Starlinger and Saedler 1972, 1976; Cohen and Kopecko 1976; Kleckner 1977; Starlinger 1977). The systematics and nomenclature of these elements have been summarized by Campbell et al. (1979).

These discrete DNA segments which range in duplex DNA length from about 750 to 80,000 nucleotide base pairs are structurally defined by repeated DNA sequences at their termini and are normally transposed intact as distinct, non-permuted units from one location to another on the same or a different molecule. No transposable element has yet been shown to replicate or to exist in a separate, physically autonomous state (i.e., they are al-

ways linearly inserted within a chromosome). As a probable consequence of their repetitious terminal sequences, these DNA
units can insert with either of two physical orientations. In
addition to sharing the common feature of terminal sequence repetition, all of these genetic elements are transposed independently of general recombination systems. Thus, these DNA
segments appear to transpose via specialized recombination
events. Transposable DNA elements have been identified, both
genetically and physically, in a large variety of different
bacterial genera, strongly suggesting their universal existence
in bacteria.

It is necessary to define certain terminology before proceeding
further. The exchange of a DNA segment between non-homologous
chromosomes or non-homologous regions of the same molecule has
been commonly termed either *transposition* or *translocation*. Though
both words seem equally appropriate, the term *translocation* has
also been used more recently by biochemists to describe the pas-
sage of molecules across a membrane or to denote the movement of
peptidyl tRNA from the "A" to the "P" site of the ribosome during
translation (Watson 1976). In order to avoid unnecessary con-
fusion, it would appear expedient to refer to the exchange of
a DNA segment between non-homologous DNA regions as *transposition*.
For this reason, I have exclusively used the term transposition
in the remainder of this review. The term *transposable elements*,
originally used to define mobile genetic elements in maize
(McClintock 1952), has been used colloquially to refer to all
transposable genetic elements in eukaryotic and prokaryotic
systems. However, in this section "transposable DNA elements"
refers to a set of structurally defined prokaryotic DNA units
that exist only in the integrated state. Transposition of a
discrete transposable DNA element, presumably directed by specialized recombination, has been termed, variously, site-specific
transposition, site-specific recombination, or simply genetic
transposition. Though transposable elements are inserted as discrete units, their insertion has been observed to occur into
many recipient chromosomal receptor sites and the term "site-
specific" now seems too constrictive. Also, it should be strongly emphasized that transposition or deletion of specific chromosomal DNA segments can occur via general recombination (i.e.,
between nearby homologous DNA regions; see Fig. 7). Perhaps the
general term "genetic transposition" should be used to refer to
the in vivo movement of DNA from one site to another regardless
of Rec-dependence. *"Specialized"* transposition might be employed
to describe those events which are mediated by specialized recombination. Finally, the endproduct of most specialized transposition events appears to be different from Rec-dependent transposition. Like the replication/transposition process of an induced Mu prophage, specialized transposition of a transposable
DNA element from one site to another does not cause loss of the
transposable element at its original locus (i.e., transposition
is not linked to precise excision). In other words, only a single
DNA strand of the element or a newly replicated transposable segment appear to be substrates for specialized transposition. In
contrast, as described in the legend to Fig. 9, general recombi-
nation-dependent deletion can occur between homologous DNA regions
bracketing any interposed sequence. Subsequent Rec-dependent in-

sertion of this deleted material could potentially occur at any chromosomal region containing significant homology to the deleted circular molecule (see Fig. 9g). Likewise, Mu-promoted transposition of chromosomal sequences appears to involve Mu-mediated deletion of the sequences, followed by Mu-promoted insertion of these sequences into a new site (see legend to Fig. 10). Therefore, specialized genetic transposition of a specific transposable element does not appear to result in loss of the transposable element at its original site whereas in Rec-dependent or Mu-promoted transposition of chromosomal sequences, the segment to be transposed is apparently first deleted from one chromosomal site and subsequently inserted elsewhere.

Transposable elements have been divided into two classes, both for historical and structural reasons. The small insertion sequence (IS) elements, which were discovered in the late 1960's and do not encode any known phenotypically identifiable proteins, represent one class (Saedler and Starlinger 1967; Jordan et al. 1968; Shapiro 1969). The second group is exemplified by the relatively large transposable elements that encode resistance to various antibiotics and which were physically identified in the mid-1970's (reviewed by Cohen 1976; Cohen and Kopecko 1976). These large elements, seemingly more complex than IS units, are comprised of a central sequence bracketed by direct or inverted DNA sequence repeats, sometimes consisting of a bona fide IS element. More recently, genes encoding toxin production and a variety of metabolic capabilities have been identified on these larger transposable elements. Although this classification scheme for transposable elements is tenuous, as you will see, it does correctly emphasize the hierarchy of structural complexity in transposable elements. Regardless of complexity, however, all of these discrete elements are responsible for a large proportion of chromosomal rearrangements in bacteria.

2. IS Elements

Analysis of the fine-structure organization of information in DNA was carried out during the 1960's through the genetic and biochemical examination of both spontaneous and induced chromosomal mutations. A considerable fraction of the spontaneous mutations studied in bacterial or phage genomes proved to be quite unusual. Unlike base substitution mutations, these novel mutations exerted a very strong polar effect on the expression of more promotor-distal genes in an operon, similar to the effect of frameshift mutations. Reversion of these unusual mutations was not enhanced by mutagens that normally cause base substitutions or frameshift mutations. Moreover, these unusual mutations reverted spontaneously to wild type, ruling out nucleotide deletion and suggesting insertion or inversion as the original alteration. Comparison, by a variety of physical measurements, of parental lambda transducing phages to mutant lambda transducing viruses containing different novel polar mutations indicated the presence of a sizable insertion in each mutant virus and essentially eliminated genetic inversion as the cause of these mutations (reviewed by Starlinger and Saedler 1976).

By DNA-RNA hybridization studies, Michaelis et al. (1969) showed that the inserted sequences in these lambda mutants were not DNA sequence duplications, but rather DNA unlike the wild-type sequences. These insertion mutations which were later measured for length in the electron microscope and for homology with one another by the DNA heteroduplex procedure (Fiandt et al. 1972; Hirsch et al. 1972; Malamy et al. 1972), are now known to comprise several distinct classes (Table 3). Therefore, these novel mutations in *E. coli* and lambda were shown to be caused by a few distinct IS elements. Starlinger and Saedler (1972, 1976) have composed excellent reviews that explore the genetic and initial physical characterization of IS elements. As discussed below, these elements have now been characterized by a variety of genetic and biochemical methods.

a) Some Genetic and Molecular Properties. The general properties of IS elements are summarized below and in Table 3. IS elements are DNA segments that range in length from 768 to about 1400 nucleotide base pairs or more and are found in bacterial, phage and plasmid chromosomes. These discrete units can be transposed in either of two physical orientations as a linear, non-permuted unit to a recipient chromosomal site on the same or a different molecule. IS elements are usually detected by the characteristic effects that they exert at new sites, i.e., abolition of the function of a gene and alteration of the expression of promotor-distal genes in an operon (i.e., polarity).

Regarding *E. coli* and its plasmids and viruses, insertion mutations occur in many different locations of these genomes (e.g., at multiple sites within different genes in the well-studied *E. coli lac* or *gal* operons). Thus, statistically speaking, the recognition sequence(s) on the recipient molecule must be fairly short, as opposed to a longer sequence which would be duplicated less often around the chromosome. For example, a specific sequence of five nucleotides should recur on a random basis about every 1000 bases. However, it should be noted that for as yet undetermined reasons, certain chromosomal regions appear to be preferred areas for insertion. For example, in separately isolated mutants, IS elements have inserted at very close (i.e., several nucleotides apart) and at identical sites within the 200 base pair *gal* operon control region (Saedler et al. 1972; Kuhn et al. 1979). It appears that IS element integration is neither entirely random like the integration of Mu phages, nor as specific as the integration of bacteriophage λ into the primary *att*B locus. However, though IS elements appear to have a preference for certain DNA regions, insertion in different mutants occurs at many nearby sites within these defined areas (i.e., receptor site clusters) and also at many other sites around the genome. Thus, unlike the case of λ which, when the primary *att*B site is deleted from the bacterial genome, integrates into several different, but specific secondary *att*B sites, IS units express what has been termed "regional" specificity, i.e., the ability to integrate as a discrete unit in many sites within a preferred short recipient DNA segment.

Table 3. Insertion sequence elements in bacteria

IS element designation	Total Tn Length (b.p.)	Length and orientation of internal terminal repeat sequences (b.p.)	Restriction enzyme suseptibility	
			Cleaved by	Not cleaved by
IS*1*	768	28 of first 34 b.p. form inverted repeat	*Alu*I *Bal*I *Hae*II, III *Hha*I *Hin*fI *Hpa*II *Hph*I *Pst*I	*Bam*HI *Bgl*I, II *Eco*RI *Hinc*II *Hin*dII, III *Hpa*I *Xba*I *Xho*I
IS*2*	∿1327	32 of first 41 b.p. form inverted repeat	*Ava*I *Bgl*I *Hae*II, III *Hin*dII, III *Hin*fI *Hpa*II *Hha*I *Mbo*I *Sma*I *Taq*I	*Eco*RI *Bam*HI
IS*3*	∿1200	?	*Hin*dIII *Pst*I	*Bam*HI *Eco*RI *Hin*dII
IS*4*	1400	16 of first 18 b.p. form inverted repeat	*Ava*I *Hin*dII	–
IS*5*	1250	<50 b.p., inverted	*Eco*RI	–
γ–δ (Tn*1000*)	∿5800	35 b.p., inverted; terminal 28 b.p. identical to Tn A termini	*Eco*RI *Bam*HI *Hin*dIII *Kpn*I *Pvu*I *Sal*I *Sma*I *Sst*I *Xho*I	*Xba*I
IS*10*	1400	10 b.p. inverted repeat located 13 b.p. in from one end	*Acc*I	–

Effect of IS element orientation on transcription	Recipient genomes carrying IS units	Length of recipient DNA duplicated at ends of IS elements (b.p.)	Pertinent references
Polar in orientation I or II	*E. coli* *S. typhimurium; Citrobacter;* Phage P1, Mu, λ, T4; Plasmids R1, R6, F, R100, Ent	9	Jordan et al. (1968) Saedler and Starlinger (1967) Starlinger and Saedler (1976) Grindley (1978) Ohtsubo and Ohtsubo (1978) Hu et al. (1975a) Shapiro (1969) Fiandt et al. (1972)
Polar and contains rho-sensitive site in I and II; Promoter function observed in variants of I or II	*E. coli;* Phage λ; Plasmids F, R6, R100	5	Saedler and Heiss (1973) Saedler et al. (1974) DeCrombrugghe et al. (1973) Ghosal and Saedler (1978) Ghosal et al. (1979a,b) Pilacinski et al. (1977) Hu et al. (1975)
Polar in I; II not yet studied	*E. coli;* Plasmid F	3 or 4	Malamy et al. (1972) Hu et al. (1975b) Deonier et al. (1979) Sommer, Cullum and Saedler (personal communication)
Polar in I & II	*E. coli* *galT*	11 or 12	Habermann et al. (1979) Pfeifer et al. (1977)
Polar in I & II	*E. coli;* Phage λ		Blattner et al. (1974) Chow and Broker (1978)
?	*E. coli;* Plasmid F	5	Guyer (1978) Broker et al. (1977a) M. Guyer (personal communication)
?	Plasmids R6, R100	9	Kleckner (1979) Kleckner and Ross (1979)

Table 3 (continued)

IS element designation	Total Tn Length (b.p.)	Length and orientation of internal terminal repeat sequences (b.p.)	Restriction enzyme suseptibility	
			Cleaved by	Not cleaved by
"ISR1"	1100	–	*Bam*HI *Hin*dIII *Pst*I	–
Unassigned	500 – 1800	–	–	–

The first five IS elements have been convincingly demonstrated to undergo specialized transposition and precise excision in *E. coli*, as has the much larger γ-δ sequence, also termed Tn*1000*. Very recently, the 1400 b.p. repeat at the ends of the Tn*10* transposon has been shown to behave as an IS element and has been named IS*10* (Kleckner and Ross 1979). Also, an IS element has been physically and genetically characterized in *Rhizobium* and has been tentatively termed "ISR1". The variety of elements that are inferred from genetic data or visualized in the electron microscope as small inverted repeat DNA sequences, but which have not been proven to transpose are listed in the "unassigned" category. The DNA sequences of IS*1* and IS*2* have been analyzed entirely (Grindley 1978; Ohtsubo and Ohtsubo 1978; Ghosal et al. 1979b; Johnsrud 1979). Total IS element length is given in nucleotide base pairs (b.p.). If known, the length of the internal short sequence that is repeated in inverse order at the termini of some IS elements is also given. The physical orientation of each IS element within a chromosome, detected by physical and genetic means, has been arbitrarily designated I or II. The effect on promoter-distal gene expression by IS elements, when inserted in an operon, is given for insertions in both orientations. The list of chromosomes known to harbor each specific IS element or that commonly acquires each element was obtained from the more detailed catalog of specific IS-promoted mutations, which was compiled by Szybalski (1977). It should be noted that not all of the given recipient chromosomes normally carry an IS element. During IS element insertion and depending upon the specific IS unit, either a 3-4, 5, 9, or 11-12 base pair recipient chromosomal sequence is duplicated in direct order such that one copy occurs at each terminus of the inserted IS unit

Effect of IS element orientation on transcription	Recipient genomes carrying IS units	Length of recipient DNA duplicated at ends of IS elements (b.p.)	Pertinent references
?	*Rhizobium lupini*	?	Puhler and Burhardt (1978)
Polar effects observed for some elements	*E. coli,* *Salmonella,* *S. dysenteriae,* *P. aeruginosa,* *P. mirabilis,* *Streptomyces;* Phage λ, P2; Plasmids R68, ColV, pAMα1	?	Starlinger (1977) Ohtsubo and Ohtsubo (1977) Yagi and Clewell (1977) Schmitt et al. (1979a) Reiss et al. (1978) Szybalski (1977)

At present, there are six named IS elements, IS1 through IS5 and IS10 as shown in Table 3, plus an assortment of unclassified "insertion" mutations that have been identified in *E. coli* and many other bacterial genera, but have not yet been shown to transpose or delete as discrete units (Starlinger and Saedler 1976; see Szybalski 1977). Very recently, a small transposable element, tentatively named "ISR1", has been identified in *Rhizobium* and has been included in this compilation. In addition, I have chosen to include with the IS elements the gamma-delta (γ-δ) sequence which is a normal constituent of the *E. coli* and F plasmid genomes. This segment is now known to transpose *recA*-independently as a discrete unit (Guyer 1978). Although it is somewhat larger than the other IS elements, the γ-δ element appears to behave similarly to them in promoting genetic rearrangements. Unlike the larger transposable elements, it is not yet known to express any phenotypic functions and, thus, bridges the gap between the "simple" IS units and the "more complex" larger transposable elements. The average nucleotide base composition of IS1 and IS2 has been determined to be about 50% G+C, comparable to that of the *E. coli* genome (Schmidt et al. 1976; Ghosal et al. 1979b; Ohtsubo and Ohtsubo 1978). Early studies, with λ transducing phage DNA carrying IS1 or IS2, employing DNA-DNA hybridizations, as well as more recent hybridization and electron microscope heteroduplex studies have shown that in *E. coli* K-12 IS1 exists as about 8 copies per genome (Saedler and Heiss 1973), IS2 as $\geq$ 7 copies per genome (Saedler and Heiss 1973; Deonier et al. 1979), IS3 as $\geq$ 5 copies per genome (Deonier et al. 1979), γδ as $\geq$ 4 copies per genome (M. Guyer, personal communication), and IS4 and IS5 as $\geq$ 1 copy each per genome (see Starlinger and Saedler 1976). First discovered in the intensively studied *E. coli* genetic system, one or more classes of the bona fide IS elements have now been found in the related genera *Salmonella*, *Citrobacter*, and *Shigella*, on the phage and plasmids of enteric bacteria, and in the gram-positive bacteria *Bacillus subtilis* (Rak, cited in Starlinger and Saedler 1976; see review by Kleckner 1977; Saedler and Ghosal 1977). Due to the conjugal or infective promiscuity of various plasmids and phage, one would expect a wide dissemination of these and other transposable elements. In fact, IS-like sequences have been detected physically and/or genetically in the diverse genera *Shigella* and *Proteus* (Ohtsubo and Ohtsubo 1977) as well as *Pseudomonas* (Jacob et al. 1977; Reiss et al. 1978), *Streptomyces* (Bibb and Hopwood 1977), *Streptococcus* (Yagi and Clewell 1977), *Staphylococcus* (Novick et al. 1979), and *Rhizobium* (Puhler and Burkardt 1978). However, the sequence relationships among these IS-like elements and the bona fide IS elements have not been established and, to date, only limited data are available on the distribution of known IS elements in bacteria.

Strong polar mutations occur spontaneously in the *E. coli* Lac or Gal operons at a frequency of 10^{-6}-10^{-7}/cell, which is probably a direct reflection of transposition or other recombinational alterations involving any transposable element normally carried by the bacterial chromosome. For this reason, it has been impossible to determine the exact transposition frequency for any specific IS element. Assuming fairly random distribution of IS-promoted events and knowing that *E. coli* contains about 3000 genes, one can say that the observed transposition frequency has been relatively low ($\sim 10^{-5}$), at best, for most IS units. However, IS element transposition is *recA*- independent and the transposition frequency almost certainly varies with the specific receptor site involved. Examination of spontaneously occuring mutations in *E. coli* and lambda phage has revealed that IS element insertion alone can cause approximately 10-15% of all mutations within a single gene (see review by Starlinger and Saedler 1972, 1976). IS1 was the most frequently identified insertion mutation. However, it is not known if the small size, high copy number per chromosome, and/or some other attribute of this element is responsible for this phenomenon. Following insertional inactivation of a gene, the function of the mutated gene can be restored at frequencies of 10^{-3}-10^{-8} depending on the IS element, its orientation, and the gene involved (Starlinger and Saedler 1976). Restoration of gene function implies that precise excision of the IS element has occurred and indicates that initial insertion usually occurs without altering the wild-type bacterial gene sequences. The excision of IS elements, known to be *recA*-independent, is probably also promoted by a specialized recombination system. Imprecise excision of an IS unit from a mutant might cause relief of polarity, but continued loss of the mutated gene function. Though such an event has not yet been demonstrated for the IS elements, imprecise excision of phage Mu and larger transposable elements does occur.

The physical aspects of IS elements, which were deduced from DNA sequence analyses, are listed in Table 3 and discussed with the larger, transposable elements in a later section. Also, current theories of IS unit transposition are discussed in the section on mechanisms of transposition.

b) Abnormal Chromosomal Rearrangements. Insertion and precise excision of IS elements appear to involve enzyme recognition of specific sequences at the IS element termini, though none of the enzymes involved has been isolated. Moreover, specialized recombination events involving these same IS element termini are seemingly responsible for chromosomal deletions and probably for duplication, inversion, transposition, and plasmid integration events. The presence of IS1 (and perhaps IS2) within an operon leads to as much as a 1000-fold increase in deletion formation frequency in this area of the chromosome, yielding deletions in one per every 10^4 cells analyzed. The deletions, which can be from several hundred to as large as 20,000 nucleotide pairs in length, generally terminate at either end of the integrated IS element and extend outward to non-randomly distributed end points within the adjacent bacterial sequences, leaving the original IS element intact (Reif and Saedler 1975, 1977; see Fig. 11). Chromosomal deletions, similar to the deletions commonly observed at the site of Mu phage integration, can also be formed during IS element insertion (see Starlinger and Saedler 1976). Recently, Nevers and Saedler (1978) have identified a function, termed *del*, which maps at ~ 61 minutes on the 100 minute *E. coli* genetic map, that is needed for high

DELETION FORMATION

Fig. 11. Deletion formation mediated by transposable elements. The *solid horizontal line* represents a portion of a chromosome containing an inserted transposable element, depicted by the *solid rectangle*. Deletion occur unidirectionally from either end of the transposable element outward to nonrandomly distributed points within the adjacent chromosomal sequences. The sequences deleted in a hypothetical deletion event are represented by each *dashed line*. Following a primary deletion event, as shown here, the transposable element is left intact and can transpose or cause secondary and tertiary deletion events

frequency IS*1*-mediated deletion formation. Cells lacking this function show a 90-99% decrease in IS*1*-mediated deletion formation. Additionally, IS*1*-promoted deletion formation at 42°C vs. 32°C is decreased to different degrees in separate mutants, suggesting that the enzyme-DNA complex, but not the enzyme itself, is temperature dependent (Reif and Saedler 1977). No other requirements of IS-mediated deletions have been identified. It is important to note that precise excision of IS*1* in a mutant lacking the *del* function is reduced ∿6-fold and may indicate some linkage between precise excision and deletion formation (Nevers and Saedler 1978).

It appears likely that IS elements mediate a variety of chromosomal rearrangements, but confirming data are presently scarce. In addition to the data presented above, IS*1* has been reported recently to mediate deletion, transposition, duplication, inversion, and plasmid cointegration events (Iida and Meyer 1979; Shapiro and MacHattie 1979). The participation of IS units as structural components of the more complex transposable elements as well as the involvement of IS elements in bacterial evolution will be discussed later.

c) *Regulation of Gene Expression.* Insertion of an IS element, beyond simply abolishing the function of the affected gene, can, depending on its orientation, affect the expression of promotor-distal genes in an operon. The mechanism(s) of polarity is not well understood. However, recent evidence suggests that there are non-random specific sites on DNA, both within and outside of genes, at which RNA polymerase and attached mRNA molecules are released from DNA through the action of a specific protein transcriptional termination factor termed rho (Roberts 1976). Current data suggest that rho initially interacts with nascent mRNA instead of directly with DNA, but that rho can not bind to actively translated mRNA regions. Successful rho attachment to the DNA-bound mRNA somehow signals the RNA polymerase to terminate transcription. Polarity, then, appears to be a composite process. In the first step, translation of the mRNA is terminated at a nonsense codon, created either by substitution or frameshift mutation or carried by an insertion element. Subsequently, rho can attach to the non-translated mRNA molecule at a specific

rho-recognition sequence that is promotor-distal to the nonsense
codon, an event which signals the RNA polymerase to terminate
transcription. This line of logic is strengthened by the finding
that the general polarity suppressor, *suA* (a mutant of the wild-
type allele that produces rho protein), can partially suppress
the polarity caused by IS*1*, IS*2*, and IS*3* in the *lac* and *gal* operons (Malamy 1970; Malamy et al. 1972; Das et al. 1976; Besemer
and Herpers 1977; Sommer, Cullum, and Saedler, personal communication). Since transcription occurs unidirectionally from a promotor sequence, only one strand of an inserted IS element is
transcribed along with the genes of an operon. IS*1* and IS*4* are
known to exert polar effects when inserted in either orientation
within an operon and these elements apparently encode nonsense
codons (Ohtsubo and Ohtsubo 1978) in both DNA strands. IS*2* exerts
polar effects in orientation I, by original definition, and acts
as a transcriptional promotor in orientation II with respect to
an operon (Saedler et al. 1974). In addition to encoding a nonsense codon in the polar orientation I, in vitro transcription
studies with λ*gal* DNA carrying IS*2* in this orientation in the
Gal operon indicate that this element also contains a rho-sensitive transcriptional termination site (Decrombrugghe et al.
1973). As discussed below, IS*2* can also act as a genetic promoter, a sequence that binds RNA polymerase and initiates transcription. When the IS*2* element is in orientation I, this element
might sometimes initiate transcription in a direction opposite
to that of the operon in which it is inserted. It is thought
that the RNA polymerase molecules initiated by the IS unit col-
lide with those polymerase molecules initiated on the opposite
DNA strand by the operon promoter, causing an additional polar
effect on operon expression. Therefore, different IS elements
appear to utilize several different mechanisms that result in
the overall decrease in expression of promoter-distal genes in
an operon (see Starlinger and Saedler 1976).

At least one bona fide IS element, IS*2*, has been found that can
positively affect gene expression. IS*2* can behave as a highly
efficient genetic promoter when inserted in orientation II as
opposed to its polar orientation (I). When inserted in orientation II within the *gal* operon control region, IS*2* has been
observed to mediate the expression of more promoter-distal
genes at a rate three fold higher than the fully induced wild-
type operon (Saedler et al. 1974). Unexpectedly, DNA sequence
analysis of one IS*2* element has revealed a rho-sensitive termination site in the strand expressed in orientation II, but no
sequence that is similar to known genetic promoters (Ghosal et
al. 1979b). Current evidence would suggest that IS*2* in orientation II does not encode a constitutively expressed genetic promoter, but that some internal sequence rearrangement may generate
promoter function (Ghosal et al. 1979b; see *bi*2 versus *int-c* in
Pilacinski et al. 1977). No other bona fide IS elements are known
to behave as genetic promoters. Thus, operon expression can be
controlled by the insertion of an IS*2* element which depending
upon its orientation can either enhance or prevent transcription
(Saedler et al. 1974). Albeit the IS*2* element has not yet been
shown to invert its orientation while remaining at the same insertion site, the unusual ability of IS*2* to control gene expres-

sion is somewhat analogous to a recently described invertible
element. Inversion of an 800 base pair IS-like segment has recently been shown to control the alternate expression of the H1
or H2 flagellar antigens in *Salmonella* (Silverman et al. 1979;
Zieg et al. 1978). In contrast, genetic variation caused by inversion of the G segment of Mu, D108, P1, or P7 viruses, as discussed earlier, appears to be a fundamentally different regulatory phenomenon. Current evidence indicates that a promoter located outside of the G segment is responsible for transcribing
essential genes within the G segment. Perhaps when the G segment is positioned in one orientation, the essential G segment
genes are expressed and viable phage are produced, and vice
versa. Alternatively, two different sets of essential genes
controlling host range may be located on different DNA strands
of the G segment so that only one set and a specific host range
is expressed for each orientation (Howe 1978). Therefore, genetic inversion, which is probably mediated by specialized recombinational processes, results in the regulation of gene expression in at least two different ways and now seems to be a
not too uncommon process.

Several independently isolated mutants containing IS*1* and IS*2*
have been used to analyze the DNA sequence of these elements
(Calos et al. 1978; Grindley 1978; Ohtsubo and Ohtsubo 1978;
Ghosal et al. 1979b; Johnsrud 1979). Although these data are
discussed in more detail later, several observations are per-
tinent here. Certain IS*2*-mediated Gal-negative mutants have
been found to revert to an unstable, intermediate level constitutive utilization of galactose. Recent analyses of these revertants show that the IS element, though remaining physically
in the polar orientation (I) with respect to the operon, now
promotes intermediate level transcription. DNA sequence studies
show that the revertant IS*2* elements, IS*2-6* and IS*2-7*, each contain a 54 or 108 base pair complex internal duplication which
was probably formed during replication (Ghosal and Saedler 1977,
1978; Ghosal et al. 1979a). Thus, variants of IS*2* in orientation
I also exhibit genetic promoter activity. Additionally, in separate studies, comparison of the DNA sequences of two different
IS*1* elements (Johnsrud 1979) or two different IS*2* elements
(Ghosal et al. 1979b) suggest that small changes in independent
isolates of IS elements do occur. Surprisingly then, although
IS elements transpose as discrete units, and elements of the
same IS class appear grossly homologous by many techniques, even
these small DNA segments appear to be constantly evolving.

3. *Phenotypically Identifiable Transposable Elements*

a) Discovery. With the widespread use of antibiotics, a phenomenal
increase in the number of bacteria resistant to antibiotics has
been observed over the past 20 years. Conjugally transferable
antibiotic resistance plasmids (R plasmids) were found to be
responsible for this rapid dissemination of resistance (Falkow
1975). Considerable genetic evidence, amassed during the 1960's,
indicated that the plasmid DNA segments encoding resistance to
one or more medically relevant antibiotics could recombine with

phage, bacterial, or other plasmid chromosomes (see reviews by Cohen and Kopecko 1976; Cohen 1976). Subsequent to the development of techniques to isolate and physically analyze entire plasmid chromosomes, a variety of discrete transposable DNA segments that specify resistance to one or more structurally distinct groups of antibiotics have been identified (Table 4). In the initial description of a large transposable element that encodes β-lactamase production (i.e., penicillin resistance phenotype), Hedges and Jacob (1974) proposed the term *transposon* to define specific DNA sequences with transposition potential. This term, now in common usage, has generally been applied to the larger transposable elements that will be discussed in this section.

b) Molecular Nature and Host Range. Transposons (abbreviated Tn's) can be defined as large DNA segments, which express a phenotypically identifiable trait(s) unrelated to their own insertion, that are capable of *recA*-independent (specialized) transposition, usually in either of two physical orientations, as a discrete non-permuted unit. As shown in Table 4, these elements range in size from 2000 to greater than 80,000 nucleotide base pairs (b.p.). In addition to causing insertional inactivation of a gene, most transposons, like IS elements, also exert polar effects on the more promoter-distal genes in an operon (see Kleckner 1977). Though the initially characterized Tn's all encoded enzymes responsible for antibiotic resistance, more recently Tn's encoding resistance to heavy metal ions (e.g., Hg^{2+}, Stanisich et al. 1977) or enzymes involved in the metabolism of lactose (Cornelis et al. 1978), raffinose (Schmitt et al. 1979b), and toluene, xylene or salicylate (Chakrabarty et al. 1978; Jacoby et al. 1978) have been identified. Furthermore, the genes for enterotoxin production (So et al. 1979) and the genes for synthesis of the K-88 bacterial surface antigen that is responsible for intestinal colonization (Schmitt et al. 1979b), previously identified on plasmids, have now been shown to exist on discrete transposable elements (Table 4). Very recently, the *his-gnd* gene sequences of the *E. coli* chromosome have been observed to transpose at a relatively high frequency (Palchaudhuri et al. 1979; Wolf 1979). Transposable DNA segments have now been identified on the plasmids or host chromosomes of these genera: *Escherichia, Citrobacter, Salmonella* (Roussel et al. 1979), *Serratia* (Hedges et al. 1977), *Proteus* (Ohtsubo and Ohtsubo 1977), *Shigella* (R100 plasmid, see Ohtsubo and Ohtsubo 1977, Kopecko et al. 1978), *Yersinia* (Cornelis et al. 1978), *Klebsiella* (Berg et al. 1975), *Pseudomonas* (Hedges and Jacob 1974; Chakrabarty et al. 1978; Jacoby et al. 1978), *Rhizobium* (Beringer et al. 1978), *Streptococcus* (Tomich et al. 1979), *Hemophilus* (Falkow et al. 1977), *Staphylococcus* (Novick et al. 1979), and *Bacillus* (see Saedler and Ghosal 1977). Also, the fact that conjugally promiscuous replicons like RP4 can transfer to and replicate in many genera not listed above would suggest that most if not all bacteria contain transposable elements.

Physical examination of all transposable elements, including IS units, mainly by electron microscope heteroduplex techniques and also by DNA sequence analyses has revealed a characteristic structure for these discrete elements (see Fig. 12). In all cases thoroughly examined, the IS or Tn unit comprises central DNA

Table 4. Phenotypically-identifiable transposable elements

Tn element designation	Tn-encoded properties	Total Tn length (b.p.)	Length and orientation of internal, terminal repeat sequences (b.p.)	Restriction enzyme susceptibility	
				Cleaved by	Not cleaved by
Tn1	Ap^R	∿ 4,600	∿ 40; inverted	BamHI HaeII,III HincII	
Tn2	Ap^R	∿ 4,600	∿ 40; inverted	Same as Tn1	EcoRI
Tn3	Ap^R	∿ 4,600	∿ 38; inverted	Same as Tn1	EcoRI
Tn401	Ap^R	∿ 4,600	∿ 40; inverted	Same as Tn1?	
Tn801 Tn802	Ap^R	∿ 4,600	∿ 40; inverted	Same as Tn1 + PstI	
Tn901	Ap^R	∿ 4,600	∿ 40; inverted	Same as Tn1	EcoRI SalI HpaI
Tn902	Ap^R	∿ 4,600	∿ 40; inverted	Same as Tn1?	
Tn1701	Ap^R	∿ 4,600	∿ 40; inverted	Same as Tn1	
Tn4	Ap^R,Sm^R, Su^R,Hg^{2+R}	∿20,500	< 140; inverted	Same as Tn1 + ?	
Tn21	Sm^R,Su^R	∿15,700	< 140; inverted	EcoRI	
Tn(Aβ)	Ap^R,Sm^R	∿14,750			
Tn(?)	Ap^R,Cm^R, Sm/Sp^R,Su^R, Tc^R	∿28,800		EcoRI + ?	
Tn5	Km^R	∿ 5,200	∿1450; inverted	HindII,III	
Tn6	Km^R	∿ 4,100			

Original chromosomal source of Tn	Length of recipient DNA duplicated at ends of Tn (b.p.)	Transposition frequency (events/cell)	Pertinent references
Pseudomonas plasmid RP4	5	10^{-2}	Hedges and Jacob (1974) Hernalsteens et al. (1977)
Salmonella plasmid RSF1030	5	–	Heffron et al. (1975) Rubens et al. (1976)
Salmonella plasmid R1-19	5	$10^{-2}-10^{-5}$	Kopecko and Cohen (1975) Kretschmer and Cohen (1977) Ohtsubo et al. (1979) Cohen et al. (1979)
Pseudomonas plasmid RP1	5?		Bennett and Richmond (1976) Grinsted et al. (1978)
Pseudomonas plasmid RP1	5?	$10^{-2}-10^{-4}$	Benedict et al. (1977)
Salmonella plasmid pRI30	5?	–	Embden et al. (1978)
E. coli phage P7	5?	–	Yun and Vapnek (1977)
Salmonella plasmid NTP1	5?	–	Yamada et al. (1979)
Salmonella plasmid R1-19	–	$10^{-6}-10^{-7}$	Kopecko and Cohen (1975) Kopecko et al. (1976)
Salmonella plasmid R100-1	–	–	Kopecko et al. (1976) Nisen et al. (1977)
Serratia plasmid R938	–	–	Hedges et al. (1977)
Salmonella ordonez	–	–	Roussel et al. (1979)
Klebsiella plasmid JR67	9	$10^{-3}-10^{-2}$	Berg et al. (1975), Berg (1977) Davies et al. (1977) Allet (1979)
E. coli plasmid JR72			Berg et al. (1975)

Table 4 (continued)

Tn element designation	Tn-encoded properties	Total Tn length (b.p.)	Length and orientation of internal, terminal repeat sequences (b.p.)	Restriction enzyme susceptibility	
				Cleaved by	Not cleaved by
Tn601	Km/Nm^R	$\sim$ 3,100	$\sim$1000; inverted	$Hind$II,III + ?	
Tn903	Km/Nm^R	$\sim$ 3,100	$\sim$1000; inverted	BpaII HaeIII HapII HgaI HhaI $Hind$III	
Tn7	Tp^R,Sm^R	$\sim$12,750	< 150; inverted	BamHI EcoRI $Hind$III	
Tn71	Tp^R,Sm^R	$\sim$12,750		Same as Tn7	
Tn72	Tp^R,Sm^R	$\sim$12,750		Same as Tn7	
Tn402	Tp^R	$\sim$ 7,500			
Tn9	Cm^R	$\sim$ 2,500	IS1; direct	BalI EcoRI + Same as IS1	
Tn1681	Heat stable enterotoxin	$\sim$ 2,060	IS1; inverted	$Hinc$II + Same as IS1	
Tn(R-det)	Cm^R,Sm^R,Su^R	$\sim$23,000	IS1; direct	–	–
Tn10	Tc^R	$\sim$ 9,300	$\sim$1400; inverted (not IS3)	AccI BamHI BglI,II EcoRI $Hind$III HpaI SstII	BalI KpnI PstI SalI XhoI XmaI

Original chromosomal source of Tn	Length of recipient DNA duplicated at ends of Tn (b.p.)	Transposition frequency (events/cell)	Pertinent references
Salmonella plasmid R6	9?	–	Davies et al. (1977)
Salmonella plasmid R6-5	9	–	Oka et al. (1978) Nomura et al. (1978)
E. coli plasmid R483	–	$5 \cdot 10^{-4}$	Barth et al. (1976) Barth and Datta (1977) Barth (personal communication)
E. coli plasmid R721		–	Barth and Datta (1977)
E. coli plasmid pBW1		–	Barth and Datta (1977)
Klebsiella plasmid R751			Shapiro and Sporn (1977)
Shigella plasmid R100	9	10^{-6}-10^{-7}	Kondo and Mitsuhashi (1964) Gottesman and Rosner (1975) MacHattie and Jackowski (1977)
E. coli plasmid ST	9		So et al. (1979)
Shigella plasmid R100-1	9?		Arber et al. (1979) Hu et al. (1975)
Shigella plasmid R100	9	10^{-6}-10^{-7}	Kleckner (1977, 1979) Kleckner et al. (1975, 1978, 1979a,b) Foster et al. (1975) Kleckner and Ross (1979)

Table 4 (continued)

Tn element designation	Tn-encoded properties	Total Tn length (b.p.)	Length and orientation of internal, terminal repeat sequences (b.p.)	Restriction enzyme susceptibility	
				Cleaved by	Not cleaved by
Tn*1721*	Tc^R	~10,700	< 38; inverted	*Eco*RI *Sma*I *Hin*dIII *Sac*II *Pst*I *Hpa*I *Sal*I	*Bam*HI *Bgl*II *Xho*I *Kpn*I
Tn*1771*	Tc^R	~10,800	< 50; inverted		
Tn*551*	Em^R	~5,200	< 100; inverted	*Bgl*II *Hpa*I	
Tn*917*	Em^R	~4,500			
Tn*501*	Hg^{2+R}	~7,800	< 150; inverted	*Eco*RI *Hin*dIII *Sal*I	*Pst*I *Sac*II *Sma*I
Tn*951*	lactose catabolism	~16,600	~100; inverted	*Bam*HI *Eco*RI *Hin*dIII *Pst*I	
Tn(Tol)	Toluene & Xylene catabolism	~52,500	?		
Tn(Raf) (Tentative)	Raffinose catabolism, H$_2$S & K88 antigen	~40,000 - 60,000	IS*1*; direct		
Tn(Ti) (Tentative)	Tumor induction	~16,500		*Sma*I *Hpa*I ?	
Tn(Sal)	Salicylate degradation	30,000			
Tn(his-gnd)	Histidine synthesis	44,000	1400; inverted		
Tn(Lac)	Lactose catabolism	80,000	IS*3*; inverted		

Original chromosomal source of Tn	Length of recipient DNA duplicated at ends of Tn (b.p.)	Transposition frequency (events/cell)	Pertinent references
E. coli plasmid pRSD1	5		Schmitt et al. (1979a,b) Mattes et al. (1979)
E. coli plasmid pFS2O2			Schöffl and Burkardt (1979) Schöffl and Puhler (1979)
Staphylococcus plasmid PI258		$10^{-4}-10^{-5}$	Novick et al. (1979)
Streptococcus plasmid pAD2			Tomich et al. (1979)
Pseudomonas plasmid pVS1		$10^{-1}-10^{-2}$	Stanisich et al. (1977) Bennett et al. (1978a)
Yersinia plasmid pGC1		10^{-4}	Cornelis et al. (1978,1979) Cornelis et al. (in preparation)
Pseudomonas Tol plasmid			Chakrabarty et al. (1978) Jacoby et al. (1978)
E. coli plasmid pRSD2			Schmitt et al. (1979b)
Agrobacterium Ti plasmids			Hernalsteens et al. (1977) Schell and van Montagu (1977) Chilton et al. (1978)
Pseudomonas Sal plasmid			Chakrabarty et al. (1978)
E. coli		$10^{-2}-10^{-4}$	Wolf (1979 and pers. comm.) Palchaudhuri et al. (1979)
E. coli phage P1*dlac*			Cornelis et al. (in preparation)

Tn numbers are those assigned by E.M. Lederberg, Plasmid Reference Center, Dept. of Medical Microbiology, Stanford University. Those Tn designations given in parentheses describe certain transposable elements for which numbers have not yet been given. The Tn-encoded properties include resistance to ampicillin (Ap^R), chloramphenicol (Cm^R), erythromycin (Em^R), kanamycin (Km^R), neomycin (Nm^R), spectinomycin (Sp^R), streptomycin (Sm^R), sulfonamide (Su^R), tetracycline (Tc^R), trimethoprim (Tp^R), and divalent mercury (Hg^{2+R}). Also, production of heat stable enterotoxin, K88 antigen, and H$_2$S, plus degradation of lactose, raffinose, toluene, xylene and salicylate are Tn-encoded properties. Recently, a segment of the *E. coli* chromosome, which encodes the *his-gnd* loci (genes for the biosynthesis of histidine and synthesis of gluconate dehydrogenase, respectively), has been shown to transpose to plasmids and phages. Also, recent evidence suggests that the tumor-inducing segment of Ti plasmids transposes to plant cells as a specific unit, but this has not been proven.

Total Tn size, as well as the length and orientation of the internal repeat sequences at Tn termini are given in nucleotide base pairs (b.p.). Tn*1, 2, 3, 401, 801, 802, 901, 902*, and *1701* appear to be homologous Ap^R elements, now collectively termed TnA. Tn*4* is a composite transposon apparently consisting of Tn*3* inserted into Tn*21* (Kopecko et al. 1976). Tn*601* and *903* are probably identical elements. Tn*7, 71*, and *72* also appear to be homologs obtained from different plasmids. Tn 1721 and 1771 appear to be identical Tc^R Tn's. Tn*9*, Tn(R-det), Tn(Raf), and Tn*1681* have direct or inverted repeats of IS*1* at their termini. The inverted sequences at the termini of Tn*10* are not IS*1, 2* or *3*, contrary to many published reports (see Kleckner and Ross 1979). A composite of published restriction endonuclease cleavage susceptibilities of various Tn's is given. However, one should be aware that there is noticeable variability in restriction patterns of similar Tn elements (Yamada et al. 1979). Since most Tn's have been isolated from plasmids, the bacterial host and/or plasmid in which the Tn was originally detected are listed. Also, the length of the recipient DNA sequence that is duplicated at the ends of a Tn element is given in b.p. A compilation of reported transposition frequencies, listed as transposition events per cell, is given, usually as a range. In most Tn identifications, transposition frequency was not assessed. Certain listed Tn properties are based on assumption (i.e., likeness to characterized Tn's) and those have been denoted by an accompanying question mark. See Campbell et al. (1979) for transposon nomenclature

sequences bracketed by either a directly or an inversely repeated terminal DNA sequence. For example, IS*2* comprises a 1250 b.p. central segment containing a 40 b.p. sequence that is repeated, imperfectly, in inverse fashion at each terminus (Table 3). Tn*3* consists of approximately a 4500 b.p. segment flanked by inverted repeats of a 38 b.p. sequence (Table 4). Though most transposable elements are structurally defined by inverted repeat DNA sequences, the Tn*9* (Cm^R), Tn(Raf), and Tn(R-det) contain direct repeats of the IS*1* sequence at their termini. However, since the ends of IS*1* contain a small inverted repeat sequence, these Tn's are actually bracketed by small inverted repeats at their termini. Thus, all IS's and Tn's for which the ends have been characterized to date, contain an inverted terminal repeat sequence. The total length of each transposable element as well as the length and orientation of the internal repeated terminal sequences, if known, are given in Tables 3 and 4 for both IS and Tn units. Examination of

TRANSPOSABLE DNA ELEMENT

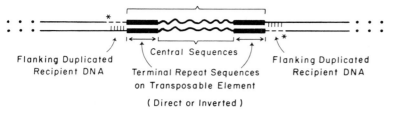

Fig. 12. The structure of transposable elements. This unscaled diagram illustrates a transposable DNA element, which is delineated by the *rectangles*
and *wavy line*, inserted within a recipient molecular DNA sequence, which is
depicted by the *thin continuous* or *broken lines*. Both IS and Tn units are
composed of a central DNA segment *(wavy line)* flanked by terminal sequences
(rectangles) that are repeated either in direct or inverted order. Apparently during insertion, staggered single-strand cleavage occurs at the sites
marked by *asterisks*, and the extended single-strand ends of the recipient
molecule are joined to the transposable element. Subsequent gap filling DNA
synthesis *(broken lines)* and ligation creates duplicated recipient molecular
sequences at the ends of the inserted Tn or IS element

the junctions of IS or Tn elements with recipient molecular DNA
has revealed the existence of directly repeated recipient molecular sequences of 3 to 4, 5, 9, or 11 to 12 b.p. at each end
of the inserted element (Fig. 12). Though the nucleotide composition of the repeated recipient sequence varies from insertion
site to insertion site for a particular element, each transposable
element is always associated with a repeated recipient DNA sequence of specific length, as shown in Tables 3 and 4 (Calos et
al. 1978; Grindley 1978; Johnsrud 1979; Ghosal et al. 1979b;
Tu and Cohen 1980). Apparently, during Tn or IS insertion an en-
zymes(s) creates a single-strand cleavage, staggered by 3 to 4,
5, 9, or 11 to 12 b.p., in each recipient DNA strand and the extended single-strand ends are joined to the transposable element,
with the complementary strand at each end being newly synthesized
(see Fig. 12; also see section on transposition mechanisms).

The large inverted repeat termini located on many transposons
make these elements easily identifiable in the electron microscope. As illustrated in Fig. 13, after denaturation and intra-
strand annealing of a DNA segment that carries a Tn which contains inversely repeated termini, one sees characteristic hairpin-loop or stem-loop structures in which the double-stranded
stem represents the reannealed, inversely repeated Tn termini.
In addition to forming characteristic stem-loop structures of
constant size, insertions of any transposon will increase the
size of the recipient molecule by a discrete length. Further
evidence for a specific transposition event can be obtained by
extensive restriction endonuclease analysis of several molecular
isolates containing the same putative Tn element, since Tn units
transpose as non-permuted DNA segments and consequently retain
the same restriction sites.

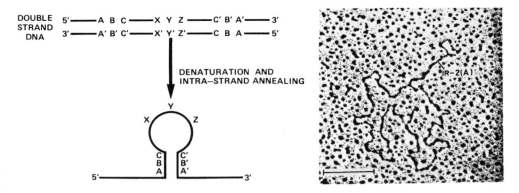

Fig. 13. Characteristic intra-strand structure of transposons with inversely repeated termini. *Top left:* A double-stranded DNA segment containing a transposon with complementary termini. Following denaturation and intra-strand annealing, hairpin-loop or stem-loop structures are observed. *Bottom left:* Single-strand molecule containing a double-stranded stalk or stem, comprising the Tn termini. The central sequences of the Tn unit (XYZ) are enclosed in the single-strand loop at the top of the stem. *Right:* Electron micrograph of a self-annealed single-strand of a small plasmid containing the Tn*601*/Tn*903* element. The double-strand stem represents the 1000 base pair inverted repeat sequences at the termini of this element

Any inserted element can be directly visualized in the electron microscope following heteroduplex formation, a technique in which single-strands of a parental molecule are allowed to re-anneal with complementary strands of an identical derivative molecule that contains the inserted element (see Fig. 14). The interaction of two linear DNA single-strands to form a double-stranded segment requires axial rotation of one strand around the other in order to form the DNA helix. Because of structural constraints then, two entirely complementary, but covalently sealed circular single-strands can not form a complete duplex molecule, but rather end up forming a molecule consisting of a mixture of duplex and single-strand regions. This is exactly what occurs when two entirely complementary DNA strands, in each of which the inverted repeat termini of a transposon have already intra-strand annealed (as shown in Fig. 13b), attempt to reanneal with each other. The sequences in the circular loop region of each strand are structurally constrained (i.e., can not undergo axial rotation) and can only form a partially duplex structure (termed underwound loop). The observation of underwound loops can be diagnostic of short inverted duplications on Tn elements as well as of the presence of new transposable elements (see Broker et al. 1977b).

c) The Transposition Process - A Genetic Definition. Transposition of Tn's occurs at frequencies, which range from 10^{-7}-10^{-1} events/cell, that probably depend upon: (1) the bacterial host (e.g., *E. coli* strain AB1157 and its derivatives decrease the normal Tn transposition frequency observed in other *E. coli* K-12 strains; Hedges et al. 1977); (2) the recipient chromosomal sites (Kleckner

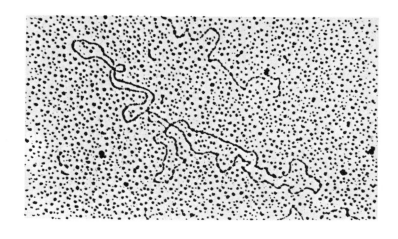

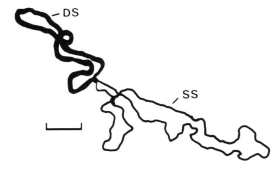

Fig. 14. DNA heteroduplex molecule. Single-strand DNA appears as a thin, uneven line in the electron micrograph, compared to the thicker double-stranded portion of the molecule. In this example, the 9,200 base pair (9.2 kilobase pair, kb) pSC101 plasmid has reannealed with the pSC120 plasmid. pSC120 is a recombinant plasmid made by inserting the 20.5 kb Tn4 element into pSC101. The tracing below shows a double-stranded (DS) circle that represents all of the pSC101 sequences, with one large single-strand (SS) insertion loop which represents the Tn4 element. Although not labeled, two small inverted repeat sequences are shown. One occurs at the termini of Tn4 and the other represents the ends of the Tn3 element that is located in this composite transposon (for more details see Kopecko et al. 1976). The *bar scale* represents 0.25 micrometer

et al. 1979b; Tu and Cohen 1980); and (3) the Tn element itself (see Table 4). Under what appear to be fairly optimal conditions Tn5, Tn501, or the TnA elements have been observed to transpose at frequencies of one transposition event per every 10 to 1000 cells (Bennett et al. 1977a; Davies et al. 1977; Grinsted et al. 1978). In contrast, other elements like Tn10 transpose less frequently (10^{-7}–10^{-6}) events/cell; Foster 1977; Kleckner 1977). Insertion in either orientation occurs with apparently equal frequency for all Tn's (Rubens et al. 1976; Kleckner 1978; Cornelis et al. 1979), with the exception of Tn7 which has the same orientation in 36 separate insertions within RP4 (Barth and Grinter 1977).

Transposition of discrete, non-permuted units implies that specific sequences at the ends of Tn elements are enzymatically recognized during insertion. Recent DNA sequence analyses of various inserted Tn's or IS units shows that for each element the same terminal nucleotide at each end of the element always forms the junction with recipient DNA (Calos et al. 1978; Grindley 1978; Ohtsubo and Ohtsubo 1978; Kleckner 1979; Tu and Cohen 1980). Available information on the recipient chromosomal recognition sites, however, is not easy to interpret. The large distribution of possible insertion sites that have been observed in the *E. coli* chromosome for the representative Tn*5* or Tη*10* elements would suggest that there is little specificity involved at the site of insertion. Like phage Mu, Tn*5* or Tn*10* insertions cause new nutritional requirements in 1-2% of the bacterial chromosomes into which these elements transpose. On the surface, these data would suggest that there is a short, three to five bp recognition sequence that is distributed randomly and often throughout most recipient chromosomes. However, despite the capability of these and other transposons to insert into many different loci, the preponderance of transposon insertions has been found to occur at preferred areas of the recipient chromosomes (termed "regional" or "local" specificity). For example, Kleckner et al. (1979b) recently found that out of 131 independently isolated Tn*10* insertions within the 10,000 bp *S. typhimurium* histidine *(his)* operon, 50 insertions occurred within a single 30 bp region of the *hisG* gene. Foster (1977) previously reported that 18 of 21 independently isolated Tn*10* insertions within the 3520 bp *E. coli lacZ* gene occurred in a single, small 175 bp region. These are examples of extremely "localized" insertion specificity in which Tn insertion occurs at non-identical, nearby sites and sometimes precisely at the same locus (Kleckner 1979). It is very interesting to note that Johnsrud et al. (1978) have found that 29 of 50 Tn*9* insertions within the *E. coli lacZ* gene map in the same "recombinational hotspot" that Foster observed for Tn*10* insertions in *lacZ*, suggesting that, at least, these different Tn's respond to the same recipient recognition sequence.

Very early reports on TnA insertion specificity (Kopecko et al. 1976; Rubens et al. 1976) as well as more recent findings for TnA, Tn*7*, and Tn*501* elements (Barth and Grinter 1977; Kretschmer and Cohen 1977; Grinsted et al. 1978; Tu and Cohen 1980) indicate that these Tn's do not insert at random, but that a large proportion of insertions occur in not-so-compact, recipient genomic regions of 500-1000 bp in length (i.e., "regional" specificity). If a frequently occurring recipient genome recognition sequence for Tn insertion was present, one would *not* expect clustered insertion sites at nearby loci, but randomly distributed insertions, with separate transposition events occasionally occurring at the same recognition site. Therefore, the non-random, "localized" or "regional" specificity of Tn insertions would argue against a randomly distributed recipient, 3-5 bp recognition sequence. The data of Tu and Cohen (1980) suggest that the "regional" specificity for Tn*3* insertion is due to recipient DNA A+T richness plus homology with the ends of Tn*3*. However, Grinsted et al. (1978) have convincingly established that a

500 bp DNA segment while present in one plasmid behaves as a "recombinationally hyperactive" recipient site for transposition, but the same DNA segment when located adjacent to different sequences in a derivative plasmid does not. Thus, a short, recipient recognition sequence is not sufficient, in itself, for transposition. Furthermore, DNA sequence analyses of various Tn- or IS-recipient DNA recombinational junctions have not revealed any such common recipient DNA recognition sequence (Grindley 1978; Johnsrud et al. 1978; Oka et al. 1978; Ghosal et al. 1979b; Kleckner 1979). Rather, it appears likely that a more complex recipient chromosomal recognition sequence is involved and that the point of cleavage of recipient DNA (i.e., the Tn insertion site) occurs some distance (up to 500 or 1000 bp) from the recognition site. Type I restriction endonucleases, which exhibit this pattern of behavior, or some similar enzyme may be responsible for the observed preferred areas of Tn insertion, as discussed in the section on mechanisms of transposition.

Little is factually known about the mechanism of Tn or IS element transposition except that it occurs independently of homologous recombination systems. Considerable genetic evidence indicates that transposition to a new site does not cause loss of the element at the original locus (Bennett et al. 1977a; Shapiro 1979). As a matter of fact, transposition of any Tn unit is detected 10^2-10^5 times more often than loss of the Tn element by precise excision. This evidence would imply that an obligatory and integrally linked replication/transposition event, like that discussed for Mu phage, occurs with transposable elements. Only a single-strand template or, perhaps, a newly replicated duplex copy of the transposable element would be inserted into the recipient site. Although the energy, enzyme, and structural requirements of transposition are unknown, for the most part, limited data obtained with Tn*3* would suggest that in *E. coli* transposition efficiency at 37°C is only 10% of that observed at 32°C and transposition does not occur at 45°C (Kretschmer and Cohen 1977). The requirements for DNA, RNA, and/or protein synthesis in transposition have not been established nor have the effects of temperature on the transposition of other transposons. As mentioned previously, the *him* and *hip* genes, which probably encode proteins that are common to several different DNA metabolic processes, affect the frequencies of transposition and excision of transposable elements, but the nature of these affects is not understood (Miller and Friedman 1977; H. Miller, personal communication). Some mechanism apparently exists that controls the frequency of transposition since this frequency appears to reach saturating levels after a period of time. For example, to date the transposition frequency is always 10^2-10^5 higher for Tn*3* than Tn*10* (Kleckner 1977; Grinsted et al. 1978). In addition, Bennett et al. (1977b, 1978b) have observed that the presence on a recipient plasmid of Tn*10* or a TnA derivative that is mutated to β-lactamase non-production *(bla⁻)* decreases the transposition frequency of a second TnA element to that plasmid, but not to other plasmids in the same cell. Therefore, the presence of a Tn element may, in some cases, suppress subsequent transposition to the carrier plasmid (i.e., it exerts a *cis*-acting suppressive effect). On the other hand, it should be noted that plasmids carrying two TnA elements have been physically identified, a fact which obsures the importance of the *cis*-acting suppressive effect (Bennett et al. 1978; Holmans et al. 1978).

Considerable effort has been applied to isolating mutants of transposons in order to see if any Tn sites or functions are necessary for transposition. To recap the conclusions before presenting the data, it appears that the inversely repeated sequences at the Tn termini are necessary for enzymatic recognition during transposition. Furthermore, at least Tn*3* and Tn*5* encode a protein(s) that is involved in their respective transposition. Heffron, Falkow, and co-workers have used a variety of novel techniques to generate addition/deletion mutations within the Tn*3* element (Heffron et al. 1977, 1978; Gill et al. 1978). By complementation of the transposition-deficient mutated Tn*3* elements with a *bla⁻* Tn*3* (i.e., phenotypically ampicillin sensitive), three classes of defective Tn*3* elements have been detected. Non-complementable mutants that contain a deletion of one terminal inverted repeat sequence demonstrate that the terminal sequences are a structural requirement for transposition. Similar conclusions were obtained by studying deletions of Tn*5* (Davies et al. 1977). Secondly, mutants obtained in approximately one-half of the Tn element could be complemented to transpose at 20% of the normal frequency. These mutants define a *trans*-acting function (RNA or protein) that is necessary for transposition. The third class of mutants occurs in a region surrounding the single *Bam* HI endonuclease cleavage site on Tn*3* and affects both the frequency and type of transposition event. When small insertions are biochemically spliced into this latter region, the transposition frequency is increased tenfold, compared to wild-type Tn*3*, and about 30% of these transposition events are abnormal, i.e., cause the insertion of the entire donor plasmid into the recipient replicon, as shown in Fig. 15c (Heffron et al. 1978; Heffron, personal communication). In the presence of a wild-type Tn*3* that is *bla⁻*, these latter insertion mutants of Tn*3* transpose as a discrete transposon greater than 99.9% of the time. It appears that the wild-type Tn*3* makes some *trans*-acting function that changes the transposition event back to normal Tn unit transposition (see Fig. 15d). In contrast, Tn*3* mutants that contain deletions of this region which affects the quality and quantity of transposition events are not observed to transpose unless complemented. Complementation of these mutants by a *bla⁻* Tn*3* restores the transposition frequency to about 20% of normal levels, but all of these Tn*3* deletion mutants form cointegrate structures upon transpositon, i.e., the entire donor plasmid, flanked by direct repeats of the mutant Tn*3* element, is inserted into the recipient replicon (Gill et al. 1978; see Fig. 15c). The basic interpretations of these data are illustrated in Fig. 15. In addition to the specific terminal sequences needed for transposition, Tn*3* encodes the production of two *trans*- acting proteins (a recently identified 110,000 mol. wt. protein (transposase or recombinase?) and a 19,000-20,000 mol. wt. "regulator" protein; Chou et al. 1979; Dougan et al. 1979; S. Cohen, F. Heffron, personal communication)

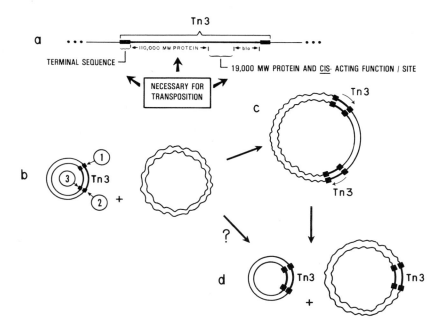

Fig. 15. Molecular and genetic aspects of Tn transposition. (a) Diagramatic depiction of the Tn*3* transposon showing the terminal repeat sequences essential for transposition. In addition, genetic and biochemical studies have revealed the Tn sequences coding for β-lactamase *(bla)* and functions/sites needed for transposition, as shown. The 110,000 mol. wt. protein may be a recombination enzyme, and the 19,000 mo. wt. protein and adjacent *cis*-acting function/site appear to regulate the frequency and type of transposition event as shown below (Gill et al. 1978; Heffron et al. 1978; Heffron, personal communcation). (b) Small donor plasmid carrying the Tn*3* transposon and a larger recipient chromosome, represented by a *circular, squiggled line.* (c) The results of studies with mutant transposons suggest that transposition occurs by cleaving the donor molecule at points labeled *1* and *3* in (b) followed by insertion of the entire donor plasmid flanked by direct repeats of the transposon into the recipient molecule. (d) Almost simultaneous processing is thought to occur generating independent donor and recipient replicons each containing one copy of the Tn unit. The regulator protein and *cis*-acting function/site, shown above, appear to be necessary for this normal processing. In the absence of either regulator function or site, the transposition event stops at the intermediate cointegrate stage (also see Meyer et al. 1979; Shapiro 1979). For ease of illustration, the recipient DNA sequences known to be directly repeated at the insertion site have not been drawn.

Current evidence does not eliminate the possibility that normal transposition occurs by initial cleavage of the Tn unit at only one end (labeled *1* in step b), followed by single-strand Tn transfer to the recipient molecule, simultaneous complementary strand synthesis on one or both molecules, cleavage of the other end of the Tn unit (at point *2* in step b), repair synthesis and ligation (see Grindley and Sherratt 1979) to give the molecule depicted in (d). The cointegrate structure (c) may be an aberrant recombinational product caused by lack of the "regulator" function/site and/or some other property (see the text) and a *cis*-acting function/site located near the gene for "regulator" protein (see Fig. 15a). Cohen and coworkers have recently demonstrated that the 20,000 mol. wt. regulator protein serves as a repressor molecule that effectively controls a bidirectional genetic transcription unit which includes the 110,000 mol. wt. transposase and production of the repressor protein itself (Chow et al. 1979; Chow, Lemaux, Casadaban and Cohen, submitted for publication). Usually, transposition involves the transient formation of cointegrate structures initially proposed by Kopecko and Cohen (1975) and shown in Fig. 15, which are processed into donor and recipient replicons, each carrying one copy of the transposable element (Fig. 15d). The Tn*3* mutants deleted for the "regulator" protein are also missing the *cis*-acting function/site, both of which are thought to direct the processing of the cointegrate structure (Gill et al. 1978). Even during complementation, all transposition events involving these Tn*3* deletion mutants remain abnormal because the *cis*-acting function/site is missing (Arthur and Sherratt 1979). These interpretations are supported by the results of similar studies with mutants of the Tn*5* element, which also suggest that a *trans*-acting function necessary for processing the normally transient cointegrate, recombinational intermediate is encoded within the Tn sequences (Meyer et al. 1979).

Based on our limited genetic knowledge of transposition, the above interpretations appear quite reasonable. However, other explanations are also plausible. For example, the normal transposition process may not involve the formation of a transient cointegrate intermediate structure (Fig. 15c), but instead may mediate the transfer of a single-strand of the transposon to a new site via the model of Grindley and Sherratt (1979) which is discussed in a later section and in the legend to Fig. 15. The regulator protein and *cis*-acting function/site might normally limit initial enzymatic cleavage to one end of the element, as described in the legend to Fig. 15. A defective regulator protein and/or lack of the *cis*-acting function/site might result in loss of control of the enzymatic cleavage and, consequently, the aberrant formation of a cointegrate structure. Cleavage of the transposon at each end but on opposite strands, and insertion into a recipient site which could occur by the model of Shapiro (1979; discussed in a later section), would result in the formation of an aberrant cointegrate structure. In the past year there were several other reports of transposon-mediated plasmid cointegration as depicted in Fig. 15c. When harbored on a multicopy colE$_1$ plasmid derivative, either Tn*9* or the enterotoxin transposon usually causes, during transposition, the cointegration of the entire donor plasmid bracketed by direct repeats of the Tn element (see So et al. 1979). However, these transposons have not been mutated and usually transpose from a larger plasmid or phage as a discrete element. One wonders if transposon-mediated, plasmid cointegrate formation is enhanced in these small multicopy plasmids because of some inherent property such as their rapid replication. More information on transposition of wild-type and mutant Tn's from large and small plasmids is obviously needed. Additionally, conditional mutants of Tn's and complementation studies among different Tn's might reveal new insight into the transposition process.

To summarize briefly the above information, genetic and physical studies indicate that Tn elements normally transpose as discrete units. It is generally assumed that the very termini of all transposons are essential structures for transposition, but their necessary, minimal length is not known. Thus far, only Tn*3* and Tn*5* have been shown to encode *trans*-acting functions essential to transposition. The transposition process may involve the initial formation of a cointegrate structure followed normally by its resolution into donor and recipient molecules, each carrying one copy of the Tn element.

Reversion of Tn-induced mutations, which is generally associated with loss of the Tn unit, is assumed to be due to precise excision of the Tn element and occurs for all transposons at frequencies ranging from 10^{-9}-10^{-6} (Kleckner et al. 1975; Berg 1977; Foster 1977). Since Tn transposition does not result in loss of the Tn unit at the original site and vice versa (Bennett et al. 1977a; Kleckner 1977), it is probable that Tn excision and transposition occur by different processes (discussed in the section on transposition mechanisms). In addition to undergoing transposition and precise excision events, most Tn's can mediate the rearrangement of nearby chromosomal sequences as discussed below.

d) Aberrant Chromosomal Rearrangements. While searching for reversion of Tn-induced mutations, imprecise excision events were detected at a frequency of 10^{-6}-10^{-4} events per cell by either polarity relief in the *lac* operon (Berg 1977; Foster 1977) or loss of Tn-encoded antibiotic resistance (Kleckner et al. 1979a; Ross et al. 1979b). Imprecise excision generally removes some of the Tn unit including drug resistance (e.g., tetracycline resistance) and, depending upon the Tn element, occurs 10-1000 times more frequently than precise excision (Kleckner 1977). Though known to occur independently of general recombination like precise excision, the mechanisms responsible for imprecise excision are not entirely understood. Physical examination of 26 imprecisely excised Tn*10* elements has revealed several different types of imprecise excision events (Ross et al. 1979a,b). Restriction enzyme analyses revealed that in ∿50% of these isolates, only 50-100 bp of Tn *10* sequences remain. DNA sequence analysis of two of these Tn*10* elements has shown that exactly 50 bp of Tn*10* remain in each case. Moreover, the deletion event appears to have occurred between short A+T rich regions internal to and directly repeated at each Tn terminus. Another group of imprecisely excised Tn*10* elements (8 of the original 26) was observed to contain simple internal Tn*10* deletions that uniquely had one deletion end point in very close proximity to either end of the 1400 bp repeat at either Tn*10* terminus. A third group of imprecisely excised Tn*10* elements (6 of 26) contain deletions of all of the Tn*10* sequences internal to the 1400 bp inverted terminal sequences plus concomitant inversion of adjacent sequences. As depicted in Fig. 16, Ross et al. (1979a) have proposed that the internal ends of the 1400 bp Tn termini recombine with some adjacent site, always resulting in concomitant deletion and inversion events.

INVERSION

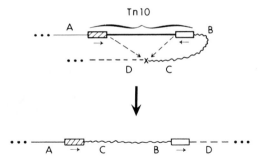

Fig. 16. Transposon-promoted inversions during imprecise Tn unit excision. The Tn*10* transposon is shown here inserted within a chromosome which is represented by *solid, squiggled*, and *broken lines*. Chromosomal regions *A* through *D* are so labeled. The 1400 bp repeats on Tn*10* are shown as *rect-angles* and their orientation is indicated by the *arrows* below each. The inside end of each 1400 bp repeat is proposed to recombine (see *dashed arrows*) with some adjacent sequence, denoted here by an *X*. This recombi-nation event results in the loss of the Tn*10* sequences normally contained within the terminal 1400 bp repeat sequences, as well as the inversion of one 1400 bp repeat and some adjacent chromosomal material (for details see Ross et al. 1979a)

Tn*3* has been found to generate, *recA*-independently, deletions which apparently occur from either terminus of the inserted transposon and extend outward to non-random points within ad-jacent chromosomal sequences, but always leave the Tn element functionally intact (Nisen et al. 1977). Similar results have been reported for Tn*10* (Noel and Ames 1978). Furthermore, in both studies recipient chromosomal regions that appeared as "recombinationally hyperactive" sites for Tn insertion also were found to act as preferred deletion end points, suggesting that transposition and deletion processes are similar.

Tn*9*, Tn*3* and the enterotoxin transposon have been observed under certain conditions to mediate the cointegration of two circular genomes, as mentioned in the previous section. Also, transposon-mediated chromosomal integration of phage λ has been detected (Davies et al. 1977; MacHattie and Shapiro 1978). In addition, Tn*10* has been used experimentally to generate duplications of chromosomal regions as well as to fuse unrelated chromosomal sequences (Kleckner et al. 1977). Though only limited data are presently available, it appears likely that most transposons will be found to generate all of the chromosomal rearrangements that are promoted by phage Mu, as discussed before.

4. Mechanisms of Transposition – Current Theories

DNA sequence analyses of many of the recipient chromosomal sites
into which Mu phage, IS elements, or transposons insert seem to
bear no great similarity to one another (Allet 1979; Ghosal et
al. 1979b). This fact separates, on a mechanistic level, these
specialized recombination systems from those of temperate bacteriophages, like λ, which appear to promote phage insertion into only one or a few short, but highly homologous regions of a
chromosome. Therefore, though similar in the ability to promote
chromosomal rearrangements in the absence of general recombination and without extended sequence homology, specialized recombination systems do display differences. The similarities among
transposition events of IS- or Tn-units and Mu phage suggest that
these different elements transpose by fundamentally related processes, as detailed below. However, the existence of several
different specialized transposition processes which result in
the same end product cannot be ruled out at the present time.

a) Essential Features of Specialized Transposition. Genetic and molecular studies of the specific transposition of Mu phage and
transposable elements (Tn, IS units) have revealed several general properties among these events. Transposition of an element
from one chromosomal site to another does not appear to result
in loss of the element at the original locus (Bennett et al.
1977a; Bukhari 1977; Shapiro 1979), a fact that ensures the involvement of replication in this event. Furthermore, insertion
of a transposable element or Mu phage into a chromosome results
in the direct duplication of 3-4, 5, 9, or 11-12 bp of recipient
DNA at the insertion site (see Calos et al. 1978; Grindley 1978;
Allet 1979; Ghosal et al. 1979b). The recipient DNA repeat sequences bracketing any specific element (e.g., IS1) are always
the same length but usually vary in nucleotide composition. Data
obtained with a derivative Tn3, Tn9, or Tn10 element, experimentally constructed so as to contain non-identical DNA flanking
each Tn terminus, indicate that this repeated DNA is not essential for the transposition event (Johnsrud et al. 1978; Kleckner
1979). However, most transposable elements are inserted in one
of two possible orientations at virtually any chromosomal site
(Bukhari 1977; Kleckner 1977). Thus, the recombination process
must involve recognition of the specific Tn or Mu termini. Although independent insertions of some Tn elements have been
found at precisely the same locus (Kleckner 1979; Tu and Cohen
1980), most Tn insertions occur in preferred chromosomal regions,
whereas Mu phage insertions appear entirely non-specific with
respect to the insertion site (Kleckner et al. 1979b; Ljungquist
et al. 1979). Based on the above observations of the structural
consequences of specialized transposition, several models have
been proposed to explain the transposition process.

b) Single-strand Transfer Model. Ljungquist and Bukhari (1977) have
provided evidence which suggests that Mu transposition follows
or occurs concomitant with Mu specific replication (Bukhari
1977). Additionally, data obtained from transposition studies
of TnA elements suggested to Bennett et al. (1977a) that transposition might involve single-strand transfer of the element

and complementary strand synthesis in the donor and recipient molecules. Grindley and Sherratt (1979) have recently described a model for single-strand transfer of any transposable unit. This simple model accomodates transposition without consequent deletion of the donor transposable element and allows for duplication of recipient sequences at the insertion site, as summarized in Fig. 17. Accordingly, one enzymatic activity is responsible for making staggered cuts in the recipient DNA (Fig. 17a). The observations that all transposable DNA units have 3-4, 5, 9, or 11-12 bp repeats of recipient DNA at insertion sites suggest that four separate, probably host-determined, factors provide this target site endonucleolytic nicking activity. A second component of this proposed reaction is an activity that recognizes, specifically cleaves, and transfers one end of a single-strand of the transposable unit to the appropriate site on the nicked recipient DNA (Fig. 17b). This second enzymatic activity may be specified by each transposable unit, as the evidence indicates for Mu, Tn3, and Tn5 (see Grindley and Sherratt 1979). Each transposable unit-specific enzyme apparently interacts with only one of the four common target site nicking proteins. Following ligation of one strand of the transposable unit to a 5' end of the recipient DNA (Fig. 17b), replication proceeds in the recipient molecule by copying the displaced transposed segment. Complementary strand synthesis on the donor molecule is not proposed by this model, presumably because no available primer exists. When complementary strand synthesis is completed in the recipient molecule, the end of the newly synthesized transposable element is ligated to the free 5' end of the recipient (Fig. 17c). Subsequently, the displaced donor strand is retransferred back to the donor molecule and complementary strand synthesis occurs on the recipient strand from the remaining free 3'-OH end that was initially created by the target site nicking activity (Fig. 17d). The result is that both strands of the transposable unit are conserved in the donor molecule and a newly replicated transposable element exists in the recipient. Interruptions in the proposed process could generate partial or complete semiconservatively replicated transposable units. Also, an aborted transposition attempt via this model to a site adjacent to the original transposable unit could generate deletions with one end point at the transposable unit (see Grindley and Sherratt 1979). By this proposal, specialized transposition is a separate process from precise excision which must result in the double-stranded removal of one flanking recipient repeat sequence in addition to the entire transposable segment. Precise excision may involve *RecA*-independent recombination between the short recipient DNA repeat sequences that flank transposed elements (see section on *recA*-independent systems for recombination). A less likely and yet untested alternative is that transposition is a non-reciprocal exchange that usually results in loss of the donor molecule (Bennett et al. 1977a; Bukhari 1977).

c) The Fusion Model for Transposition. Over the past few years there were repeated observations in which transposable units caused the fusion of two replicons, where only one of the replicons initially contained the element (Faelen et al. 1975; Heffron et al. 1978; Shapiro 1979). Although the model presented above can be adapted to creating replicon fusions (see Grindley and

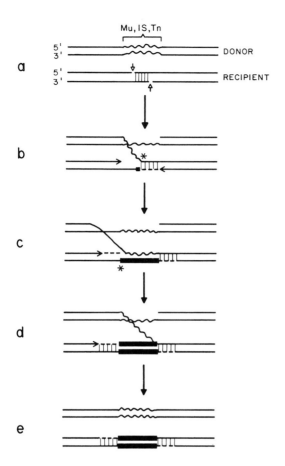

Fig. 17. A single-strand transfer model for transposition. Donor and recipient double-stranded DNA regions are represented by *pairs of horizontal
lines*. The discrete transposable DNA unit is depicted by *squiggled lines*.
Newly replicated chromosomal DNA is represented by *dashed lines* while newly
synthesized transposable segment sequences are depicted by a *filled rectangle*. (a) Transposition is initiated by an enzyme complex that causes two
opposing single-strand cleavages separated by 3-4, 5, 9, or 11-12 bp, depending upon the transposable unit. (b) Subsequently, a second enzymatic
component recognizes, cleaves, and transfers one end of the transposable
segment to the recipient molecule, where it is ligated to a free 5'-end
(see *asterisk*). As the recipient molecule is opened to accomodate the displaced strand, complementary strand synthesis occurs on the lower strand
from the free 3'-end (see *dashed line*). (c) Complementary strand synthesis
continues in the recipient molecule until the entire transposable element
has been copied, at which point the bottom recipient strand is ligated
(asterisk). Because of lack of a primer, no complementary strand synthesis
has occurred in the donor molecule. (d) Thus, the displaced strand is retransferred to the donor molecule and complementary strand synthesis occurs
on the top recipient strand from the free 3'-end. (e) Following ligation
of the ends of the transposable segment to the adjacent chromosomal DNA,
the donor transposable unit is completely conserved. For more details
of this proposed model, see Grindley and Sherratt (1979)

Sherratt 1979), Shapiro (1979) has proposed a different approach to the transposition event. According to this scheme, a circular donor molecule containing one copy of a transposable unit is co-integrated into the recipient molecule. Subsequently, the donor molecule is deleted along with a hybrid copy of the transposable unit while leaving another copy of the transposed segment flanked by duplicated recipient sequences. The details of this proposal are given in Fig. 18 (Shapiro 1979). A significant facet of this model, cointegrate formation/replicon fusion, is a normal inter-mediate structure of the transposition process between inter-acting circular molecules and can be observed whenever the final reciprocal exchange event is blocked. Furthermore, this model can produce deletion and inversion events during transposition if the donor and target DNA regions are on the same molecule (for details see Shapiro 1979). More recently, a very similar transposition model was proposed by Arthur and Sherratt (1979).

d) Other Aspects of Transposition. Do transposable elements exist, as transposition intermediates, in the autonomous, nonreplicating circular state? By biochemically splicing a known small replicon into the central sequences of a transposable element, Cohen et al. (1979) have isolated a self-replicating transposition "inter-mediate". Although one wonders if this structure is merely a pro-duct of precise excision, it should prove valuable in further defining the transposition process. Similarly, further study of the recently isolated single-strand phages carrying transposons may aid our understanding of transposition (Nomura et al. 1978; Ray and Kook 1978).

The actual recognition sites in transposons or in recipient DNA for the transposition-enzyme complex have not been deciphered, but it is noteworthy that the ends of transposable elements are palindromic and A+T rich. Palindromic DNA sequences are known to be protein interaction sites for a variety of enzymes and repres-sor molecules; and A+T rich regions are susceptible to "localized denaturation", perhaps prior to cleavage (see Vogel 1977). It seems likely that the inverted repeat sequences within and at the ends of the termini of transposable segments serve as recog-nition sequences. Comparison among the ends of various trans-posable elements has revealed that the ends of Tn*3* and γ-δ (both of which produce 5 bp direct repetitions; M. Guyer and N. Grind-ley, personal communication) or the ends of Tn*10*, Tn*9*/IS*1*, and Tn*903* (all of which produce 9 bp direct repeats; Kleckner 1979) share significant segments of homology. In addition, Kleckner (1979) has reported very limited evidence for sequence homology between a site internal to Tn*10* and a region in the recipient molecule, but some distance from the insertion site, that helps align the incoming Tn*10* element with respect to its insertion site. Alternatively, and more likely, the non-random clustering of transposon insertion sites suggests that the enzyme component that makes the 3-4, 5, 9, or 11-12 bp staggered cut in the reci-pient sequence is similar to type I restriction endonucleases, which are known to cleave DNA at a distance from the actual re-cognition site (Rosamond et al. 1979). Finally, one wonders if the apparent regulation of transposition frequency is not a re-sult of modification of the recipient DNA recognition sequences.

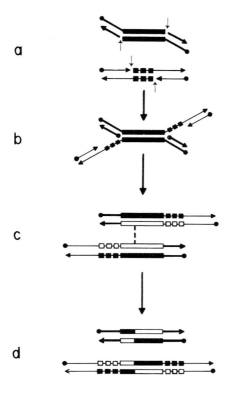

Fig. 18. Fusion model of transposition. Double-stranded donor *(thick lines)*
and recipient *(thin lines)* DNA regions are shown horizontally. Only the interacting portions of these DNA regions are illustrated; they may exist on
the same or different molecules. The transposable DNA unit is represented
by a *rectangle*, while the target site is depicted by *small boxes*. (a) Transposition is initiated by four single-strand cleavages (shown by small *vertical arrows*), one at each end of the target site and transposable DNA segment. The two pairs of cleavages must have opposite polarities. *Arrowheads*
and *dots* on these strands represent 3'-OH and 5'-PO_4 ends, respectively.
(b) Subsequently, the donor and recipient strands are joined in a chi-shaped
structure by ligation at points marked by small arrows. Replication from the
free 3'-end is thought to produce two duplex DNA regions, each carrying a
semi-conservatively replicated transposable unit, as shown in (c). Newly
synthesized DNA is represented by *open blocks* and *rectangles*. If both initial interacting DNA regions represented different circular molecules, this
step would result in a figure eight or replicon fusion structure. A final
reciprocal exchange is proposed to take place between the repeated transposable units (see *vertical dashed line*), generating separate donor and
recipient molecules, each carrying one hybrid copy of the transposable unit.
(d) In addition, the recipient molecule contains direct repeats of the target sequence flanking the newly inserted DNA segment. For details see
Shapiro (1979)

V. Bacterial System(s) for *recA*-independent Recombination

The spontaneous occurrence of chromosomal deletion or duplication events in the absence of known general recombination systems has been recognized for some years (reviewed by Franklin 1971). The discovery of transposable DNA units and our newly acquired knowledge of how Mu phage and IS elements create chromosomal rearrangements can explain the formation of virtually any chromosomal reshuffling. Despite these remarkable findings, there are still many spontaneous deletion and duplication events that do not appear to involve either general recombination or any of the known specialized recombination systems. The recent advances in nucleotide sequencing techniques have allowed for a long-awaited examination of the nucleotide sequence relationships among the ends of these aberrant duplication and deletion mutations.

Evidence obtained from nucleotide sequence studies of *lacI* gene mutants indicates that DNA segments located between identically oriented repeats of 5 or 8 bp can be deleted. The deletion always removes one repeat sequence and all of the intervening sequences. Also, duplication of a DNA segment that lies between small, identically oriented repeats has been observed (see Farabaugh et al. 1978). However, it is not known if these sequence-specific chromosomal alterations occur *recA*-independently. In a separate study, nearly precise excision of Tn*10*, which occurs in the absence of Rec ability, has been observed to result apparently from recombination between 24 bp repeat sequences that are located at each Tn*10* terminus (Ross et al. 1979b). Although all of these rearrangements could have occurred during replication or repair by mispairing events, one can also speculate that an uncharacterized bacterial system(s) for genetic exchanges at short, repeat DNA sequences might be responsible. As mentioned previously, precise excision by cleavage at the specific termini of a transposable element or Mu prophage and subsequent closure of the chromosome would leave one extra 3-4, 5, 9, or 11-12 bp copy of the recipient site sequence. Perhaps some system like that discussed above is involved in the deletion of Tn or IS units so that one copy of the repeated recipient site sequence and the entire intervening sequence (i.e., the Tn or IS unit) are deleted. Such a system is not entirely without precedent. The integration and excision of lambda phage occurs between 15 bp common core repeat sequences, one each in *att*P and *att*B. The sequences surrounding the core region are involved in the specificity and requirements of the reaction. Likewise, the chromosomal sequences surrounding any repeat sequence on a molecule may affect its reactivity/involvement in recombinational exchanges. Though seemingly logical, these suggestions are very speculative.

An unusual recombination pathway in *E. coli* has recently been detected during the study of recombination between different, genetically marked λ phage. Neither host Rec nor λ Red, Int or Der pathways of recombination appear to be responsible. In *recA* cells, these recombination events occur at a frequency of ∿10% of that seen in Rec[+] hosts and appear to involve exchanges between homologous DNA regions. Furthermore, these events require RNA polymerase and probably active transcription; this has been termed the "Rpo pathway". It has been proposed that some local change (e.g., unwinding) of DNA structure caused by transcription is required for this process (Ikeda and Kobayashi 1977). It is not yet known if such a process is involved in specialized transposition or the precise excision events discussed above. It is interesting to note, however, that the Ohtsubos have found sequence homolgy between the ends of IS*1* and known genetic promoters (i.e., RNA polymerase binding sites) and have proposed that RNA polymerase and perhaps the Rpo pathway are involved in Tn or IS unit transposition and/or precise excision (Ohtsubo and Ohtsubo 1978).

D. Effect of Transposable Elements and Viruses on Bacterial Evolution

Transposable genetic elements appear to be normal constituents of bacterial, plasmid, and phage chromosomes. Conjugative plasmids allow for the rapid intercellular dissemination of genetic information and have been identified in many diverse bacterial species (Reanney 1976; Kopecko et al. 1979). Considering the fact that conjugally promiscuous plasmids can transfer between quite different bacterial genera, it appears reasonable to suggest that transposable genetic elements exist universally in bacteria.

Temperate viruses and transposable elements (IS, Tn units) cause mutations and mediate macro-evolutionary chromosomal rearrangements. In addition, some transposable elements are involved in the regulation of gene expression. Specific examples of how transposable DNA units affect bacterial evolution have been given in the previous sections and have been extensively reviewed (Cohen 1976; Cohen and Kopecko 1976; Starlinger and Saedler 1976; Bukhari et al. 1977; Kleckner 1977; Starlinger 1977; Schwesinger 1977). This section is intended to focus on several specific points that have not been discussed elsewhere.

I. The Chromosome - Constancy in the Face of Change

IS*1* through IS*5* and γ-δ, from data presented above, exist as 8, 7, 5, 1, 1, and 4 copies per genome, respectively, which represents 1% of the *E. coli* K-12 chromosome. In fact, many segments of the *E. coli* K-12 chromosome that are bordered by inverted repeat DNA segments of 750 bp or larger have been visualized in the electron microscope (Chow 1977; Ohtsubo and Ohtsubo 1977). Since the chromosome is circular, all genes lie between one or more sets of inversely repeated DNA segments. The known locations of IS elements on the *E. coli* K-12 chromosome and the F plasmid are shown in Figs. 19 and 20. As discussed below, recombination between IS elements carried on plasmids (e.g., the F plasmid) and chromosomally borne IS units results in the chromosomal integration of the plasmid (Davidson et al. 1974).

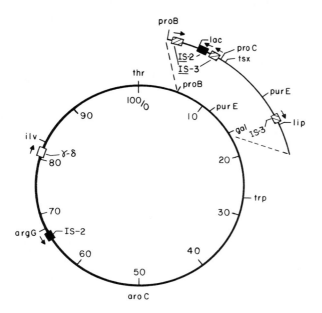

Fig. 19. Insertion sequences act as recombinational "hotspots" for Hfr formation. The circular linkage map of *E. coli* K-12 is schematically divided into 10 min length segments and the chromosomal locations of various genes and insertion sequences are shown. The genetic markers listed include those that affect the synthesis of threonine *(thr)*, proline *(proB, proC)*, adenine *(purE)*, tryptophan *(trp)*, chorismic acid *(aroC)*, arginine *(argG)*, and iso-leucine-valine *(ilv)*; those that affect utilization of lactose *(lac)* or galactose *(gal)*; and those that affect the requirement for lipoate *(lip)* or the resistance to phage T6 *(tsx)* (see Bachmann et al. 1976). Through an intensive study of the molecular relationships among F and several F' plasmids, N. Davidson and co-workers (1974) have deduced the identities and physical orientations of the insertion sequence elements that are actively involved in Hfr formation (indicated by *shaded blocks* in the figure) at six different locations on the *E. coli* genome. Although the *E. coli* K-12 chromosome was estimated, by DNA-DNA hybridization studies, to carry seven copies of IS*2*, the locations of only two such sequences are known. To create an Hfr strain the autonomous F plasmid apparently integrates into the chromosome after complementary pairing between an insertion sequence region in the F plasmid and a homologous sequence in the chromosome. Thus, Hfr polarity would be a direct consequence of the orientations of the homologous sequences on each parental molecule. Twenty-seven different Hfr strains are thought to have been formed by F integration at one of the sequences mapped above. It is likely that other known Hfr strains were constructed by F integration at other insertion sequences located on the chromosome which have not yet been mapped

The frequency of spontaneously formed duplication in the *E. coli* chromosome has been reported to be 10^{-4}-10^{-7}, while spontaneous deletion formation has been estimated to be 10^{-6}-10^{-9}, similar to point mutations (see Starlinger 1977). Non-randomness of end points has been noted for both duplication (Starlinger 1977) and

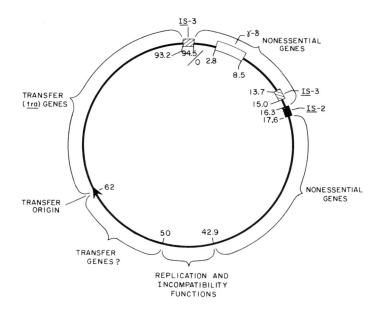

Fig. 20. Structural map of the F (sex) plasmid. In this scaled map, the locations of various functions, constructed from published data, are physically positioned through the use of a kilobase (kb) coordinate mapping system
devised by N. Davidson and co-workers (1974). The map order of 12 genes that
determine the necessary functions for conjugal transfer *(tra)* plus two associated regulatory genes is known, and these genes occupy the last one-
third of the total 94.5 kb pair length of the F plasmid. The origin of conjugal transfer has recently been mapped accurately at kilobase coordinate
62 and the region between kilobase coordinates 50 and 62 may contain other
transfer-associated functions (M. Guyer, personal communication). By the
use of restriction enzyme procedures, construction and study of large deletion mutants of F has led to the conclusion that all functions necessary
for self-replication and incompatibility (a locus responsible for the inability of two similar plasmids to coexist within the same cell) are encoded
on a small 7.1 kb long DNA segment. The regions labeled as non-essential
genes have been genetically deleted without observed consequences on either
F self-replication or transfer ability. Additionally, locations are given
for IS*2*, IS*3*, and the gamma-delta sequences, all of which are separate integration sites for Hfr formation, and which also participate in F plasmid
recombinational rearrangements

deletion mutations (Farabaugh et al. 1978). Spontaneous transposition and inversion events have also been observed in *E. coli*
and their instability suggests the presence of terminal repeat
sequences which subsequently undergo *rec*-dependent reversal of
the original event (Starlinger 1977). Although some of these
events would likely be IS-mediated, recent evidence indicates
that deletions and duplications can occur via a bacterial *recA*-
independent recombination system(s) which recognizes similarly
oriented, short repeat DNA sequences (discussed above). In addition, other as yet uncharacterized recombination systems are

probably also involved. Much ado has been made about the Rec-independence of specialized recombination events and it should be emphasized that *rec*-dependent recombination between these transposable elements (i.e., mobile regions of homology) probably is more important evolutionarily.

It is obvious that chromosomal rearrangements occur quite often in bacteria, but why bacteria apparently remain genetically stable is not so obvious. Evidently, most genetic alterations, even those affecting only a few nucleotides, are not evolutionarily advantageous. Perhaps each small region of the chromosome has been and is being constantly exposed to different surrounding sequences, always striving for the most stable arrangement for that particular environment. Thus, when mutations occur but the environment remains essentially unchanged, the mutations would be selected against. As the environment and, consequently, cellular requirements change, various other chromosomal rearrangements are selected for and the chromosome would remain in a quasistable state while the individual nucleotide regions seek their most stable interrelationships for the new environment. Observations made with cloned eukaryotic DNA in *E. coli* cells (Cohen et al. 1978) or results of analyses of single plasmids isolated from different bacterial hosts (Causey and Brown 1978) support this conjecture.

II. Plasmids - Recombinational Assemblages of Transposable Units

Early electron microscope heteroduplex analyses of the nucleotide sequence relationships among F and related R plasmids indicated that these plasmids are largely homologous. However, areas of homology among these molecules terminated at identical points in several different plasmids, hinting at the modular nature of plasmid evolution (Sharp et al. 1973; Cohen and Kopecko 1976). Cataloging of the properties of many transposons, detected on plasmids obtained from a large variety of different bacteria, has clearly demonstrated the fact that plasmids are evolving through a process of exchange of discrete transposable units. For instance, knowing the location of several transposons on the R1 plasmid, Kopecko et al. (1976) deduced from the earlier data of Sharp et al. (1973) that the R100, R6, and R1 plasmids, though obtained from bacteria originally isolated in Japan, Germany, and England, respectively, carry some of the same transposable elements (see Fig. 21). Moreover, the finding that the shared transposons were located at the same location in each R plasmid further suggested that these plasmids have evolved from a common ancestral plasmid. In another example, Tn*1*, Tn*2*, Tn*3*, Tn*801*, Tn*802*, and Tn*1701* are members of a class of transposons, collectively called TnA elements, that encode ampicillin resistance. These highly homologous TnA elements, detected initially on different plasmids isolated from *Pseudomonas aeruginosa* or *E. coli*, have more recently been detected in plasmids from *Haemophilus* and *Neisseria* (Falkow et al. 1977). Thus, transposons have played a major role in the dissemination of drug resistance genes and other medically relevant determinants (e.g., enterotoxin or K88 antigen synthesis) among bacteria.

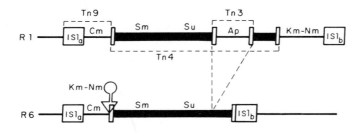

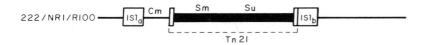

Fig. 21. Relationship of r-determinant regions on R1, R6, and 222/NR1/R100 plasmids. Only the r-determinant region of these plasmids is depicted; direct repeats of IS1 were found to bracket the r-det region (Hu et al. 1975; Ptashne and Cohen 1975). R1, R6 and 222/NR1/R100 are plasmids that were obtained from different bacteria isolated initially in England, Germany, and Japan, respectively. The same transposable sequence encoding chloramphenicol resistance (Cm) is located in all three plasmids as well as the larger transposon encoding streptomycin (Sm) and sulfonamide (Su) resistance. However, in the R1 plasmid the Tn3 (Ap) element is inserted within the Sm, Su transposon and now comprises the composite transposable element, Tn4. The R1 plasmid also carries an additional segment which encodes resistance to kanamycin-neomycin (Km-Nm) and which is deleted spontaneously at a high frequency, but has not been shown to transpose (Kopecko and Cohen 1975; Kopecko et al. 1976). Finally, the R6 plasmid carries an inserted transposable kanamycin/neomycin resistance gene segment located between the Cm and the Sm, Su transposons (see Kopecko et al. 1976, 1978)

Susskind and Botstein (1978) recently reviewed considerable data which indicated that P22 and the lambdoid viruses share some identical functional DNA segments. These authors suggest that these viruses are comprised of a set of interchangeable modular units, but the mechanism of interchange is unknown.

Many plasmids carry, in addition to transposons, IS elements which allow for chromosomal integration of plasmids. For example, during Hfr formation the F plasmid, shown in Figure 20, integrates into the chromosome via an homologous IS segment shared with the chromosome (Fig. 19; Davidson et al. 1974). Plasmid-borne IS units are also used to amplify genes located between two similar IS elements (Cohen 1976; Yagi and Clewell 1977; Schmitt et al. 1979a).

III. Evolution of Transposable Elements

From an evolutionary viewpoint it seems instructive to mention that although IS elements can transpose as independent units, it appears that two nearby, identical IS elements flanking any

sequence can form a transposon (MacHattie and Shapiro 1978;
So et al. 1979; M. Guyer and J.L. Rosner, personal communication). Considering the recent observation that the *his-gnd* region
of the *E. coli* chromosome is a transposon, is every gene on the
chromosome located on a transposable segment at one time or
another when bracketed by IS elements?

There does appear to be a hierarchy of transposable elements
beginning with IS units and ending with the most complex unit,
Mu phage. It is possible that transposons containing long inverted repeat termini (e.g., Tn*10*) which are not susceptible
to *recA*-dependent deletion, are more highly evolved than Tn*9*,
for instance, which is flanked by direct repeats of IS*1*. Furthermore, Tn units encoding functions necessary for their own
transposition would seem more advanced than units that use
host-encoded transposition machinery.

The origins of transposons are unknown, but the similarity of
the terminal sequence, as determined by sequence analysis and
discussed earlier, of some transposable elements would suggest
the existence of four evolutionarily separate classes of transposons (those with 3-4, 5, 9, or 11-12 bp repeats). The minimal
length of a transposable segment is not known but one might
assume that it could be as short as two adjacent recognition/
cleavage sites, which might be as short as 4-8 bp each. However, this seems unlikely because one would then expect there
to be innumerable IS units and, instead, only a few classes
have been detected.

E. Use of Transposable Elements as Experimental Tools

Two recent extensive and masterfully composed reports on the
practical in vivo employment of transposons and Mu phage in experimental genetic manipulations are available (Faelen and
Toussaint 1976; Kleckner et al. 1977). However, a brief listing
of the experimental uses of Tn elements has been included here
for general information and to stimulate interest in these experimental tools. The ease of selection for a drug resistance
phenotype, the ability of transposons to insert at many chromosomal sites at a relatively high frequency and, in addition, to
generate various chromosomal rearrangements make Tn elements
very useful in the laboratory. Transposons can be introduced
into a bacterial cell via infection with defective transducing
phages carrying a Tn element (Kleckner et al. 1975), or by conjugation or transformation of plasmid vectors containing Tn's.
These vectors can be eliminated by a variety of procedures, e.g.,
conditionally lethal mutations or conditions (Kretschmer and
Cohen 1977). Mutations in virtually any chromosomal site can be
obtained by simultaneous selection for the transposon phenotype
(e.g., tetracycline resistance) and loss of the vector. Subsequently, isolated colonies can be replica-plated on appropriate
media to obtain the desired mutants. Transposon-induced mutations can be polar, allowing for the location of genes in an

operon, as well as for genetic mapping of mutants with pheno-
types that are not scorable (Kleckner et al. 1978). Furthermore,
these remarkable Tn-induced mutations are revertible so that
the original phenotype can be restored.

Tn units, inserted into sites immediately adjacent to the trait(s)
of interest, can be used to manipulate a particular trait by
chromosomal duplication or deletion, as described below. In addition, a non-scorable trait can be molecularly cloned by selec-
ting for an easily identifiable adjacent drug resistance transposon. Transposons can also be used in genetic engineering to
generate known restriction endonuclease cleavage sites in specific chromosomal regions. Furthermore, inserted Tn elements can
mediate deletions from the element outward to various chromosomal
sites. Alternatively, two identical Tn elements inserted at nearby chromosomal sites can generate specific deletions of the interposed sequences. By similar techniques, Tn*10* has been used to
generate chromosomal duplications with predetermined end points
(Kleckner et al. 1977) and probably could be used to promote inversions of specific chromosomal genes. It should be emphasized
that fusion of nearby, but unrelated DNA sequences (e.g., two
different operons) can be constructed by deleting the sequences
located between Tn units inserted within each operon, followed
by selection for precise excision of the remaining Tn element.

As noted in previous sections, Tn units can promote the random
chromosomal integration of both plasmids and phage, a condition
which is conducive to the subsequent formation of a variety of
novel specialized transducing phages and plasmids carrying
various chromosomal segments (e.g., F-prime plasmids). Moreover,
recombination between a Tn unit located on a conjugative plasmid, like F, and an identical Tn unit inserted at a known site
on the bacterial chromosome will lead to the construction of Hfr
strains with predetermined transfer origins and orientations.
Transposition of Tn elements to plasmids that are phenotypically
cryptic or not easily scorable offers new possibilities for the
manipulation and study of these elements. In addition to transferring traits of interest through the above manipulations, it
appears that one can construct new transposable elements by inserting known IS or Tn units to either side of the DNA segment
of interest (McHattie and Shapiro 1978; Shapiro and MacHattie
1979; M. Guyer and J.L. Rosner, personal communication) or by
the molecular cloning of a DNA segment into the middle of a
characterized transposon (Goebel et al. 1977; Heffron et al.
1978; J. Manis and B. Kline, personal communication). The usefulness of Tn units as experimental tools is not limited to *E.
coli* by any means, as Tn elements can be transferred to a wide
variety of bacterial genera by conjugally promiscuous plasmids
such as RP4. In addition, the Tn elements in gram-positive bacteria should prove to be just as useful experimentally. Finally,
one should be aware that Mu phage can mediate all of the events
listed above and may be useful under conditions in which Tn units
cannot be employed.

F. Specialized Recombination in Eukaryotic Cells: A Prologue

A rapidly increasing body of evidence indicates that specialized recombination systems are not unique to prokaryotes, but rather are found in a wide range of eukaryotic organisms as well (see Bukhari et al. 1977). From the results of classic genetic studies of maize, McClintock (1957, 1965) has described distinct transposable genetic "controlling elements" that are capable of affecting the expression of various genes. Like IS segments, insertion of a controlling element into a gene can cause inactivation of that gene; restoration of gene activity occurs following excision of the controlling element. In addition, major chromosomal rearrangements (inversion, deletions, duplications) are often found in association with loci carrying these elements. Recently, Nevers and Saedler (1977) have composed an excellent summary of controlling elements in maize and, based on known properties of bacterial IS units, offered an elegantly simple model to explain their behavior (see also Peterson 1977). Furthermore, considerable genetic and cytological data obtained with *Drosophila* suggest the presence of transposable "IS-like" elements that are capable of causing mutations and site-specific deletions (Green 1977). In addition to mediating host chromosomal integration of various eukaryotic viruses, it seems reasonable to assume that specialized recombination systems are involved in transposition of the controlling elements in maize and the IS-like mutants in *Drosophila*, as well as in mediating various chromosomal reshufflings associated with these mobile DNA units.

Several examples of potential specialized recombination systems have been observed in yeast. The alternating and exclusive expression of one of two mating types in *Saccharomyces cerevisiae* has been hypothesized to occur via exchange of specific DNA segments (the cassette model; Hicks et al. 1977). A similar system for mating-type interconversion seemingly exists for *Schizosaccharomyces pombe* (Egel 1977). With the use of relatively new molecular cloning systems, Cameron et al. (1979) recently have physically identified transposon-like elements in the DNA of *S. cerevisiae*. One element, TY1, is 5.6 kb in length and is flanked by a 0.25 kb direct repeat, termed delta. Hybridization studies show that TY1 is present as 35 copies per haploid genome (i.e., 2% of the total haploid DNA content), whereas delta, which is not always associated with TY1, exists as 100 copies per haploid cell. Furthermore, both TY1 transposition and linked chromosomal alterations have been observed. Although speculative, middle repetitive DNA, like TY1, which may be involved in gene regulation in eukaryotes, could be transposed via specialized recombination processes.

Finally, numerous eukaryotic genes have been identified which are interrupted within the coding regions by intervening sequences. These intervening sequences are found in the primary transcript of the gene, but not in the functional mRNA, i.e., they are spliced out and the resulting ends of the RNA molecule are rejoined (Darnell 1978; Knapp et al. 1978; Tilghman et al.

1978). Are there specialized recombination systems for exchange between RNA molecules?

Acknowledgements. I am indebted to my wife, Patricia Guerry-Kopecko, who helped with the editing, writing, and final compilation of material and without whose help and encouragement I could not have written this review. I appreciate the helpful advice of P. Barth, A. Campbell, R. Deonier, N. Grindley, M. Guyer, M. Howe, A. Puhler, H. Saedler, F. Stahl, W. Szybalski and A. Toussaint and thank the numerous co-workers who sent me reprints and preprints of their work, especially P. Bennett, D. Botstein, S. Cohen, M. Howe, N. Kleckner, H. Nash, E. Ohtsubo, H. Saedler, R. Schmitt, W. Szybalski, and A. Toussaint. Finally, I thank L. Baron and J. Wohlhieter for their patience and encouragement and for reviewing the manuscript.

This review was completed in July 1979.

References

Allet, B.: Nucleotide sequences at the ends of bacteriophage Mu DNA. Nature (London) 274, 553-558 (1978)

Allet, B.: Mu insertion duplicates a 5 base pair sequence at the host inserted site. Cell 16, 123-129 (1979)

Arber, W., Iida, S., Jutle, H., Caspers, P., Meyer, J., Hanni, C.: Rearrangements of genetic material in *E. coli* as observed on the bacteriophage P1 plasmid. Cold Spring Harbor Symp. Quant. Biol. 43, 1197-1208 (1979)

Arthur, A., Sherratt, D.: Dissection of the transposition process: a transposon-encoded site-specific recombination system. Mol. Gen. Genet. 175, 267-274 (1979)

Bachmann, B.J., Low, K.B., Taylor, A.L.: Recalibrated linkage map of *E. coli* K-12. Bacteriol. Rev. 40, 116-167 (1976)

Barth, P.T., Datta, N.: Two naturally occurring transposons indistinguishable from Tn 7. J. Gen. Microbiol. 102, 129-134 (1977)

Barth, P.T., Grinter, N.J.: Map of plasmid RP4 derived by insertion of transposon C. J. Mol. Biol. 113, 455-474 (1977)

Barth, P.T., Datta, N., Hedges, R.W., Grinter, N.J.: Transposition of a deoxyribonucleic acid sequence encoding trimethoprim and streptomycin resistances from R483 to other replicons. J. Bacteriol. 125, 800-810 (1976)

Bellett, A.J.D., Busse, H.G., Baldwin, R.L.: Tandem genetic duplications in a derivative of phage lambda. In: The Bacteriophage Lambda (ed. A.D. Hershey), pp. 501-513. New York: Cold Spring Harbor Laboratories 1971

Benedict, M., Fennewald, M., Shapiro, J.: Transposition of a beta-lactamase locus from RP1 into *Pseudomonas putida* degradative plasmids. J. Bacteriol. 129, 809-814 (1977)

Bennett, P. M., Richmond, M.H.: Translocation of a discrete piece of deoxyribonucleic acid carrying an *amp* gene between replicons in *Escherichia coli*. J. Bacteriol. 126, 1-6 (1976)

Bennett, P.M., Grinsted, J., Richmond, M.H.: Transposition of Tn A does not generate deletions. Mol. Gen. Genet. 154, 205-211 (1977a)

Bennett, P.M., Robinson, M.K., Richmond, M.H.: Limitations on the transposition of Tn A. In: Topics in Infectious Diseases, Vol. 2: R Factors: Their Properties and Possible Control (eds. J. Drews, G. Högenauer), pp. 81-99. Berlin, Heidelberg, New York: Springer 1977b

Bennett, P.M., Grinsted, J., Choi, C.L., Richmond, M.H.: Characterization of Tn 501, a transposon determing resistance to mercuric ions. Mol. Gen. Genet. 159, 101-106 (1978a)

Bennett, P.M., Robinson, M.K., Richmond, M.H.: Self-limitation of multiple transposition of Tn A. In: Microbiology-1978 (ed. D. Schlessinger), pp. 16-18. Washington, D.C.: American Society for Microbiology 1978b

Berg, D.: Genetic evidence for two types of gene arrangements in new λ*dv* plasmid mutants. J. Mol. Biol. 86, 59-68 (1974)

Berg, D., Davies, J., Allet, B., Rochaix, J.D.: Transposition of R factor genes to bacteriophage λ. Proc. Natl. Acad. Sci. USA 72, 3628-3632 (1975)

Berg, D.E.: Insertion and excision of the transposable kanamycin resistance determinant Tn 5. In: DNA Insertion Elements, Plasmids and Episomes (eds. A. Bukhari, J. Shapiro, S. Adhya), pp. 205-212. New York: Cold Spring Harbor Laboratory 1977

Beringer, J.E., Beynon, J.L., Buchanan-Wollaston, A.V., Johnson, A.W.B.: Transfer of the drug resistance transposon Tn 5 to *Rhizobium*. Nature (London) 276, 633-634 (1978)

Besemer, J., Herpers, M.: Suppression of polarity of insertion mutations within the *gal* operon of *E. coli*. Mol. Gen. Genet. 151, 295-304 (1977)

Bibb, M.J., Hopwood, D.A.: Genetic and physical characterization of a *Streptomyces coelicolor* sex factor. In: Microbiology-1978 (ed. D. Schlessinger), pp. 139-141. Washington D.C.: American Society for Microbiology 1978

Blattner, F.R., Fiandt, M., Hass, K.K., Twose, P.A., Szybalski, W.: Deletions and insertions in the immunity region of coliphage lambda: revised measurements of the promoter-startpoint distance. Virology 62, 458-471 (1974)

Broker, T.R., Chow, L.T., Soll, L.: The *E. coli* gamma-delta recombination sequence is flanked by inverted duplications. In: DNA Insertion Elements, Plasmids and Episomes (eds. A. Bukhari, J. Shapiro, S. Adhya), pp. 575-580. New York: Cold Spring Harbor Laboratories 1977a

Broker, T.R., Soll, L., Chow, L.T.: Underwound loops in self-renatured DNA can be diagnostic of inverted duplications and translocated sequences J. Mol. Biol. 113, 579-598 (1977b)

Bukhari, A.I.: Reversal of mutator phage Mu integration. J. Mol. Biol. 96, 87-99 (1975)

Bukhari, A.I.: Bacteriophage Mu as a transposition element. Annu. Rev. Genet. 10, 389-412 (1976)

Bukhari, A.I.: Transposition of DNA sequences. In: Genetic Interaction and Gene Transfer (ed. C.W. Anderson), Vol. 29, pp. 218-232. New York: Brookhaven Symposium 1977

Bukhari, A.I., Shapiro, J., Adhya, S., (eds.): DNA Insertion Elements, Plasmids and Episomes. New York: Cold Spring Harbor Laboratories 1977

Calos, M.P., Johnsrud, L., Miller, J.H.: DNA sequence at the integration sites of the insertion element IS1. Cell 13, 411-418 (1978)

Cameron, J.R., Loh, E.Y., Davis, R.W.: Evidence for transposition of dispersed repetitive DNA families in yeast. Cell 16, 739-751 (1979)

Campbell, A.: Distribution of genetic types of transducing phages. Genetics 48, 409-421 (1963)

Campbell, A.: Episomes. New York: Harper and Row 1969

Campbell, A.: How viruses insert their DNA into the DNA of the host cell. Sci. Am. 235, 103-113 (1976)

Campbell, A., Berg, D.E., Botstein, D., Lederberg, E.M., Novick, R.P., Starlinger, P., Szybalski, W.: Nomenclature of transposable elements in prokaryotes. Gene 5, 197-206 (1979)

Casadaban, M.J., Silhavy, T.J., Berman, M.L., Shuman, H.A., Sarthy, A.V., Beckwith, J.R.: Construction and use of gene fusions directed by bacteriophage Mu insertions. In: DNA Insertion Elements, Plasmids, and Episomes (eds. A. Bukhari, J. Shapiro, S. Adhya), pp. 531-535. New York: Cold Spring Harbor Laboratories 1977

Causey, S.C., Brown, L.R.: Transconjugant analysis: limitations on the use of sequence-specific endonucleases for plasmid identification. J. Bacteriol. 135, 1070-1079 (1978)

Chakrabarty, A.M., Friello, D.A., Bopp, L.H.: Transposition of plasmid DNA segments specifying hydrocarbon degradation and their expression in various microorganisms. Proc. Natl. Acad. Sci. (USA) 75, 3109-3112 (1978)

Chattoraj, D., Crasemann, J., Dower, N., Faulds, D., Faulds, P., Malone, R., Stahl, F., Stahl, M.: Chi. Cold Spring Harbor Symp. Quant. Biol. 43, 1063-1068 (1979)

Chilton, M -D., Montoya, A.L., Merlo, D.J., Drummond, M.H., Nutter, R., Gordon, M.P., Nester, E.W.: Restriction endonuclease mapping of a plasmid that confers oncogenicity upon *Agrobacterium tumefaciens* strain B6-806. Plasmid 1, 254-269 (1978)

Chou, J., Casadaban, M.J., Lemaux, P.G., Cohen, S.N.: Identification and characterization of a self-regulated repressor of translocation of the Tn3 element. Proc. Natl. Acad. Sci. USA 76, 4020-4024 (1979)

Chow, L.T.: Sequence arrangements of the *Escherichia coli* chromosome and of putative insertion sequences, as revealed by electron microscopic hetero-duplex studies. J. Mol. Biol. 113, 611-621 (1977)

Chow, L.T., Broker, T.R.: Adjacent insertion sequences IS2 and IS5 in phage Mu mutants and an IS5 in a lambda *darg* phage. J. Bacteriol. 133, 1427-1436 (1978)

Chow, L.T., Bukhari, A.I.: Bacteriophage Mu genome: structural studies on Mu DNA and Mu mutants carrying insertions. In: DNA Insertion Elements, Plasmids and Episomes (eds. A.I. Bukhari, J. Shapiro, S. Adhya), pp. 295-306. New York: Cold Spring Harbor Laboratories 1977

Chow, L.T., Davidson, N., Berg, D.: Electron microscope study of the structures of λdv DNAs. J. Mol. Biol. 86, 69-89 (1974)

Chow, L.T., Kahmann, R., Kamp, D.: Electron microscopic characterization of DNAs of non-defective deletion mutants of bacteriophage Mu. J. Mol. Biol. 113, 591-609 (1977)

Chow, L.T., Broker, T.R., Kahmann, R., Kamp, D.: Comparison of the G DNA inversion in bacteriophages Mu, P1 and P7. In: Microbiology-1978 (ed. D. Schlessinger), pp. 55-56. Washington D.C.: American Society for Microbiology 1978

Clark, A.J.: Recombination-deficient mutants of *E. coli* and other bacteria. Annu. Rev. Genet. 7, 67-86 (1973)

Clark, A.J.: Progress toward a metabolic interpretation of genetic recombination of *Escherichia coli* and bacteriophage lambda. Genetics 78, 259-271 (1974)

Cohen, S.N.: Transposable genetic elements and plasmid evolution. Nature (London) 263, 731-738 (1976)

Cohen, S.N., Kopecko, D.J.: Structural evolution of bacterial plasmids: role of translocating genetic elements and DNA sequence insertions. Fed. Proc. 35, 2031-2036 (1976)

Cohen, S.N., Brevet, J., Cabello, F., Chang, A., Chow, J., Kopecko, D.J., Kretschmer, P.J., Nisen, P., Timmis, K.: Macro- and micro-evolution of bacterial plasmids. In: Microbiology-1978 (ed. D. Schlessinger), pp. 217-220. Washington: American Society for Microbiology 1978

Cohen, S.N., Casadaban, M.J., Chou, J., Tu, C.-P.D.: Studies of the specificity and control of transposition of the Tn3 element. Cold Spring Harbor Symp. Quant. Biol. 43, 1247-1255 (1979)

Cornelis, G., Ghosal, D., Saedler, H.: Tn 951: a new transposon carrying a lactose operon. Mol. Gen. Genet. 160, 215-224 (1978)

Cornelis, G., Ghosal, D., Saedler, H.: Multiple integration sites for the lactose transposon Tn 951 on plasmid RP1 and establishment of a coordinate system for Tn 951. Mol. Gen. Genet. 168, 61-67 (1979)

Couturier, M.: The integration and excision of bacteriophage Mu. Cell 7, 155-163 (1976)

Daniell, E., Abelson, J.: *lac* messenger RNA in *lacZ* gene mutants of *Escherichia coli* caused by insertion of bacteriophage Mu. J. Mol. Biol. 76, 319-322 (1973)

Daniell, E., Abelson, J., Kim, J.S., Davidson, N.: Heteroduplex structures of bacteriophage Mu DNA. Virology 51, 237-239 (1973a)

Daniell, E., Boram, W., Abelson, J.: Genetic mapping of the inversion loop in bacteriophage Mu DNA. Proc. Natl. Acad. Sci. USA 70, 2153-2156 (1973b)

Darnell, J.E., Jr.: Implications of RNA-RNA splicing in evolution of eukaryotic cells. Science 202, 1257-1260 (1978)

Das, A., Court, D., Adhya, S.: Isolation and characterization of conditional lethal mutants of *E. coli* defective in transcription termination factor rho. Proc. Natl. Acad. Sci. USA 73, 1959-1963 (1976)

Davidson, N., Deonier, R.C., Hu, S., Ohtsubo, E.: Electron microscope studies of sequence relations among plasmids of *Escherichia coli*. X. Deoxyribonucleic acid sequence organization of F and F-primes, and the sequence involved in Hfr formation. In: Microbiology-1974 (ed. D. Schlessinger), pp. 56-65. Washington D.C.: American Society for Microbiology 1974

Davies, J., Berg, D., Jorgensen, R., Fiandt, M., Huang, T.-S. R., Courvalin, P., Schloff, J.: Transposable neomycin phosphotransferase. In: Topics in Infectious Diseases, Vol. 2. R Factors: Their Properties and Possible Control (eds. J. Drew, G. Högenauer), pp. 101-110. Berlin, Heidelberg, New York: Springer 1977

Davis, R.W., Parkinson, J.S.: Deletion mutants of bacteriophage lambda. III. Physical structure of *att*. J. Mol. Biol. 56, 403-423 (1971)

Davis, R.W., Simon, M., Davidson, N.: Electron microscope heteroduplex methods for mapping regions of base sequence homology in nucleic acids. In: Methods in Enzymology (eds. L. Grossman, K. Moldave), Vol. 21, pp. 413-428. New York: Academic Press 1971

Decrombrugghe, B., Adhya, S., Gottesman, M., Pastan, I.: Effect of rho on transcription of bacterial operons. Nature New Biol. 241, 260-264 (1973)

Demerec, M., Adelberg, E.A., Clark, A.J., Hartman, P.E.: A proposal for a uniform nomenclature in bacterial genetics. Genetics 54, 61-76 (1966)

Denarie, J., Rosenberg, C., Bergeron, B., Boucher, C., Michel, M., Barate De Bertalmio, M.: Potential of RP4: Mu plasmids for in vivo genetic engineering of gram-negative bacteria. In: DNA Insertions, Plasmids and Episomes (eds. A. Bukhari, J. Shapiro, S. Adhya), pp. 507-520. New York: Cold Spring Harbor Laboratories 1977

Deonier, R.C., Hadley, R.G.: Distribution of inverted IS-length sequences in the *E. coli* K-12 genome. Nature (London) 264, 191-193 (1976)

Deonier, R.C., Hadley, R.G., Hu, M.: Enumeration and identification of IS3 elements in *Escherichia coli* strains. J. Bacteriol. 137, 1421-1424 (1979)

Dobzhansky, T.: Evolution, Genetics, and Man. New York: Wiley 1955

Dougan, G., Saul, M., Twigg, A., Gill, R., Sherratt, D.: Polypeptides expressed in *Escherichia coli* K-12 minicells by transposition elements Tn 1 and Tn 3. J. Bacteriol. 138, 48-54 (1979)

Echols, H., Gingery, R., Moore, L.: Integrative recombination function of phage λ: evidence for a site-specific recombination enzyme. J. Mol. Biol. 34, 251-260 (1968)

Egel, R.: "Flip-flop" control and transposition of mating type genes in fission yeast. In: DNA Insertion Elements, Plasmids and Episomes (eds. A. Bukhari, J. Shapiro, S. Adhya), pp. 447-544. New York, Cold Spring Harbor Laboratories 1977

Embden, J.D.A. van, Veltkamp, E., Stuitje, T., Andreoli, P.M., Nijkamp, H.J.J.: Integration of a transposable DNA sequence which mediates ampicillin resistance into Clo DF13 plasmid DNA: determination of the site and orientation of TnA insertions. Plasmid 1, 204-217 (1978)

Emmons, S.W., Maccosham, V., Baldwin, R.L.: Tandem genetic duplications in phage lambda. III. The frequency of duplication mutants in two derivatives of phage lambda is independent of known recombination systems. J. Mol. Biol. 91, 133-146 (1975)

Engler, J., Inman, R.B.: Site-specific recombination in bacteriophage lambda. J. Mol. Biol. 113, 385-400 (1977)

Faelen, M., Toussaint, A.: Bacteriophage Mu-1: a tool to transpose and to localize bacterial genes. J. Mol. Biol. 104, 525-539 (1976)

Faelen, M., Toussaint, A., De Lafonteyne, J.: Model for the enhancement of λ-gal integration into partially induced Mu-1 lysogens. J. Bacteriol. 121, 873-882 (1975)

Faelen, M., Toussaint, A., Van Montagu, M., Van den Elsacker, S., Engler, G., Schell, J.: In vivo genetic engineering: the Mu-mediated transposition of chromosomal DNA segments onto transmissible plasmids. In: DNA Insertions, Plasmids, and Episomes (eds. A. Bukhari, J. Shapiro, S. Adhya), pp. 521-530. New York: Cold Spring Harbor Laboratories 1977

Faelen, M., Huisman, O., Toussaint, A.: Involvement of phage Mu-1 early functions in Mu-mediated chromosomal rearrangements. Nature (London) 271, 580-582 (1978)

Falkow, S.: Infectious Multiple Drug Resistance. London: Pion 1975

Falkow, S., Elwell, L.P., Roberts, M., Heffron, F., Gill, R.: The transposition of ampicillin resistance: nature of ampicillin resistant Haemophilus influenza and Neisseria gonorrhea. In: Teopics in Infectious Diseases, Vol. 2: R Factors: Their Properties and Possible Control (eds. J. Drews, G. Högenauer), pp. 115-125. Berlin, Heidelberg, New York: Springer 1977

Farabaugh, P.J., Schmeissner, U., Hofer, M., Miller, J.H.: Genetic studies of the lac repressor. VII. On the molecular nature of spontaneous hotspots in the lacI gene of Escherichia coli. J. Mol. Biol. 126, 847-863 (1978)

Fiandt, M., Szybalski, W., Małamy, M.H.: Polar mutations in lac, gal, and phage λ consist of a few IS-DNA sequences inserted with either orientation. Mol. Gen. Genet. 119, 223-231 (1972)

Foster, T.J.: Insertion of the tetracycline resistance translocation unit Tn 10 in the lac operon of Escherichia coli K-12. Mol. Gen. Genet. 154, 305-309 (1977)

Foster, T.J., Howe, T.G.B., Richmond, K.M.V.: Translocation of the tetracycline resistance determinant from R100-1 to the Escherichia coli chromosome. J. Bacteriol. 124, 1153-1158 (1975)

Franklin, N.C.: Illegitimate recombination. In: The Bacteriophage Lambda (ed. A. D. Hershey), pp. 175-194. New York: Cold Spring Harbor Laboratories 1971

Ghosal, D., Saedler, H.: Isolation of the mini-insertions IS-6 and IS-7 of Escherichia coli. Mol. Gen. Genet. 158, 123-128 (1977).

Ghosal, D., Saedler, H.: The DNA sequence of IS6 and its relation to the sequence of IS2. Nature (London) 275, 611-617 (1978)

Ghosal, D., Gross, J., Saedler, H.: The DNA sequence of IS2-7 and generation of mini-insertions by replication of IS2 sequences. Cold Spring Harbor Sym. Quant. Biol. 43, 1193-1196 (1979a)

Ghosal, D., Sommer, H., Saedler, H.: Nucleotide sequence of the transposable DNA element IS2. Nucleic Acids. Res. 6, 1111-1122 (1979b)

Gill, R., Heffron, F., Dougan, G., Falkow, S.: Analysis of sequences transposed by complementation of two classes of transposition-deficient mutants of Tn 3. J. Bacteriol. 136, 742-756 (1978)

Goebel, W., Lindenmaier, W., Pfeifer, F., Schrempf, H., Schelle, B.: Transposition and insertion of intact, deleted and enlarged ampicillin transposon Tn3 from mini-R1 (Rsc) plasmids into transfer factors. Mol. Gen. Genet. 157, 119-129 (1977)

Gottesman, M.E.: The integration and excision of bacteriophage lambda. Cell 1, 69-72 (1974)

Gottesman, M., Rosner, J.L.: Acquisition of a determinant for chloramphenicol resistance by coliphage lambda. Proc. Natl. Acad. Sci. USA 72, 5041-5045 (1975)

Green, M.M.: The case for DNA insertion mutations in *Drosophila*. In: DNA Insertion Elements, Plasmids and Episomes (eds. A. Bukhari, J. Shapiro, S. Adhya), pp. 437-445. New York: Cold Spring Harbor Laboratory 1977

Grindley, N.D.F.: IS1 insertion generates duplication of nine base pair sequence at its target site. Cell 13, 419-426 (1978)

Grindley, N.D.F., Sherratt, D.: Sequence analysis at IS1 insertion sites: models for transposition. Cold Spring Harbor Symp. Quant. Biol. 43, 1257-1261 (1979)

Grinsted, J., Bennett, P.M., Higginson, S., Richmond, M.H.: Regional preference of insertion of Tn501 and Tn802 into RP1 and its derivatives. Mol. Gen. Genet. 166, 313-320 (1978)

Guarneros, G., Echols, H.: Thermal asymmetry of site-specific recombination by bacteriophage λ. Virology 52, 30-38 (1973)

Guyer, M.S.: The gamma-delta sequence of F is an insertion sequence. J. Mol. Biol. 126, 347-365 (1978)

Habermann, P., Klaer, R., Kühn, S., Starlinger, P.: IS4 is found between 11 or 12 base pair duplications. Mol. Gen. Genet. 175, 369-373 (1979)

Hayes, W.: The Genetics of Bacteria and their Viruses. New York: Wiley 1968

Hedges, R.W., Jacob, A.E.: Transposition of ampicillin resistance from RP4 to other replicons. Mol. Gen. Genet. 132, 31-40 (1974)

Hedges, R.W., Matthew, M., Smith, D.I., Cresswell, J.M., Jacob, A.E.: Properties of a transposon conferring resistance to penicillin and streptomycin. Gene 1, 241-253 (1977)

Heffron, F., Rubens, C., Falkow, S.: Translocation of a plasmid DNA sequence which mediates ampicillin resistance: molecular nature and specificity of insertion. Proc. Natl. Acad. Sci. USA 72, 3623-3627 (1975)

Heffron, F., Bedinger, P., Champoux, J.J., Falkow, S.: Deletions affecting the transposition of an antibiotic resistance gene. Proc. Natl. Acad. Sci. USA 74, 702-706 (1977)

Heffron, F., So, M., McCarthy, B.J.: In vitro mutagenesis of a circular DNA molecule by using synthetic restriction sites. Proc. Natl. Acad. Sci. USA 75, 6012-6016 (1978)

Hernalsteens, J.P., Villarroel-Mandiola, R., Van Montagu, M., Schell, J.: Transposition of Tn1 to a broad host-range drug resistance plasmid. In: DNA Insertion Elements, Plasmids and Episomes (eds. A. Bukhari, J. Shapiro, S. Adhya), pp. 179-183. New York: Cold Spring Harbor Laboratories 1977

Hershey, A.D.(ed.): The Bacteriophage Lambda. New York: Cold Spring Harbor Laboratories 1971a

Hershey, A.D.: Persistent heterozygotes in phage T4. Carnegie Inst. Wash. Yearb. 69, 717-722 (1971b)

Hicks, J.B., Strathern, J.N., Herskowitz, I.: The cassette model of mating type interconversion. In: DNA Insertion Elements, Plasmids and Episomes (eds. A. Bukhari, J. Shapiro, S. Adhya), pp. 457-462. New York: Cold Spring Harbor Laboratory 1977

Hirsch, H.J., Starlinger, P., Brachet, P.: Two kinds of insertions in bacterial genes. Mol. Gen. Genet. 119, 191-206 (1972)

Holmans, P., Anderson, G.C., Clowes, R.C.: Tn A-directed deletions and translocations within the R6K plasmid. In: Microbiology-1978 (ed. D. Schlessinger), pp. 38-41. Washington D.C.: American Society for Microbiology 1978

Howe, M.M.: Invertible DNA sequences. Nature (London) 271, 608-610 (1978)

Howe, M., Bade, E.: Molecular biology of bacteriophage Mu. Science 190, 624-632 (1975)

Hsu, M.-T., Davidson, N.: Electron microscope heteroduplex study of the heterogeneity of Mu phage and prophage DNA. Virology 58, 229-239 (1974)

Hu, S., Ohtsubo, E., Davidson, N., Saedler, H.: Electron microscope heteroduplex studies of sequence relations among bacterial plasmids. XII. Identification and mapping of the insertion sequences IS1 and IS2 in F and R plasmids. J. Bacteriol. 122, 764-775 (1975a)

Hu, S., Ptashne, K., Cohen, S.N., Davidson, N.: The αβ-sequence of F is IS3. J. Bacteriol. 123, 687-692 (1975b)

Hull, R.H., Gill, G.S., Curtis, R. III: Genetic characterization of Mu-like bacteriophage D108. J. Virol. 27, 513-518 (1978)

Iida, S., Meyer, J.: Involvement of insertion sequences in the formation of hybrid phages between phage P1 and R plasmid NR1. Abstr. Annu. Meet. Am. Soc. Microbiol., p. 139 (1979)

Ikeda, H., Kobayashi, I.: Involvement of DNA-dependent RNA polymerase in a recA-independent pathway of genetic recombination in Escherichia coli. Proc. Natl. Acad. Sci. USA 74, 3932-3936 (1977)

Inman, R.B., Schnos, M.: Partial denaturation of thymine and 5-bromouracil containing λ DNA in alkali. J. Mol. Biol. 49, 93-98 (1970)

Jacob, A.E., Cresswell, J.M., Hedges, R.W.: Molecular characterization of the P group plasmid R68 and variants with enhanced chromosome mobilizing ability. Fed. Eur. Microbiol. Soc. Lett. 1, 71-74 (1977)

Jacoby, G.A., Rogers, J.E., Jacob, A.E., Hedges, R.W.: Transposition of Pseudomonas toluene-degrading genes and expression in Escherichia coli. Nature (London) 274, 179-180 (1978)

Johnsrud, L.: DNA sequence of the transposable element IS1. Mol. Gen. Genet. 169, 213-218 (1979)

Johnsrud, L., Calos, M.P., Miller, J.H.: The transposon Tn9 generates a 9 base pair repeated sequence during integration. Cell 15, 1209-1219 (1978)

Jordan, E., Saedler, H., Starlinger, P.: 0° - and strong polar mutations in the gal operon are insertions. Mol. Gen. Genet. 102, 353-363 (1968)

Kahmann, R., Kamp, D., Zipser, D.: Mapping of restriction sites in Mu DNA. In: DNA Insertion Elements, Plasmids and Episomes (eds. A. Bukhari, J. Shapiro, S. Adhya), pp. 335-339. New York: Cold Spring Harbor Laboratory 1977

Kamp, D., Kahmann, R., Zipser, D., Broker, T.R., Chow, L.T.: Inversion of the G DNA segment of phage Mu controls phage infectivity. Nature (London) 271, 577-580 (1978)

Kamp, D., Chow, L.T., Broker, T.R., Kwoh, D., Zipser, D., Kahmann, R.: Site-specific recombination in phage Mu. Cold Spring Harbor Symp. Quant. Biol. 43, 1159-1167 (1979)

Kikuchi, Y., Nash, H.A.: The bacteriophage λ int gene product. A filter assay for genetic recombination, purification of int and specific binding to DNA. J. Biol. Chem. 253, 7149-7157 (1978)

Kleckner, N.: Translocatable elements in prokaryotes. Cell 11, 11-23 (1977)

Kleckner, N.: DNA sequence analysis of Tn 10 insertions: origin and role of 9 base pair flanking repetitions during Tn10 translocation. Cell 16, 711-720 (1979)

Kleckner, N., Ross, D.G.: Translocation and other recombination events involving the tetracycline resistance element Tn10. Cold Spring Harbor Symp. Quant. Biol. 43, 1233-1246 (1979)

Kleckner, N., Chan, R.K., Tye, B.-K., Botstein, D.: Mutagenesis by insertion of a drug resistance element carrying an inverted repetition. J. Mol. Biol. 97, 561-575 (1975)

Kleckner, N., Roth, J., Botstein, D.: Genetic engineering in vivo using translocatable drug-resistance elements. J. Mol. Biol. 116, 125-159 (1977)

Kleckner, N., Barker, D.F., Ross, D.G., Botstein, D., Swan, J., Zabeau, M.: Properties of the translocatable tetracycline-resistance element Tn 10 in *Escherichia coli* and bacteriophage lambda. Genetics 90, 427-461 (1978)

Kleckner, N., Reichardt, K., Botstein, D.: Inversions and deletions of the Salmonella chromosome generated by the translocatable tetracycline resistance element Tn10. J. Mol. Biol. 127, 89-115 (1979a)

Kleckner, N., Reichardt, K., Botstein, D.: Preferred sites for the insertion of the translocatable tetracycline-resistance element Tn10. J. Mol. Biol. (1979b)

Knapp, G., Beckmann, J.S., Fuhrman, P.F., Abelson, J.: Transcription and processing of intervening sequences in yeast tRNA genes. Cell 14, 221-236 (1978)

Kondo, E., Mitsuhashi, S.: Drug resistance of enteric bacteria. IV. Active transducing bacteriophage P1 Cm produced by combination of F factor with bacteriophage P1. J. Bacteriol. 88, 1266-1276 (1964)

Kopecko, D.J., Cohen, S.N.: Site-specific *recA*-independent recombination between bacterial plasmids: involvement of palindromes at the recombinational loci. Proc. Natl. Acad. Sci. USA 72, 1371-1377 (1975)

Kopecko, D.J., Brevet, J., Cohen, S.N.: Involvement of multiple translocating DNA segments and recombinational hotspots in the structural evolution of bacterial plasmids. J. Mol. Biol. 108, 333-360 (1976)

Kopecko, D.J., Brevet, J., Cohen, S.N., Nisen, P.D., Zabielski, J.: Involvement of the termini of translocating DNA segments as recombinational hot spots in the structural evolution of plasmids. In: Microbiology-1978 (ed. D. Schlessinger), pp. 25-28. Washington D.C.: American Society for Microbiology 1978

Kopecko, D.J., Johnson, E.M., Baron, L.S.: Genetic and molecular aspects of bacterial heredity. In: Burrow's Textbook of Microbiology, 21st ed. (ed. W. Freeman). Philadelphia: Saunders 1979

Kornberg, A.: DNA Synthesis. San Francisco: Freeman 1974

Kretschmer, P.J., Cohen, S.N.: Selected translocation of plasmid genes: frequency and regional specificity of translocation of the Tn 3 element. J. Bacteriol. 130, 888-899 (1977)

Kuhn, S., Fritz, H.-J., Starlinger, P.: Close vicinity of IS1 integration sites in the leader sequence of the *gal* operon of *E. coli*. Mol. Gen. Genet. 167, 235-241 (1979)

Landy, A., Ross, W.: Viral integration and excision: structure of the lambda *att* sites. Science 197, 1147-1160 (1977)

Lewin, B.: Gene Expression - 3. New York: Wiley 1977

Little, J.W., Gottesman, M.: Defective lambda particles whose DNA carries only a single cohesive end. In: The Bacteriophage Lambda (ed. A.D. Hershey), pp. 371-394. New York: Cold Spring Harbor Laboratory 1971

Ljungquist, E., Bukhari, A.I.: State of prophage Mu DNA upon induction. Proc. Natl. Acad. Sci. USA 74, 3143-3147 (1977)

Ljungquist, E., Khatoon. H., Du Bow, M., Ambrosio, L., De Bruijn, F., Bukhari, A.I.: Integration of bacteriophage Mu DNA. Cold Spring Harbor Symp. Quant. Biol. 43, 1151-1158 (1979)

Luria, S.E., Darnell, J.E., Baltimore, D., Campbell, A.: General Virology 3rd ed. New York: Wiley 1978

McHattie, L.A., Jackowski, J.B.: Physical structure and deletion effects of the chloramphenicol resistance element Tn 9 in phage lambda. In: DNA Insertion Elements, Plasmids and Episomes (eds. A. Bukhari, J. Shapiro, S. Adhya), pp. 219-228. New York: Cold Spring Harbor Laboratory 1977

MacHattie, L.A., Shapiro, J.A.: Chromosomal integration of phage λ by means of a DNA insertion element. Proc. Natl. Acad. Sci. USA 75, 1490-1494 (1978)

Mahajan, S.K., Datta, A.R.: Mechanisms of recombination by the Rec BC and the Rec F pathways following conjugation in *Escherichia coli* K-12. Mol. Gen. Genet. 169, 67-78 (1979)

Malamy, M.H.: Some properties of insertion mutations in the *lac* operon. In: The Lactose Operon (eds. J.R. Beckwith, D. Zipser), pp. 359-373. New York: Cold Spring Harbor Laboratory 1970

Malamy, M.H., Fiandt, M., Szybalski, W.: Electron microscopy of polar insertions in the *lac* operon of *Escherichia coli*. Mol. Gen. Genet. 119, 207-222 (1972)

Malone, R.E., Chattoraj, D.K., Foulds, D.H., Stahl, M.M., Stahl, F.W.: Hotspots for generalized recombination in the *Escherichia coli* chromosome. J. Mol. Biol. 121, 473-491 (1978)

Martuscelli, J., Taylor, A.L., Cummings, D., Chapman, V., Delong, S., Canedo, L.: Electron microscope evidence for linear insertion of bacteriophage Mu-1 in lysogenic bacteria. J. Virol. 8, 551-563 (1971)

Matsubara, K., Kaiser, A.D.: λdv: an autonomously replicating DNA fragment. Cold Spring Harbor Symp. Quant. Biol. 33, 769-775 (1968)

Matsubara, K., Otsuji, Y.: Preparation of plasmids from lambdoid phages and studies on their incompatibilities. Plasmid 1, 284-296 (1978)

Mattes, R., Burkardt, H.J., Schmitt, R.: Repetition of tetracycline resistance determinant genes on R plasmid pRSD1 in *Escherichia coli*. Mol. Gen. Genet. 168, 173-184 (1979)

Maxam, A., Gilbert, W.: A new method for sequencing DNA. Proc. Natl. Acad. Sci. USA 74, 560-564 (1977)

McClintock, B.: Chromosome organization and gene expression. Cold Spring Harbor Symp. Quant. Biol. 16, 13-47 (1952)

McClintock, B.: Controlling elements and the gene. Cold Spring Harbor Symp. Quant. Biol. 21, 197-226 (1957)

McClintock, B.: The control of gene action in maize. Brookhaven Symp. Biol. 18, 162-184 (1965)

McEntee, K., Epstein, W.: Isolation and characterization of specialized transducing bacteriophages for the *recA* gene of *Escherichia coli*. Virology 77, 306-318 (1977)

Meselson, M., Weigle, J.: Chromosome breakage accompanying genetic recombination in bacteriophage. Proc. Natl. Acad. Sci. USA 47, 857-868 (1961)

Meyer, R., Boch, G., Shapiro, J.: Transposition of DNA inserted into deletions of the Tn 5 kanamycin resistance element. Mol. Gen. Genet. 171, 7-13 (1979)

Michaelis, G., Saedler, H., Venkov, P., Starlinger, P.: Two insertions in the galactose operon having different sizes but homologous DNA sequences. Mol. Gen. Genet. 104, 371-377 (1969)

Miller, H.I., Friedman, D.I.: Isolation of *Escherichia coli* mutants unable to support lambda integrative recombination. In: Insertion Elements, Plasmids and Episomes (eds. A. Bukhari, J. Shapiro, S. Adhya), pp. 349-356. New York: Cold Spring Harbor Laboratory 1977

Mizuuchi, K., Gellert, M., Nash, H.A.: Involvement of supertwisted DNA in integrative recombination of bacteriophage λ. J. Mol. Biol. 121, 375-392 (1978)

Nash, H.A.: Integration and excision of bacteriophage lambda. Curr. Top. Microbiol. Immunol. 78, 171-199 (1977)

Nash, H.A., Kikuchi, Y., Mizuuchi, K., Gellert, M.: Integrative recombination of bacteriophage lambda: genetics and biochemistry. In: Integration and Excision of DNA Molecules (eds. P. Hofschneider, P. Starlinger), pp. 21-27. Berlin, Heidelberg, New York: Springer 1978

Nevers, P., Saedler, H.: Transposable genetic elements as agents of gene instability and chromosomal rearrangements. Nature (London) 268, 109-115 (1977)

Nevers, P., Saedler, H.: Mapping and characterization of an *E. coli* mutant defective in IS1-mediated deletion formation. Mol. Gen. Genet. 160, 209-214 (1978)

Nisen, P.D., Kopecko, D.J., Chou, J., Cohen, S.N.: Site-specific deletions occurring adjacent to the termini of a transposable ampicillin resistance element (Tn 3). J. Mol. Biol. 117, 975-998 (1977)

Noel, K.D., Ames, G.F.-L.: Evidence for a common mechanism for the insertion of the Tn 10 transposon and for the generation of Tn 10-stimulated deletions. Mol. Gen. Genet. 166, 217-223 (1978)

Nomura, N., Yamagishi, H., Oka, A.: Isolation and characterization of transducing coliphage fd carrying a kanamycin resistance gene. Gene 3, 39-51 (1978)

Novick, R.P., Clowes, R.C., Cohen, S.N., Curtiss III, R., Datta, N., Falkow, S.: Uniform nomenclature for bacterial plasmids: a proposal. Bacteriol. Rev. 40, 168-189 (1976)

Novick, R.P., Edelman, I., Schwesinger, M.D., Gruss, A.D., Swanson, E.C., Pattee, P.A.: Genetic translocation in *Staphylococcus aureus*. Proc. Natl. Acad. Sci. USA 76, 400-404 (1979)

O'Day, K.J., Schultz, D.W., Howe, M.M.: Search for integration-deficient mutants of bacteriophage Mu. In: Microbiology-1978 (ed. D. Schlessinger), pp. 48-51. Washington D.C.: American Society for Microbiology 1978

Ohtsubo, H., Ohtsubo, E.: Repeated DNA sequences in plasmids, phages and bacterial chromosomes. In: DNA Insertion Elements, Plasmids and Episomes (eds. A. Bukhari, J. Shapiro, S. Adhya), pp. 49-63. New York: Cold Spring Harbor Laboratory 1977

Ohtsubo, H., Ohtsubo, E.: Nucleotide sequence of an insertion element IS1. Proc. Natl. Acad. Sci. USA 75, 615-619 (1978)

Ohtsubo, H., Ohmori, H., Ohtsubo, E.: Nucleotide sequence analysis of Tn3 (Ap): implications for insertion and deletion. Cold Spring Harbor Symp. Quant. Biol. 43, 1267-1277 (1979)

Oka, A., Nomura, N., Sugimoto, K., Sugisaki, H., Takanami, M.: Nucleotide sequence at the insertion sites of a kanamycin transposon. Nature (London) 276, 845-847 (1978)

Palchaudhuri, S., Goldberg, S., Lawrence, M.: Transposon-like behavior of *E. coli* histidine genes. Abstr. Annu. Meet. Am. Soc. Microbiol., p. 140 (1979)

Parkinson, J.S.: Deletion mutants of bacteriophage lambda. II. Genetic properties of *att*-defective mutants. J. Mol. Biol. 56, 385-401 (1971)

Parkinson, J.S., Huskey, R.J.: Deletion mutants of bacteriophage lambda. I. Isolation and initial characterization. J. Mol. Biol. 56, 369-384 (1971)

Peterson, P.A.: The position hypothesis for controlling elements in maize. In: DNA Insertion Elements, Plasmids and Episomes (eds. A. Bukhari, J. Shapiro, S. Adhya), pp. 429-435. New York: Cold Spring Harbor Laboratory 1977

Pfeifer, D., Habermann, P., Kubai-Maroni, D.: Specific sites for integration of IS elements within the transferase gene of the *gal* operon of *E. coli* K-12. In: DNA Insertion Elements, Plasmids, and Episomes (eds. A. Bukhari, J. Shapiro, S. Adhya), pp. 31-36. New York: Cold Spring Harbor Laboratory 1977

Pilacinski, W., Mosharrafa, E., Edmundson, R., Zissler, J., Fiandt, M., Szybalski, W.: Insertion sequence IS2 associated with *int*-constitutive mutants of bacteriophage lambda. Gene 2, 61-74 (1977)

Potter, H., Dressler, D.: DNA recombination: in vivo and in vitro studies. Cold Spring Harbor Symp. Quant. Biol. 43, 969-985 (1979)

Ptashne, K., Cohen, S.N.: Occurrence of insertion sequence (IS) regions on plasmid deoxyribonucleic acid as direct and inverted nucleotide sequence duplications. J. Bacteriol. 122, 776-781 (1975)

Puhler, A., Burkardt, H.-J.: Fertility Inhibition in *Rhizobium lupini* by the resistance plasmid RP4. Mol. Gen. Genet. 162, 163-171 (1978)

Radding, C.M.: Molecular mechanisms in genetic recombination. Annu. Rev. Genet. 7, 87-111 (1973)

Ray, D.S., Kook, K.: Insertion of the Tn 3 transposon into the genome of the single-stranded DNA phage M13. Gene 4, 109-119 (1978)

Reanney, D.: Extrachromosomal elements as possible agents of adaptation and development. Bacteriol. Rev. 40, 552-590 (1976)

Reif, H.J., Saedler, H.: IS1 is involved in deletion formation in the *gal* region of *E. coli* K-12. Mol. Gen. Genet. 137, 17-28 (1975)

Reif, H.J., Saedler, H.: Chromosomal rearrangements in the *gal* region of *E. coli* K-12 after integration of IS1. In: DNA Insertion Elements, Plasmids and Episomes (eds. A. Bukhari, J. Shapiro, S. Adhya), pp. 81-91. New York: Cold Spring Harbor Laboratory 1977

Reiss, G., Burkardt, H., Puhler, A.: Molecular characterization of R68-45, a plasmid with chromosomal donor ability. Hoppeseylers Z. Physiol. Chem. 359, 1139 (1978)

Roberts, J.W.: Transcription termination and its control in *E. coli*. In: RNA Polymerase (eds. R. Losick, M. Chamberlin), pp. 247-271. New York: Cold Spring Harbor Laboratory 1976

Roberts, J.W., Roberts, C.W., Craig, N.L., Phizicky, E.M.: Activity of the *E. coli recA* gene product. Cold Spring Harbor Symp. Quant. Biol. 43, 917-920 (1979)

Rosamond, J., Endlich, B., Linn, S.: Electron microscopic studies of the mechanism of action of the restriction endonuclease of *Escherichia coli* B. J. Mol. Biol. 129, 619-635 (1979)

Ross, D.G., Swan, J., Kleckner, N.: Physical structure of Tn 10-promoted deletions and inversion: role of 1400 base pair inverted repetitions. Cell 16, 721-731 (1979a)

Ross, D.G., Swan, J., Kleckner, N.: Nearly precise excision: a new type of DNA alteration associated with the translocatable element Tn 10. Cell 16, 733-738 (1979b)

Roussel, A., Carlier, C.A., Gerband, C., Chabbert, Y.A., Croissant, O., Blangy, D.: Reversible translocation of antibiotic resistance determinants in *Salmonella ordonez*. Mol. Gen. Genet. 169, 13-25 (1979)

Rubens, C., Heffron, F., Falkow, S.: Transposition of a plasmid deoxyribonucleic acid sequence that mediates ampicillin resistance: independence from host *rec* functions and orientation of insertion. J. Bacteriol. 128, 425-434 (1976)

Saedler, H.: Implications for the evolution of the chromosome and some plasmids. In: DNA Insertion Elements, Plasmids and Episomes (eds. A. Bukhari, J. Shapiro, S. Adhya), pp. 65-72. New York: Cold Spring Harbor Laboratory 1977

Saedler, H., Ghosal, D.: Properties of DNA insertion elements in *E. coli*. In: Topics in Infectious Diseases, Vol. 2: R Factors: Their Properties and Possible Control (eds. J. Drews, G. Högenauer), pp. 131-140. Berlin, Heidelberg, New York: Springer 1977

Saedler, H., Heiss, B.: Multiple copies of the insertion DNA sequences IS1 and IS2 in the chromosome of *E. coli* K-12. Mol. Gen. Genet. $\underline{122}$, 267-277 (1973)

Saedler, H., Starlinger, P.: O^O-mutations in the galactose operon in *E. coli*. I. Genetic characterization. Mol. Gen. Genet. $\underline{100}$, 178-189 (1967)

Saedler, H., Reif, H. J., Hu, S., Davidson, N.: IS2, a genetic element for turn-off and turn-on of gene activity in *E. coli*. Mol. Gen. Genet. $\underline{132}$, 265-289 (1974)

Saedler, H., Besemer, J., Kemper, B., Rosenwirth, B., Starlinger, P.: Insertion mutations in the control region of the *gal* operon in *E. coli*. I. Biological characterization of the mutations. Mol. Gen. Genet. $\underline{115}$, 258-265 (1972)

Schell, J., Van Montagu, M.: Transfer, maintenance, and expression of bacterial Ti-plasmid DNA in plant cells transformed with *Agrobacterium tumefaciens*. In: Genetic Interaction and Gene Transfer (ed. C.W. Anderson), Vol. 29, pp. 36-49. New York: Brookhaven Symposium 1977

Schmidt, F., Besemer, J., Starlinger, P.: The isolation of IS1 and IS2 DNA. Mol. Gen. Genet. $\underline{145}$, 145-154 (1976)

Schmitt, R., Bernhard, E., Mattes, R.: Characterization of Tn1721, a new transposon containing tetracycline resistance genes capable of amplification. Mol. Gen. Genet. $\underline{172}$, 53-65 (1979a)

Schmitt, R., Mattes, R., Schmid, K., Altenbuchner, J.: Raf-plasmids in strains of *Escherichia coli* and their possible role in enteropathogeny. In: Plasmids of Medical, Environmental and Commercial Importance (eds. K. Timmis, A. Puhler), pp. 199-210. Amsterdam-New York: Elsevier/North Holland 1979b

Schöffl, F., Puhler, A.: Intramolecular amplification of the tetracycline resistance determinant of transposon Tn1771 in *Escherichia coli*. Genet. Res. Camb. $\underline{33}$, 253-262 (1979)

Schöffl, F., Burkardt, H.J.: Intramolecular amplification of the tetracycline resistance determinant of transposon Tn 1771 in *Escherichia coli*. In: Plasmids of Medical, Environmental and Commercial Importance (eds. K. Timmis, A. Puhler), pp. 211-223. Amsterdam-New York: Elsevier/North Holland 1979

Schwesinger, M.: Additive recombination in bacteria. Microbiol. Rev. $\underline{41}$, 872-902 (1977)

Shapiro, J.A.: Mutations caused by the insertion of genetic material into the galactose operon of *Escherichia coli*. J. Mol. Biol. $\underline{40}$, 93-105 (1969)

Shapiro, J.A.: Molecular model for the transposition and replication of bacteriophage Mu and other transposable elements. Proc. Natl. Acad. Sci. USA $\underline{76}$, 1933-1937 (1979)

Shapiro, J.A., MacHattie, L.A.: Integration and excision of prophage λ mediated by the IS1 element. Cold Spring Harbor Symp. Quant. Biol. $\underline{43}$, 1135-1142 (1979)

Shapiro, J.A., Sporn, P.: Tn 402: a new transposable element determining trimethoprin resistance that inserts in bacteriophage lambda. J. Bacteriol. $\underline{129}$, 1632-1635 (1977)

Sharp, P., Cohen, S.N., Davidson, N.: Electron microscope heteroduplex studies of sequence relations among plasmids of *Escherichia coli*. II. Structure of drug resistance (R) factors and F factors. J. Mol. Biol. $\underline{75}$, 235-255 (1973)

Shibata, T., Das Gupta, C., Cunningham, R.P., Radding, C.M.: Purified *Escherichia coli recA* protein catalyzes homologous pairing of super-helical DNA and single-stranded fragments. Proc. Natl. Acad. Sci. USA 76, 1638-1642 (1979)

Shimada, K., Weisberg, R.A., Gottesman, M.E.: Prophage lambda at unusual chromosomal locations. I. Location of the secondary attachment sites and the properties of the lysogens. J. Mol. Biol. 63, 483-503 (1972)

Silverman, M., Zieg, J., Hilmen, M., Simon, M.: Phase variation in *Salmonella*: genetic analysis of a recombinational switch. Proc. Natl. Acad. Sci. USA 76, 391-395 (1979)

So, M., Heffron, F., McCarthy, B.J.: The *E. coli* gene encoding heat stable toxin is a bacterial transposon flanked by inverted repeats of IS1. Nature (London) 277, 453-456 (1979)

Stahl, F.W., Crasemann, J.N., Stahl, M.M.: Rec-mediated recombinational hot spot activity in phage lambda. III. Chi mutations are site mutations stimulating rec-mediated recombination. J. Mol. Biol. 94, 203-212 (1975)

Stanisich, V.A., Bennett, P.M., Richmond, M.H.: Characterization of a translocation unit encoding resistance to mercuric ions that occurs on a nonconjugative plasmid in *Pseudomonas aeruginosa*. J. Bacteriol. 129, 1227-1233 (1977)

Starlinger, P.: DNA rearrangements in prokaryotes. Annu. Rev. Genet. 11, 103-126 (1977)

Starlinger, P., Saedler, H.: Insertion mutations in microorganisms. Biochimie 54, 177-185 (1972)

Starlinger, P., Saedler, H.: IS-elements in microoragnisms. Curr. Top. Microbiol. Immunol. 75, 111-152 (1976)

Susskind, M.M., Botstein, D.: Molecular genetics of bacteriophage P22. Microbiol. Rev. 42, 385-413 (1978)

Szybalski, E.H., Szybalski, W.: A comprehensive molecular map of bacterio-phage lambda. Gene 7, 217-270 (1979)

Szybalski, W.: IS elements in *Escherichia coli*, plasmids, and bacterio-phages. In: DNA Insertion Elements, Plasmids, and Episomes (eds. A. Buk-hari, J. Shapiro, S. Adhya), pp. 583-590. New York: Cold Spring Harbor Laboratory 1977

Taylor, A.L.: Bacteriophage-induced mutation in *Escherichia coli*. Proc. Natl. Acad. Sci. USA 50, 1043-1051 (1963)

Tilghman, S.M., Curtis, P.J., Tiemeier, D.C., Leder, P., Weissman, C.: The intervening sequence of a mouse β-globin gene is transcribed within the 15S β-globin mRNA precursor. Proc. Natl. Acad. Sci. USA 75, 1309-1313 (1978)

Tomich, P.K., An, F.Y., Clewell, D.B.: A transposon (Tn917) in *Streptococcus faecalis* that exhibits enhanced transposition during induction of drug resistance. Cold Spring Harbor Symp. Quant. Biol. 43, 1217-1221 (1979)

Toussaint, A.: Insertion of phage Mu-1 within prophage λ: a new approach for studying the control of the late functions in bacteriophage λ. Mol. Gen. Genet. 106, 89-92 (1969)

Toussaint, A., Faelen, M.: The dependence of temperate phage Mu-1 upon replication functions of *E. coli* K-12. Mol. Gen. Genet. 131, 209-214 1974)

Toussaint, A., Faelen, M., Bukhari, A.I.: Mu-mediated illegitimate recombi-nation as an integral part of the Mu life cycle. In: DNA Insertion Ele-ments, Plasmids and Episomes (eds. A.I. Bukhari, J. Shapiro, S. Adhya), pp. 275-285. New York: Cold Spring Harbor Laboratory 1977

Tu, C.-P. D., Cohen, S.N.: Translocation specificity of the Tn3 element: characterization of sites of multiple insertions. Cell 19, 151-160 (1980)

Vogel, H.J. (ed.): Nucleic Acid-Protein Recognition. New York: Academic Press 1977

Waggoner, B.T., Pato, M.L., Taylor, A.L.: Characterization of covalently closed circular DNA molecules isolated after bacteriophage Mu induction. In: DNA Insertion Elements, Plasmids and Episomes (eds. A.I. Bukhari, J. Shapiro, S. Adhya), pp. 263-274. New York: Cold Spring Harbor Laboratory 1977

Watson, J.D.: Molecular Biology of the Gene. 3rd edn. Menlo Park: Benjamin 1976

Weil, J., Signer, E.R.: Recombination in bacteriophage λ. II. Site-specific recombination promoted by the integration system. J. Mol. Biol. 34, 273-279 (1968)

Weisberg, R.A., Adhya, S.: Illegitimate recombination in bacteria and bacteriophage. Annu. Rev. Genet. 11, 451-473 (1977)

Weisberg, R.A., Gottesman, S., Gottesman, M.E.: Bacteriophage lambda: the lysogenic pathway. In: Comprehensive Virology, (eds. H. Frankel-Conrat, R. Wagner), Vol. III, pp. 197-258. New York: Plenum Press 1977

Westmoreland, B., Szybalski, W., Ris, H.: Mapping of deletions and substitutions in heteroduplex DNA molecules of bacteriophage lambda by electron microscopy. Science 163, 1343-1348 (1969)

Wolf, R.E.: Evolution of a transposon carrying the *gnd-his* region of the *Escherichia coli* chromosome. Abstr. Annu. Meet. Am. Soc. Microbiol., p. 140. Washington D.C.: American Society for Microbiology 1979

Yagi, Y., Clewell, D.B.: Identification and characterization of a small sequence located at two sites on the amplifiable tetracycline resistance plasmid paMα1 in *Streptococcus faecalis*. J. Bacteriol. 129, 400-406 (1977)

Yamada, Y., Calame, K.L., Grindley, J.N., Nakada, D.: Location of an ampicillin resistance transposon, Tn 1701, in a group of small, non-transferring plasmids. J. Bacteriol. 137, 990-999 (1979)

Yun, T., Vapnek, D.: Structure and location of antibiotic resistance determinats in bacteriophages P1Cm and P7 (ϕ Amp). In: DNA Insertion Elements, Plasmids, and Episomes (eds. A.I. Bukhari, J. Shapiro, S. Adhya), pp. 229-234. New York: Cold Spring Harbor Laboratory 1977

Zieg, J., Hilmen, M., Simon, M.: Regulation of gene expression by site-specific inversion. Cell 15, 237-244 (1978)

Viroids

T. O. Diener and R. A. Owens

A. Introduction

Viroids are nucleic acid species of relatively low molecular
weight (ca. 10^5 daltons) that are present in certain organisms
afflicted with specific maladies. They are not detectable in
healthy individuals of the same species, but, when introduced
into such individuals, they replicate autonomously and cause the
appearance of the characteristic disease syndrome (Diener 1979).
Unlike viral nucleic acids, viroids are not encapsidated. Viroids are highly resistant to heat, as well as to ultraviolet
and ionizing radiation. Known viroids are single-stranded molecules that contain extensive regions of intramolecular comple-
mentarity and exist as covalently closed circular structures.

So far, viroids are definitely known to exist only in higher
plants and to consist of RNA. It has been suggested, however,
that a group of animal or human diseases, the subacute spongi-
form encephalopathies (Gajdusek 1977), may be caused by nucleic
acid molecules with properties similar to those of the plant viroids (Diener 1972b). Recent evidence with one of these diseases,
scrapie of sheep, suggests that this concept may be correct.
Scrapie-specific low molecular weight DNA (Adams 1972; Hunter
et al. 1973) and a DNA component essential for the expression
of scrapie infectivity (Marsh et al. 1978) have been identified,
suggesting that viroids may occur in animals as well as in plants.

The first viroid was discovered in attempts to isolate and characterize the agent of the potato spindle tuber disease, which
for many years had been assumed to be of viral etiology. Diener
and Raymer (1967) reported that the transmissible agent of this
disease is a free RNA and that no viral nucleoprotein particles
(virions) are detectable in infected tissue. Later, sedimentation and gel electrophoretic analyses conclusively demonstrated
that the infectious RNA has a low molecular weight (Diener 1971b)
and that, therefore, the agent drastically differs from conventional viruses. Because of this basic difference, the term viroid
was introduced to denote pathogenic nucleic acids with properties
similar to those of the potato spindle tuber agent (PSTV) (Diener
1971b). Five additional plant diseases - citrus exocortis (Semancik and Weathers 1972), chrysanthemum stunt (Diener and Lawson
1973), cucumber pale fruit (Van Dorst and Peters 1974), chrysanthemum chlorotic mottle (Romaine and Horst 1975), and hop stunt
(Sasaki and Shikata 1977) - are now known to be caused by viroids;

and a sixth, coconut cadang-cadang, most likely also has viroid etiology (Randles et al. 1976, 1977).

In this report, recent evidence on the molecular structure of viroids, on possible mechanisms of viroid replication, and on viroid-host cell interactions are discussed.

B. Physical and Chemical Properties of Viroids

I. Molecular Weight

The importance of molecular weight determination in the elucidation of viroid structure has long been recognized. Difficulties in securing quantities of viroids sufficient for conventional biochemical analyses and purifying them from host RNAs complicated early efforts to measure viroid molecular weights. Recently, these difficulties have been overcome, and accurate molecular weight determinations have been published.

1. Comparative Methods Using Biological Activity

Early attempts to determine the molecular weight of viroids had to depend upon detection of the RNA by virtue of its biological activity. Although it had been known for some time that PSTV sediments more slowly than single-stranded viral RNA molecules (Diener and Raymer 1967), no conclusions concerning its molecular weight could be made since the conformation of the RNA was unknown. A principle elaborated by Loening (1967), however, was thought to allow estimation of a molecular weight based on biological activity alone. Loening reasoned that since the secondary structure of an RNA species has opposite effects on the sedimentation properties and electrophoretic mobility in polyacrylamide gels, a combination of these two analytical methods should be useful to distinguish structural differences from molecular weight differences. Biological activity is the only parameter necessary for evaluation of results.

Combined sedimentation and gel electrophoretic analyses under non-denaturing conditions showed that infectious PSTV has a very low molecular weight, 5×10^4 daltons (Diener 1971b). This conclusion was confirmed by the ability of PSTV to penetrate small pore size polyacrylamide gels which exclude high molecular weight RNAs (Diener and Smith 1971). Diener and Lawson (1973) have shown chrysanthemum stunt viroid (CSV) to be a low molecular weight RNA similar to, but distinct from, PSTV by combined sedimentation-electrophoretic analysis. Singh and Clark (1971), and Sänger (1972) have reported similar molecular weight estimates for PSTV and citrus exocortis viroid (CEV), respectively. Semancik and Weathers (1972), however, estimated the molecular weight of CEV to be 1.25×10^5.

Although these determinations conclusively demonstrated that known viroids are low molecular weight RNAs, the molecular weight estimates must be regarded with caution for two reasons: (1) The electrophoretic mobility of an RNA depends not only on its molecular weight (more correctly, its molecular volume) but also on its particular secondary and tertiary structure (Boedtker 1971). As shown below, viroids differ greatly in their structure from the RNA standards used. (2) Comparison of results with PSTV and CEV suggest that isolated viroid RNA may contain aggregates.

Electrophoresis of PSTV under non-denaturing conditions in 3%, 5%, and 7.5% polyacrylamide gels yielded a paucidisperse infectivity distribution with apparent molecular weight values of 5×10^4 and 1×10^5 (Diener 1971b). Because of this 1:2 ratio of apparent molecular weights and the fact that all of the PSTV enters a 20% polyacrylamide gel and migrates with an apparent molecular weight of 5×10^4 (Diener and Smith 1971), native PSTV may occur in several states of aggregation. Electrophoresis in small pore size polyacrylamide gels may cause disaggregation. A somewhat similar phenomenon was reported by Semancik et al. (1973b) for CEV. At gel concentrations of less than 8-10%, CEV has an apparent molecular weight of approximately 1×10^5, whereas at higher gel concentration, the apparent molecular weight decreases to approximately 5×10^4. These authors accepted the 1×10^5 value as correct and attributed the behavior of CEV in high percentage gels to "its unusual structural properties".

2. *Comparative Methods Using Purified Viroids*

Although the experiments described above demonstrate that viroids are low molecular weight RNAs, further characterization required isolation of the RNAs as physically recognizable entities. Diener (1972a) made the first steps toward the purification of PSTV. Figure 1 compares the UV absorption profiles of low molecular weight RNA preparations isolated from uninfected control (panel A) and PSTV-infected (panel B) tomato seedlings fractionated by polyacrylamide gel electrophoresis. A prominent component (marked II) is found only in the RNA prepared from PSTV-infected seedlings, and infectivity coincides with this component. These observations allowed purification of PSTV in amounts sufficient for biophysical and biochemical analyses.

Boedtker (1971) has proposed that treatment with formaldehyde at $63°C$ completely denatures RNA. When the electrophoretic mobilities of PSTV and several standard RNAs are compared after formaldehyde treatment, a molecular weight of $7.5-8.5 \times 10^4$ was estimated (Diener and Smith 1973). This estimate agrees closely with a value of 8.9×10^4 calculated from the contour length of PSTV spread for electron microscopy from a solution containing 8 M urea (Sogo et al. 1973).

Semancik et al. (1973a) obtained identical molecular weight estimates for PSTV and CEV by gel electrophoresis under non-denaturing conditions (1.25×10^5). Dickson et al. (1975) have analyzed PSTV and CEV using gel electrophoresis under both denaturing and non-denaturing conditions; identical mobilities were only observed in 98% formamide - 10% polyacrylamide gel. A comparative study of the disease-specific low molecular weight RNA of

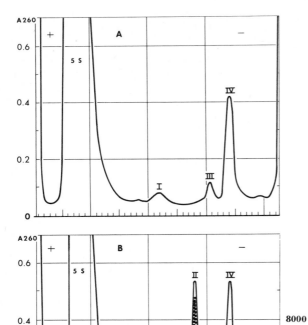

Fig. 1. (A) UV absorption profile of an RNA preparation from
healthy tomato leaves after
electrophoresis in a 20% polyacrylamide gel. (B) UV absorption (———) and infectivity
distribution (----) of an RNA
preparation from PSTV-infected
tomato leaves after electrophoresis in a 20% polyacrylamide gel. 5 S, 5 S ribosomal
RNA; I, III, IV, unidentified
minor components of cellular
RNA; II, PSTV; A_{260}, absorbance
at 260 nm. Electrophoretic
migration from right to left.
(From Diener 1972a)

cadang-cadang (ccRNA-1) and PSTV in denaturing formamide-poly-
acrylamide gels showed that ccRNA-1 is smaller than PSTV, and its
molecular weight was estimated to be 6.3 to 7.3 x 10^4 (Randles
et al. 1976).

Sensitivity to irradiation with both UV light and ionizing radiation has also been used as a comparative method for estimating
viroid molecular weights (target volumes). Diener et al. (1974)
found that 70-90 times more 254 nm UV irradiation is required
to inactivate PSTV than to inactivate tobacco ringspot virus.
Semancik et al. (1973b) compared the relative sensitivities of
CEV and TMV to ionizing radiation and calculated a molecular
weight of 1.1 x 10^5 for CEV.

3. Independent Methods Using Purified Viroids

Sänger et al (1976) have determined the molecular weights of
three viroids by sedimentation equilibrium under denaturing conditions. Molecular weights of 119 ± 4 x 10^3, 127 ± 4 x 10^3, and
110 ± 5 x 10^3 were reported for CEV, PSTV, and cucumber pale

fruit viroid (CPFV), respectively. Since the complete primary sequence of PSTV has recently been established (Gross et al. 1978), the molecular weight of PSTV (sodium salt) can be calculated to 123,337.

II. Conformation

The molecular structure of viroids remained enigmatic for a number of years because they display properties typical of double-stranded RNA in some analytical systems and properties typical of single-stranded RNA in other systems. Recent determination of the complete primary sequence of PSTV and the thermodynamic parameters of five different highly purified viroids have indicated that viroids have a common secondary structure (Gross et al. 1978; Langowski et al. 1978). Native viroids have an extended, rod-like structure containing a series of double helical segments interrupted by internal loops.

1. Chromatographic Behavior and Thermal Denaturation

The elution pattern of PSTV from columns of methylated serum albumin suggests double-strandedness, while the elution pattern from CF-11 cellulose columns is consistent with both single- and double-stranded molecules (Diener and Raymer 1969). Most PSTV elutes from hydroxylapatite at phosphate buffer concentrations lower than expected of a double-stranded RNA (Diener 1971c; Lewandowski et al. 1971). Similar behavior has been reported for CEV (Singh and Sänger 1976). T_m values of 50-58°C and total hyperchromicities of 10-22% have been measured for PSTV, CEV, CPFV, and cc-RNA 1 under ionic conditions equivalent to 0.01-0.1 x SSC (Diener 1972a; Semancik et al. 1975; Randles 1975; Sänger et al. 1976). Thermal denaturation of viroids occurs at lower temperatures and over a wider temperature range than expected for double-stranded RNAs of similar base compositions (Fig. 2).

These data suggest that viroids contain extensive secondary structure consisting of alternating single-stranded and base-

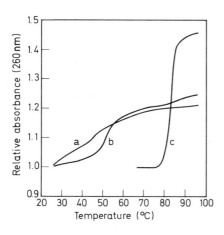

Fig. 2. Thermal denaturation profiles of yeast tRNA (a), PSTV (b), and a mixture of cucumber mosaic virus double-stranded RNAs 1, 2, and 3 (c) in 0.1 x SSC containing 0.1 mM EDTA. The cucumber mosaic virus double-stranded RNAs were a gift from Dr. J.M. Kaper, Plant Virology Laboratory, PPI, SEA, U.S.D.A., Beltsville, MD 20705

paired regions. Finally, immunological tests made with antisera that react specifically with double-stranded RNA gave no evidence for the presence of double-stranded RNA in highly infectious PSTV preparations (Stollar and Diener 1971).

2. *Electron Microscopy*

Sogo et al. (1973) processed purified preparations of PSTV for electron microscopy by the protein monolayer spreading technique of Kleinschmidt and Zahn (1959). When PSTV solutions containing 8 M urea were spread onto a hypophase of distilled water, individual short strands of relatively uniform length (50 nm) were visible. These strands were ribonuclease-sensitive and thicker than single-stranded RNA.

Two subsequent studies have examined both non-denatured and denatured viroids by electron microscopy. The most important result of these studies is the discovery that viroid RNAs can exist as covalently closed circular molecules. The relative proportions of circular and linear molecules observed in the two studies were different, and the significance of this variation is presently unclear. Viroids are the only type of RNA known to occur naturally as covalently closed circular molecules.

Sänger et al. (1976) examined non-denatured and denatured preparations of CPFV, PSTV, and CEV, using benzyldimethylakylammonium chloride rather than cytochrome C, as the film-forming surfactant. Examination of undenatured CPFV revealed rod-like or dumbbell-shaped molecules with asymmetric terminal knots; the average particle length was 35 nm. Denatured preparations of CPFV (heated for 5 min at 60°C in the presence of 80% formamide, 4.8 M urea, 0.4% formaldehyde, and 25 µg/ml benzyldimethylalkylammonium chloride) contained a variety of structures ranging from rod-like molecules to fully denatured single-stranded circles. Nicked molecules (0.5-1%) were also present in the preparations. Treatment with proteinase K or Pronase did not alter the structures observed.

McClements and Kaesberg (1977) used the spreading technique developed by Inman and Schnös (1970) to examine non-denatured and denatured PSTV preparations. Non-denatured molecules resembled double-stranded rods with a mean length of 50 ± 10 nm. Partially denaturing (23-51% formamide at room temperature) and completely denaturing conditions (heating for 15 min at 65°C with 1.4 M formaldehyde followed by addition of formamide to 45%) led to the appearance of two single-stranded components. Completely denatured circular molecules (~20% of total molecules) had a mean contour length of 140 ± 10 nm, while single-stranded linear molecules (~70% of total molecules) had a mean contour length of 110 ± 10 nm.

Owens et al. (1977) have separated PSTV preparations into two fractions by polyacrylamide gel electrophoresis in the presence of formamide and urea. One fraction contains predominantly circular molecules, the second fraction contains almost exclusively linear molecules. The purity of the fractions was estimated by electron microscopy of formaldehyde-denatured molecules. Both linear and circular molecules were found to be infectious.

3. *Thermodynamic Studies*

The thermal denaturation and electron microscopic studies allow a qualitative description of viroid structure: viroids exist as single-stranded, covalently closed circular RNA molecules whose rod-like conformation results from extensive base-pairing. Low field NMR spectroscopy (Semancik et al. 1975) indicates that the double-stranded regions of CEV are enriched in GC base-pairs — 70-80% compared to an overall base composition of 58.8% G+C. This qualitative description of viroid conformation has served as the basis for a series of recent quantitative thermodynamic studies.

Henco et al. (1977) used UV absorption melting analysis and temperature jump methods under one set of ionic conditions (10 mM sodium cacodylate, 1 mM EDTA, pH 6.8) to obtain quantitative thermodynamic and kinetic data for CPFV and CEV. Two relaxation processes were resolved; one in the sec range, and one in the msec range. The data suggest that CEV and CPFV contain (1) an uninterrupted double helix of 52 base-pairs (G-C content = 72%), and (2) several short double helical stretches containing a total of 15-25 base-pairs (G-C content = 50%). These two parts were postulated to fulfill different functions in replication and pathogenicity: the uninterrupted double helix representing the rigid core of the molecule, while the short double helical regions could dissociate under physiological conditions.

This model has been refined by Langowski et al. (1978) who measured the thermodynamic parameters of five different viroids (CEV, PSTV, CPFV, chrysanthemum stunt viroid, and chrysanthemum chlorotic mottle viroid) and presented a detailed statistical thermodynamic treatment of their denaturation. This treatment does not require the use of an all-or-none approximation where only the native and fully denatured states are considered and is in accordance with both the dependence of T_m values upon ionic strength and the measured apparent reaction enthalpies. Although the cooperativity of viroid melting is as high as that of authentic double-stranded RNA, the T_m values are more than 30° lower than those of double-stranded RNA with the same G-C content. A model where viroids exist in their native conformation as extended rod-like structures containing a series of double helical sections and internal loops is in complete accord with experimental data. The highly cooperative melting is at least partially explained by the circular RNA structure. The major part (250-300 of the approximately 350 nucleotides present) of each viroid is required to interpret the thermodynamic behavior. Parallel microcalorimetric studies (Klump et al. 1978) are in accordance with this model and, in addition, indicate that the native conformations of CEV and CPFV contain approximately 85 base-pairs.

III. Primary and Secondary Structure of PSTV

The complete primary sequence of PSTV has recently been established (Gross et al. 1978). The sequencing strategy employed a
combination of conventional (complete and partial pancreatic and
T_1 ribonuclease digestions) and newer, rapid (base-specific cleavage of long 5'-terminal labeled oligonucleotides followed by polyacrylamide gel fractionation) procedures. The slow rate of viroid replication required that the RNA be labeled in vitro using
$[\gamma-^{32}P]$ATP and T_4 bacteriophage-induced 5'-polynucleotide kinase
from *E. coli*. PSTV was shown to be covalently closed, single-stranded RNA containing 359 ribonucleotides (73 AMP, 77 UMP, 101 GMP,
and 108 CMP); no modified nucleotides occur at the 5' terminus
of any fragment sequenced.

Figure 3 shows the secondary structure model for PSTV proposed
by Gross et al. (1978). The complete primary sequence has been
arranged to maximize the number of base-pairs found, and then
the model refined to reflect (1) location of sites of preferential nuclease cleavage, and (2) sensitivity of cytidine residues
to modification by sodium bisulfite (Domdey et al. 1978). A total of 122 base-pairs (73 G-C, 38 A-U, and 11 G-U pairs) are present; the serial arrangement of double helical sections and internal loops is consistent with the independent thermodynamic
measurements described in the previous section.

This secondary structure model is also consistent with the results of *E. coli* RNase III digestion of PSTV. Under conditions
that allow the enzyme to demonstrate its full capacity to cleave
RNA, neither the electrophoretic mobility (Dickson 1976) nor the
infectivity of PSTV (Diener, unpublished) is affected. RNase III
requires either an extended region of perfect RNA·RNA duplex
(approximately 25 base-pairs) or a highly specialized RNA sequence (Robertson and Hunter 1975) to cleave RNA.

C. Replication of Viroids

In contrast to the detailed knowledge of viroid nucleic acid sequence and structure just summarized, comparatively little has
been definitely established concerning viroid replication and
pathogenesis. Consideration of the biological properties of infectious viroid RNA and what is known about the replication of
viroid RNA per se identifies both differences and similarities
between viroids and conventional RNA viruses.

I. Viroids as Messenger RNAs

Several lines of evidence suggest that the viroid disease process is initiated by direct interaction of the viroid RNA with
the host and not by a viroid-coded polypeptide. Although known
viroids could code for a polypeptide with a molecular weight of
~10,000, PSTV is neither translated nor aminoacylated by a varie-

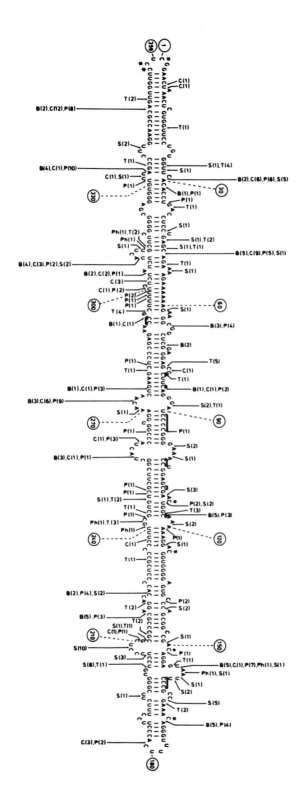

Fig. 3. Proposed secondary structure of PSTV. This structure was derived by arranging the primary sequence of PSTV to maximize intramolecular base-pairing and subsequent consideration of the observed nuclease and bisulphite sensitivities of PSTV. Arrows indicate sites easily cleaved during controlled nuclease digestion; only those sites whose location has been precisely determined were considered. The length of the arrows is related to the number of cleavages observed at the respective sites, and the numbers at the arrows indicate how often the cleavage was observed. Single letter codes identify the different nucleases used: B, bacterial alkaline phosphatase from Boehringer containing an unknown nuclease; C, pancreatic RNAse containing ε-carboxymethyllysine at position 41; T, RNAse T_1; P, unmodified pancreatic RNAse; S, nuclease from *Staphylococcus aureus*; Ph, RNAse Phy I from *Physarum polycephalum*. Heavy bars indicate resistance of cytosine residues to bisulphite modification, and asterisks indicate cytosine residues converted to uracil residues by bisulphite treatment. (From Gross et al. 1978)

ty of in vitro protein synthesis systems (Davies et al. 1974; Hall et al. 1974). Furthermore, CEV was not translated in *Xeno-pus laevis* oocytes, even after in vitro polyadenylation, and did not interfere with the translation of endogenous mRNAs (Semancik et al. 1977). Viroids also lack detectable mRNA activity in vivo. Zaitlin and Hariharasubramanian (1972) compared proteins synthesized in PSTV-infected and healthy tomato seedlings by a double-label protocol and could not detect a viroid-specific polypeptide of ~10,000 molecular weight. Although Conejero and Semancik (1977) found that the synthesis of certain polypeptides was enhanced in CEV-infected *Gynura*, no viroid-specific proteins could be detected.

Although these results suggest that neither infectious viroid "plus" strands nor any complementary "minus" strands that may be synthesized during replication function as mRNA, two additional facts must be considered. (1) The one-dimensional SDS-polyacrylamide gel electrophoresis systems used in all these studies lack resolution for the study of low molecular weight polypeptides expected to be viroid-specific. (2) Determination of the primary sequence of PSTV allows the sequence of a unit length complementary strand to be predicted; at least three potential translation products can be identified. Two factors, their apparent "single-stranded" character and presence in the cytoplasm, suggest that RNA molecules complementary to CEV may be translated.

II. Fundamental Characteristics of Viroid Replication

A number of viroids (PSTV, CEV, CSV, and CPFV) have overlapping host ranges and induce similar symptoms in some hosts (discussed by Diener and Hadidi 1977). The induction of similar symptoms in several hosts by PSTV and CEV and their identical mobilities in 5% polyacrylamide gels led some workers to conclude that the two viroids were independent isolates of the same pathogen (Singh et al. 1972; Semancik et al. 1973a; Singh and Clark 1973). Significant differences in the RNA fingerprints of PSTV, CEV, and CSV (Dickson et al. 1975; Gross et al. 1977), however, support the concept of individual viroid species.

Although the replication of viroids is apparently completely dependent upon host-coded proteins, the sequence of the progeny appears, as expected, to be determined by the sequence of the infecting viroid. Two different techniques, RNA fingerprinting and DNA·RNA hybridization, have failed to detect host-specific changes in viroid sequence. Dickson et al. (1978) analyzed PSTV and CEV, each purified from both tomato and *Gynura*, by two-dimensional RNA fingerprinting after in vitro iodination with ^{125}I and digestion with ribonuclease T_1. Because each viroid retained its distinctive pattern of cytosine-containing oligonucleotides with only minor variations, they concluded that viroid nucleotide sequences are principally determined by the infecting viroid and not the host. Owens et al. (1978) compared the thermal denaturation profiles of RNA·DNA hybrids containing PSTV "type" strain grown in four different hosts (tomato, tobacco, *Gynura*, and chrysanthemum) and DNA complementary to PSTV "type"

strain isolated from tomato. No significant differences in the
thermal stabilities of the hybrids were detected. Since a sequence difference of 1.5% will reduce the T_m of an RNA·DNA hybrid
approximately 1° (McCarthy and Church 1970), a change in only
6-10 of the 359 nucleotides in PSTV would have been detectable.

A third and most important characteristic of viroid replication
is its autonomy; i. e., viroids do not require the presence of
a helper virus for replication (Diener 1971b). Although PSTV and
the satellite RNAs associated with several RNA plant viruses have
similar molecular weights, no helper virus can be detected in uninoculated Rutgers tomato seedlings. Although simultaneous transfer of PSTV and a helper virus could occur during inoculation
with crude extracts, PSTV can be transmitted with low molecular
weight RNA fractions eluted from polyacrylamide gels. Since all
Rutgers tomato seedlings are susceptible to PSTV, the putative
helper virus would have to be seed-transmitted at 100% efficiency. Similar undetected helper viruses would have to be universally present in apparently normal plants of a number of solanaceous
hosts of PSTV. In summary, it is very unlikely that helper viruses are involved in viroid replication.

III. Subcellular Location of Viroid Accumulation

Knowledge of the subcellular location of viroids is important for
two reasons: (1) It was valuable in early attempts to characterize and purify PSTV; (2) it continues to indicate the various
possible levels of viroid-host interaction (replication and pathogenesis).

Bioassays of subcellular fractions prepared from PSTV-infected
tomato tissue demonstrated that only the tissue debris and nuclear fractions contain appreciable infectivity (Diener 1971a).
The chloroplast, mitochondrial, ribosomal, and supernatant fractions contained only traces of infectivity. Most infectivity was
chromatin-associated and could be extracted as free RNA with phosphate buffer. The significant infectivity associated with the
tissue debris probably results from incomplete extraction of nuclei. This association of PSTV with chromatin may explain its heterogeneous sedimentation properties in crude extracts, because
no evidence for the existence of stable, sedimentable virions in
PSTV-infected tissue could be obtained. Sänger (1972) has reported that CEV is also located in the nuclear fraction in close association with the chromatin. Comparison of nuclei isolated from
uninoculated and PSTV-infected tomato leaf tissue showed that
nuclei from infected tissue were larger and had increased RNA/
DNA and protein/DNA ratios (Takahashi and Diener 1975).

After examining the subcellular distribution of CEV in *Gynura aurantiaca* DC with techniques expressly designed to minimize organelle
damage, Semancik et al. (1976) proposed a distinction between the
site of CEV synthesis in the nucleus and the site of "pathologi-
cal lesion" (or "accumulation") at the plasma membrane. Although
a majority of recoverable infectivity was usually recovered in
the tissue debris and nuclear fractions, significant infectivity

was present in a membrane fraction identified as plasma membrane vesicles. Failure to detect CEV in nucleic acids isolated from purified nuclei by electrophoretic analysis was cited as supporting a second subcellular location for CEV, but corresponding electrophoretic analyses of nucleic acid prepared from membrane-rich fractions were not reported. Although the subcellular sites of viroid synthesis and pathogenic activity may be different, presently available evidence is inconclusive.

IV. Complementary Nucleic Acid Sequences

Theoretically possible schemes for viroid replication involve either DNA- or RNA-directed RNA synthesis. A DNA-directed mechanism requires the presence of DNA sequences complementary to the entire viroid to be present in infected tissues; an RNA-directed mechanism requires the presence of complementary RNA sequences. Evidence for the existence of both complementary DNA and RNA sequences has been presented.

Semancik and Geelen (1975) originally reported that DNA sequences complementary to ^{125}I-CEV are present in CEV-infected tomato and *Gynura* but absent from uninoculated plants. There are basically two reasons why these results have not been accepted. (1) The experimental results can be criticized on technical grounds. The "DNA-rich" preparations contained significant amounts of low molecular weight RNA; pretreatment with pancreatic RNAse or alkali before hybridization would have eliminated possible complementary RNA sequences. The presence of complementary RNA sequences would allow the formation of RNA-RNA duplexes as well as DNA·RNA hybrids, and the methods used to characterize the presumed DNA· RNA hybrids were not rigorous. (2) Subsequent reports by Hadidi et al. (1976) and even the same laboratory (Grill and Semancik 1978) have not confirmed these results. Hadidi et al. (1976) found that although alkali-treated DNA from several solanaceous hosts of PSTV contained infrequent sequences complementary to at least 60% of PSTV, the levels of these sequences in tomato and *Gynura* DNA did not increase after PSTV infection.

In contrast to the results discussed above, Grill and Semancik (1978) have recently reported that RNA sequences complementary to CEV can be detected in CEV-infected, but not uninoculated *Gynura*. Treatment of nucleic acid preparations with pancreatic RNAse or NaOH, but not pancreatic DNAse, abolished their ability to hybridize with ^{125}I-CEV. Their high T_m values, sharp melting profile, and insensitivity to RNAse H indicate that these hybrids are well-matched RNA·RNA duplexes.

Three facts about these CEV-complementary RNA sequences must be considered when assessing their possible role in viroid replication and/or pathogenesis. (1) They are present in both the high-speed supernatant and nuclear fractions after cell fractionation; infectious CEV is found primarily in the nucleus. (2) The complementary sequences found in the high-speed supernatant (cytoplasm) seem to be essentially single-stranded molecules (insoluble in 2 M LiCl), while those in the nucleus are apparently

more heterogeneous (present in both the precipitate and supernatant after 2 M LiCl fractionation). (3) The presence of detectable levels of complementary RNA sequences in the 2 M LiCl-soluble nuclear RNA fraction suggests that comparatively large amounts
of complementary RNA must be present. If the level of complementary sequences were comparable to those present in TMV infection
where the ratio of "minus" to "plus" strands never exceeds 1:13
(Kielland-Brandt and Nilsson-Tillgren 1973), hybridization of
^{125}I-CEV to complementary RNA sequences in the 2 M LiCl-soluble
fraction would be inhibited by the unlabeled CEV present.

V. Mechanism of Viroid Replication

Viroid replication has been studied in leaf slices prepared from
PSTV-infected tomato (Diener and Smith 1975), nuclei isolated
from PSTV-infected tomato leaves (Takahashi and Diener 1975),
protoplasts isolated from leaves of CEV-infected tomato (Mühlbach et al. 1977), and tomato protoplasts infected in vitro with
CPFV (Mühlbach and Sänger 1977; Sänger, personal communication).
Actinomycin D inhibited the incorporation of radioactive nucleotide precursors into viroid RNA in all systems. Intracellular
concentrations of α-amanitin sufficient to inhibit tomato DNA-
dependent RNA polymerase II inhibited CPFV replication in tomato
protoplasts, but did not inhibit either TMV RNA replication or
the RNA polymerase III-dependent synthesis of host 5 S and tRNA
(Sänger, personal cummunication). Viroid RNA was detected by its
characteristic mobility during polyacrylamide gel electrophoresis in all studies.

Actinomycin D is widely used as a specific inhibitor of DNA-directed RNA synthesis and neither binds to double-stranded RNA
(Haselkorn 1964) nor inhibits TMV replication after the infection has been established (Dawson and Schlegel 1976). The results
discussed above suggest that either the synthesis of progeny viroid molecules is DNA-directed or, more likely, that either a
host-coded RNA primer or rapidly turning-over host protein(s) is
required for RNA-directed viroid replication.

Bouloy et al. (1978) have recently shown that in vitro synthesis
of influenza virus "plus" strands by the virion-associated polymerase requires eukaryotic mRNA as primer. Influenza virus replication in vivo is inhibited by the presence of Actinomycin D and
α-amanitin (reviewed by Bishop 1977), but RNA synthesis in vitro
is resistant to these inhibitors. DNA-directed RNA polymerase II,
the enzyme responsible for synthesis of eukaryotic mRNAs, is completely inhibited by low levels of α-amanitin.

A possible mechanism for viroid replication, therefore, involves
RNA-directed RNA synthesis catalyzed by a polymerase that is resistant to Actinomycin D and α-amanitin but completely dependent
upon primer(s) synthesized by DNA-directed RNA polymerase II.
This mechanism is consistent with both in vivo and in vitro studies, but the available evidence does not exclude other schemes.

D. Possible Mechanisms of Pathogenesis

No evidence exists on how viroids interfere with their host's metabolism to produce the characteristic macroscopic symptoms observed in certain host species. The nuclear location and replication of viroids and their apparent inability to serve as messenger RNAs suggest, however, that these symptoms are caused by interference with gene regulation in the infected host cells (Diener 1971b).

Several viroids have an extended host range, but no damage resulting from viroid replication is discernible in a large majority of host species (Diener and Hadidi 1977). Thus, in these symptomless hosts, viroid-induced metabolic aberrations do not occur or, if they do occur, they must be harmless in the particular genetic milieu of the host.

With both PSTV (Zaitlin and Hariharasubramanian 1972) and CEV (Conejero and Semancik 1977) certain host proteins occur in larger amounts in infected than in healthy tissue. Possibly these aberrations in host protein synthesis are related to the pathogenic properties of viroids.

E. Possible Origin of Viroids

The finding of sequences complementary to PSTV in the DNAs of several uninfected host species (Hadidi et al. 1976) suggests that PSTV may have originated from host genetic material. Among the species tested, several solanaceous species have DNAs with the highest affinity to PSTV; the more phylogenetically distant a plant species is from solanaceous plants, the fewer PSTV-related sequences are generally found in its DNA (Hadidi et al. 1976). This supports the idea that PSTV originated from genes normally present in certain solanaceous plant species.

References

Adams, D.H.: Studies on DNA from normal and scrapie-affected mouse brain. J. Neurochem. $\underline{19}$, 1869-1882 (1972)

Bishop, D.H.L.: Virion polymerases. In: Comprehensive Virology (eds. H. Fraenkel-Conrat, R.R. Wagner), Vol. 10, pp. 117-278. New York: Plenum Press 1977

Boedtker, H.: Conformation independent molecular weight determinations of RNA by gel electrophoresis. Biochim. Biophys. Acta $\underline{240}$, 448-453 (1971)

Bouloy, M., Plotch, S.J., Krug, R.M.: Globin mRNAs are primers for the transcription of influenza viral RNA in vitro. Proc. Natl. Acad. Sci. USA $\underline{75}$, 4886-4890 (1978)

Conejero, V., Semancik, J.S.: Exocortis viroid: Alteration in the proteins of *Gynura aurantiaca* accompanying viroid infection. Virology $\underline{77}$, 221-232 (1977)

Davies, J.W., Kaesberg, P., Diener, T.O.: Potato spindle tuber viroid. XII. An investigation of viroid RNA as a messenger for protein synthesis. Virology 61, 281-286 (1974)

Dawson, W.O., Schlegel, D.E.: The sequence of inhibition of tobacco mosaic virus synthesis by actinomycin-D, thiouracil, and cycloheximide in a synchronous infection. Phytopathology 66, 177-181 (1976)

Dickson, E.: Studies of plant viroid RNA and other RNA species of unusual function. Ph.D. Thesis, New York: The Rockefeller University 1976

Dickson, E., Prensky, W., Robertson, H.D.: Comparative studies of two viroids: Analysis of potato spindle tuber and citrus exocortis viroids by RNA fingerprinting and polyacrylamide-gel electrophoresis. Virology 68, 309-316 (1975)

Dickson, E., Diener, T.O., Robertson, H.D.: Potato spindle tuber and citrus exocortis viroids undergo no major sequence changes during replication in two different hosts. Proc. Natl. Acad. Sci. USA 75, 951-954 (1978)

Diener, T.O.: Potato spindle tuber virus: A plant virus with properties of a free nucleic acid. III. Subcellular location of PSTV-RNA and the question of whether virions exist in extracts or in situ. Virology 43, 75-89 (1971a)

Diener, T.O.: Potato spindle tuber "virus". IV. A replicating, low molecular weight RNA. Virology 45, 411-428 (1971b)

Diener, T.O.: A plant virus with properties of a free ribonucleic acid: Potato spindle tuber virus. In: Comparative Virology (eds. K. Maramorosch, E. Kurstak), pp. 433-478. New York: Academic Press 1971c

Diener, T.O.: Potato spindle tuber viroid. VIII. Correlation of infectivity with a UV-absorbing component and thermal denaturation properties of the RNA. Virology 50, 606-609 (1972a)

Diener, T.O.: Is the scrapie agent a viroid? Nature New Biol. 235, 218-219 (1972b)

Diener, T.O.: Viroids and Viroid Diseases. New York: Wiley Interscience 1979

Diener, T.O., Hadidi, A.: Viroids. In: Comprehensive Virology (eds. H. Fraenkel-Conrat, R.R. Wagner), Vol. 11, pp. 285-337. New York: Plenum Press 1977

Diener, T.O., Lawson, R.H.: Chrysanthemum stunt: A viroid disease. Virology 51, 94-101 (1973)

Diener, T.O., Raymer, W.B.: Potato spindle tuber virus: A plant virus with properties of a free nucleic acid. Science 158, 378-381 (1967)

Diener, T.O., Raymer, W.B.: Potato spindle tuber virus: A plant virus with properties of a free nucleic acid. II. Characterization and partial purification. Virology 37, 351-366 (1969)

Diener, T.O., Smith, D.R.: Potato spindle tuber viroid: VI. Monodisperse distribution after electrophoresis in 20% polyacrylamide gels. Virology 46, 498-499 (1971)

Diener, T.O., Smith, D.R.: Potato spindle tuber viroid. IX. Molecular-weight determination by gel electrophoresis of formylated RNA. Virology 53, 359-365 (1973)

Diener, T.O., Smith, D.R.: Potato spindle tuber viroid. XIII. Inhibition of replication by Actinomycin D. Virology 63, 421-427 (1975)

Diener, T.O., Schneider, I.R., Smith, D.R.: Potato spindle tuber viroid. XI. A comparison of the ultraviolet light sensitivities of PSTV, tobacco ringspot virus, and its satellite. Virology 57, 577-581 (1974)

Domdey, H., Jank, P., Sänger, H.L., Gross, H.J.: Studies on the primary and secondary structure of potato spindle tuber viroid: Products of digestion with ribonuclease A and ribonuclease T_1 and modification with bisulfite. Nucleic Acids Res. 5, 1221-1236 (1978)

Gajdusek, D.C.: Unconventional viruses and the origin and disappearance of Kuru. Science 197, 943-960 (1977)

Grill, L.K., Semancik, J.S.: RNA sequences complementary to citrus exocortis viroid in nucleic acid preparations from infected *Gynura aurantiaca*. Proc. Natl. Acad. Sci. USA 75, 896-900 (1978)

Gross, H.J., Domdey, H., Sänger, H.L.: Comparative oligonucleotide fingerprints of three plant viroids. Nucleic Acids Res. 4, 2021-2028 (1977)

Gross, H.J., Domdey, H., Lossow, C., Jank, P., Raba, M., Alberty, H., Sänger, H.L.: Nucleotide sequence and secondary structure of potato spindle tuber viroid. Nature (London) 273, 203-208 (1978)

Hadidi, A., Jones, D.M., Gillespie, D.H., Wong-Staal, F., Diener, T.O.: Hybridization of potato spindle tuber viroid to cellular DNA of normal plants. Proc. Natl. Acad. Sci. USA 73, 2453-2457 (1976)

Hall, T.C., Wepprich, R.K., Davies, J.W., Weathers, L.G., Semancik, J.S.: Functional distinctions between the ribonucleic acids from citrus exocortis viroid and plant viruses: Cell-free translation and aminoacylation reactions. Virology 61, 486-492 (1974)

Haselkorn, R.: Actinomycin D as a probe for nucleic acid secondary structure. Science 143, 682-684 (1964)

Henco, K., Riesner, D., Sänger, H.L.: Conformation of viroids. Nucleic Acids Res. 4, 177-194 (1977)

Hunter, G.D., Kimberlin, R.H., Collis, S., Millson, G.C.: Viral and non-viral properties of the scrapie agent. Ann. Clin. Res. 5, 262-267 (1973)

Inman, R.B., Schnös, M.: Partial denaturation of thymine- and 5-bromouracil-containing λ DNA in alkali. J. Mol. Biol. 49, 93-98 (1970)

Kielland-Brandt, M.C., Nilsson-Tillgren, T.: Studies on the biosynthesis of TMV. V. Determination of TMV RNA and its complementary RNA at different times after infection. Mol. Gen. Genet. 121, 229-238 (1973)

Kleinschmidt, A.K., Zahn, R.K.: Über Desoxyribonucleinsäure-Molekeln in Protein-Mischfilmen. Z. Naturforsch. B 14, 770-779 (1959)

Klump, H., Riesner, D., Sänger, H.L.: Calorimetric studies on viroids. Nucleic Acids Res. 5, 1581-1587 (1978)

Langowski, J., Henco, K., Riesner, D., Sänger, H.L.: Common structural features of different viroids: Serial arrangement of double helical sections and internal loops. Nucleic Acids Res. 5, 1589-1610 (1978)

Lewandowski, L.J., Kimball, P.C., Knight, C.A.: Separation of the infectious ribonucleic acid of potato spindle tuber virus from double-stranded ribonucleic acid of plant tissue extracts. J. Virol. 8, 809-812 (1971)

Loening, U.E.: The fractionation of high-molecular weight ribonucleic acid by polyacrylamide-gel electrophoresis. Biochem. J. 102, 251-257 (1967)

Marsh, R.F., Malone, T.G. Semancik, J.S., Lancaster, W.D., Hanson, R.P.: Evidence for an essential DNA component in the scrapie agent. Nature (London) 275, 146-147 (1978)

McCarthy, B.J., Church, R.B.: The specificity of molecular hybridization reactions. Annu. Rev. Biochem. 39, 131-150 (1970)

McClements, W.L., Kaesberg, P.: Size and secondary structure of potato spindle tuber viroid. Virology 76, 477-484 (1977)

Mühlbach, H.P., Sänger, H.L.: Multiplication of cucumber pale fruit viroid in inoculated tomato leaf protoplasts. J. Gen. Virol. 35, 377-386 (1977)

Mühlbach, H.P., Camacho-Henriquez, A., Sänger, H.L.: Isolation and properties of protoplasts from leaves of healthy and viroid-infected tomato plants. Plant Sci. Lett. 8, 183-189 (1977)

Owens, R.A., Erbe, E., Hadidi, A., Steere, R.L., Diener, T.O.: Separation and infectivity of circular and linear forms of potato spindle tuber viroid. Proc. Natl. Acad. Sci. USA 74, 3859-3863 (1977)

Owens, R.A., Smith, D.R., Diener, T.O.: Measurement of viroid sequence homology by hybridization with complementary DNA. Virology 89, 388-394 (1978)

Randles, J.W.: Association of two ribonucleic acid species with cadang-cadang disease of coconut palm. Phytopathology 65, 163-167 (1975)

Randles, J.W., Rillo, E.P., Diener, T.O.: The viroidlike structure and cellular location of anomalous RNA associated with the cadang-cadang disease. Virology 74, 128-139 (1976)

Randles, J.W., Boccardo, G., Retuerma, M.L., Rillo, E.P.: Transmission of the RNA species associated with cadang-cadang of coconut palm, and the insensitivity of the disease to antibiotics. Phytopathology 67, 1211-1216 (1977)

Robertson, H.D., Hunter, T.: Sensitive methods for the detection and characterization of double helical ribonucleic acid. J. Biol. Chem. 250, 418-425 (1975)

Romaine, C.P., Horst, R.K.: Suggested viroid etiology for chrysanthemum chlorotic mottle disease. Virology 64, 86-95 (1975)

Sänger, H.L.: An infectious and replicating RNA of low molecular weight: The agent of the exocortis disease of citrus. Adv. Biosci. 8, 103-116 (1972)

Sänger, H.L., Klotz, G., Riesner, D., Gross, H.J., Kleinschmidt, A.K.: Viroids are single-stranded covalently closed circular RNA molecules existing as highly base-paired rod-like structures. Proc. Natl. Acad. Sci. USA 73, 3852-3856 (1976)

Sasaki, M., Shikata, E.: On some properties of hop stunt disease agent, a viroid. Proc. Jpn. Acad. 53B, 109-112 (1977)

Semancik, J.S., Geelen, J.L.M.C.: Detection of DNA complementary to pathogenic viroid RNA in exocortis disease. Nature (London) 256, 753-756 (1975)

Semancik, J.S., Weathers, L.G.: Exocortis disease: Evidence for a new species of "infectious" low molecular weight RNA in plants. Nature New Biol. 237, 242-244 (1972)

Semancik, J.S., Magnuson, D.S., Weathers, L.G.: Potato spindle tuber disease produced by pathogenic RNA from citrus exocortis disease: Evidence for the identity of the causal agents. Virology 52, 292-294 (1973a)

Semancik, J.S., Morris, T.J., Weathers, L.G.: Structure and conformation of low molecular weight pathogenic RNA from exocortis disease. Virology 53, 448-456 (1973b)

Semancik, J.S., Morris, T.J., Weathers, L.G., Rodorf, B.F., Kearns, D.R.: Physical properties of a minimal infectious RNA (viroid) associated with the exocortis disease. Virology 63, 160-167 (1975)

Semancik, J.S., Tsuruda, D., Zaner, L., Geelen, J.L.M.C., Weathers, L.G.: Exocortis disease: Subcellular distribution of pathogenic (viroid) RNA. Virology 69, 669-676 (1976)

Semancik, J.S., Conejero, V., Gerhart, J.: Citrus exocortis viroid: Survey of protein synthesis in *Xenopus laevis* oocytes following addition of viroid RNA. Virology 80, 218-221 (1977)

Siegel, A., Hariharasubramanian, V.: Reproduction of small plant RNA viruses. In: Comprehensive Virology (eds. H. Fraenkel-Conrat, R.R. Wagner), Vol. 2, pp. 61-108. New York: Plenum Press 1974

Singh, A., Sänger, H.L.: Chromatographic behavior of the viroids of the exocortis disease of citrus and the spindle tuber disease of potato. Phytopathol. Z. 87, 143-160 (1976)

Singh, R.P., Clark, M.C.: Infectious low-molecular-weight ribonucleic acid. Biochem. Biophys. Res. Commun. 44, 1077-1083 (1971)

Singh, R.P., Clark, M.C.: Similarity of host response to both potato spindle tuber and citrus exocortis viruses. FAO Plant Protect. Bull. 21, 121-125 (1973)

Singh, R.P., Clark, M.C., Weathers, L.G.: Similarity of host symptoms induced by citrus exocortis and potato spindle tuber causal agents. Phytopathology 62, 790 (1972)

Sogo, J.M., Koller, T., Diener, T.O.: Potato spindle tuber viroid. X. Visualization and size determination by electron microscopy. Virology 55, 70-80 (1973)

Stollar, B.D., Diener, T.O.: Potato spindle tuber viroid. V. Failure of immunological tests to disclose double-stranded RNA or RNA-DNA hybrids. Virology 46, 168-170 (1971)

Takahashi, T., Diener, T.O.: Potato spindle tuber viroid. XIV. Replication in nuclei isolated from infected leaves. Virology 64, 106-114 (1975)

Van Dorst, H.J.M., Peters, D.: Some biological observations on pale fruit, a viroid-incited disease of cucumber. Neth. J. Plant Pathol. 80, 85-96 (1974)

Zaitlin, M., Hariharasubramanian, V.: A gel electrophoretic analysis of proteins from plants infected with tobacco mosaic and potato spindle tuber viruses. Virology 47, 296-305 (1972)

Subject Index

Progress in Molecular and Subcellular Biology

Editors: F. E. Hahn, H. Kersten,
W. Kersten, W. Szybalski
Advisors: T. T. Puck,
G. F. Springer, K. Wallenfels
Managing Editor: F. E. Hahn

Volume 1

1969. 32 figures. VII, 237 pages
ISBN 3-540-04674-7

Contents: F. E. Hahn: On Molecular Biology. – C. R. Woese: The Biological Significance of the Genetic Code. – J. Davies: Errors in Translation. – H. G. Mandel: The Incorporation of Fluorouracil into RNA and its Molecular Consequences. – R. M. Smillie, N. S. Scott: Organelle Biosynthesis: The Chloroplast. – B. W. Agranoff: Macromolecules and Brain Function – A 1969 Baedeker.

From the reviews:
"...a thoroughly enthralling collection of four excellent recollection of four excellent reviews and two brief but informative essays, which make easy reading for both specialist or student... If the future volumes of Progress in Molecular and Subcellular Biology continue at the high standard set by Vol. 1, this series will form an important part of any biochemical or molecular and cellular biology library."
Search

Volume 2

Proceedings of the Research Symposium on Complexes of Biologically Active Substances with Nucleic Acids and Their Models of Action
Held at the Water Reed Army Institute of Research, Washington, 16–19 March 1970
1971. 158 figures. IX, 400 pages
ISBN 3-540-05321-2

From the reviews:
"...The book will be of great value to the active research workers in the field of chemistry, biology and pharmacology of DNA-drug complexes. Future works along this line may furnish essential informations on the action some of the small molecules play as genetic repressors. Of great importance are experimental procedures which are presented in the articles und up-to-date references."
Acta Microbiologica Polonica

Volume 3

1973. 58 figures. VII, 251 pages
ISBN 3-540-06227-0

Contents: F. E. Hahn: Reverse Transcription and the Central Dogma. – M. J. Fournier, Jr., D. J. Brenner, B. P. Doctor: The Isolation of Genes: A Review of Advances in the Enrichment, Isolation and in vitro Synthesis of Specific Cistrons. – A. Kaji: Mechanism of Protein Synthesis and the Use of Inhibitors in the Study of Protein Synthesis. – A. B. Edmundson, M. Schiffer, K. R. Ely, M. K. Wood: Structural Features of Immunoglobulin Light Chains. – A. S. Braverman: The Thalassemia Syndromes: Genetically Determined Disorders of the Regulation of Protein Synthesis in Eukaryotic Cells. – C. A. Paoletti, G. Riou: The Mitochondrial DNA of Malignant Cells.

From the reviews:
"Some of the most exciting recent developments in molecular genetics are reviewed in the six contributions to this volume. Subjects: Reverse transcription and the central dogma; The isolation of genes; The mechanism of protein synthesis and its inhibitors; Structural features of the immunoglobulin light chains; The thalassemia syndromes: genetically determined disorders of protein synthesis; and The mitochondrial DNA of malignant cells. The result is an authoritative handbook that few geneticists and molecular biologists can afford to be without." *Quarterly Review of Biology*

Springer-Verlag
Berlin
Heidelberg
New York

Progress in Molecular and Subcellular Biology

Editors: F. E. Hahn, H. Kersten,
W. Kersten, W. Szybalski
Advisors: T. T. Puck,
G. F. Springer, K. Wallenfels
Managing Editor: F. E. Hahn

Volume 4

1976. 88 figures, 27 tables. XI, 251 pages
ISBN 3-540-07487-2

The new volume of this well-established series contains seven articles
which range from biophysical studies on the conformations of DNA,
through papers on gene expression in bacterial viruses and cellular
slime molds, to inhibitors of nucleic acid syntheses as antitumor
agents and the role of adenosine as a regulator of coronary blood
flow. This volume continues the tradition of the series to combine
carefully selected articles in the field of biomedicine with those in
which life scientists are particulary interested.

Contents: S. Bram: The Polymorphism of DNA. – H. J. Witmer: Regu-
lation of Bacteriophage T4 Gene Expression. – W. Lotz: Defective
Bacteriophages: The Phage Tail-Like Particles. – M. Sussman: The
Genesis of Multicellular Organization and the Control of Gene Ex-
pression in Dictyostelium discoideum. – D. Oesterhelt: Isoprenoids
and Bacteriorhodopsin in Halobacteria. – P. Chandra, L. K. Steel,
U. Ebener, M. Woltersdorf, H. Laube, G. Will: Inhibitors of DNA
Synthesis in RNA Tumor Viruses: Biological Implications and Their
Mode of Action. – R. A. Olsson, R. E. Patterson: Adenosine as a Physio-
logical Regulator of Coronary Blood Flow.

Volume 5

1977. 56 figures. XIII, 176 pages
ISBN 3-540-08192-5

Volume 5 of this series presents a collection of five progress reports
concerning recent developments in molecular biology outside the
classical area of molecular genetics. It continues the emphasis on
medical pharmacological aspects of molecular biology which has
characterized this series throughout its existence.

Contents: F. E. Hahn: The Double Helix Revisted: Watson and Olby. –
C. Reiss, T. Arpa-Gabarro: Thermal Transition Spectroscopy: A New
Tool for Submolecular Investigation of Biologic Macromolecules. –
I. Schuster: Interactions of Drugs with Liver Microsomes. –
W. Flamenbaum, J. H. Schwartz, R. J. Hamburger, J. S. Kaufman: The
Pathogenesis of Experimental Acute Renal Failure: The Role of
Membrane Dysfunction. – P. Gund: Three-Dimensional Pharmaco-
phoric Pattern Searching. – N. W. Gabel: Chemical Evolution: A
Terrestrial Reassessment.

Volume 6

1978. 65 figures, 59 tables. XI, 354 pages
ISBN 3-540-08588-2

Volume 6 of this series continues to report current progress in molecular genetics, molecular pharmacology, biochemistry and cell biology
of medical and general scientific interest. Of special interest is an
article on recombinant DNA research and another article on the mor-
phine receptors in the brain.

Contents: K. N. Timmis, S. N. Cohen, F. C. Cabello: DNA Cloning and
the Analysis of Plasmid Structure and Function. – H. Kleinkauf,
H. Koischwitz: Peptide Bond Formation in Non-Ribosomal
Systems. – A. Goldstein, B. M. Cox: Opiate Receptors and Their Endo-
genous Ligands (Endorphins). – S. L. Hajduk: Influence of DNA
Complexing Compounds on the Kinetoplast of Trypanosomatids. –
R. L. O'Brien, J. W. Parker, J. F. P. Dixon: Mechanisms of Lymphocyte
Transformation. – L. Ebringer: Effects of Drugs on Chloroplasts.

Springer-Verlag
Berlin
Heidelberg
New York